3 in 1
Practice Book

Practice
Reteach
Spiral Review

Grade 4

CALIFORNIA HSP Mat

Harcourt
SCHOOL PUBLISHERS

Visit *The Learning Site!*
www.harcourtschool.com

Printed in the United States of America

ISBN 13: 978-0-15-383385-4
ISBN 10: 0-15-383385-8

16 0868 15 14 13
4500434922

Contents

Key: PW Practice Workbook RW Reteach Workbook SR Spiral Review

Contents

Key: PW Practice Workbook RW Reteach Workbook SR Spiral Review

Contents

Key: PW Practice Workbook RW Reteach Workbook SR Spiral Review

Contents

Key: PW Practice Workbook RW Reteach Workbook SR Spiral Review

Name____ **Lesson 1.1**

Place Value Through Hundred Thousands

You can use a place-value chart to read and write large numbers. Each group of three digits is called a **period**. When you write a number, separate each period with a comma.

	THOUSANDS PERIOD			ONES PERIOD		
Place	Hundreds	Tens	Ones	Hundreds	Tens	Ones
Value	100,000	10,000	1,000	100	10	1
Think	100 thousands	10 thousands	1 thousand	100 ones	10 ones	1 one

To write a number in **standard form**, write the number using digits.

407,051

To write a number in **word form**, write the number in each period and follow it with the name of the period. Do not name the ones period. Remember to use commas to separate each period of numbers.

407,051 is four hundred seven thousand, fifty-one.

To write a number in **expanded form**, write the sum of the value of each digit. Do not include digits for which the value is 0.

407,051 = 400,000 + 7,000 + 50 + 1

Write each number in word form.

1. 52,321 ________

2. 300,000 + 20,000 + 4,000 + 900 ________

3. 965,143 ________

4. 500,000 + 80,000 + 7,000 + 300 + 10 + 6 ________

Write each number in standard form.

5. 90,000 + 6,000 + 200 + 80 + 1 ________

6. nine hundred thousand, thirteen ________

Write each number in expanded form.

7. 40,571 ________

8. 999,293 ________

9. six hundred thousand, ninety-four ________

10. one hunded twenty thousand, fifty-six ________

Name___________________________________ **Lesson 1.1**

Place Value Through Hundred Thousands

Write each number in two other forms.

1. 50,000 + 3,000 + 700 + 5

2. eight hundred thousand, nine hundred thirty-seven

3. 420,068

4. 78,641

Complete.

5. 290,515 = two hundred ninety ____________, five hundred fifteen = ____________ + 90,000 + ____________ + 10 + 5

6. ____________ + 10,000 + 3,000 + 100 + 80 + 9 = 413,1 ______ = four hundred thirteen thousand, one ____________ eighty-nine

Write the value of the underlined digit in each number.

7. 70<u>5</u>,239 **8.** 4<u>1</u>7,208 **9.** 914,3<u>2</u>5 **10.** 360,04<u>4</u>

Problem Solving and Test Prep

11. In 2005, there were 20,556 Bulldogs registered in the American Kennel Club. What are two ways you can represent the number?

12. In 2005, the Labrador Retriever was the most popular breed in the American Kennel Club with 137,867 registered. Write the number in two other forms.

13. What is the value of the digit 9 in 390,215?

A 900
B 9,000
C 90,000
D 900,000

14. In February, eighty-five thousand, six hundred thirteen people went to the Westminster Dog Show. What is the number in standard form?

A 850,630 **C** 850,613
B 85,630 **D** 85,613

Spiral Review

For 1–4, write the numbers in order from greatest to least.

1. 659; 671; 603 ______________________

2. 567; 312; 410 ______________________

3. 1,320; 1,412; 1,398 ______________________

4. 3,050; 3,765; 3,246 ______________________

For 8–10, use the data in the table.

Mr. Conrad's students voted for their favorite colors.

Color	Votes
Black	2
Blue	8
Green	5
Red	7
Yellow	4

8. Which color had the fewest votes? __________

9. How many votes were there for red? __________

10. How many more students liked blue than green? ______________

For 5–7, find the area of each figure. Write the answer in square units.

5. ______________________ 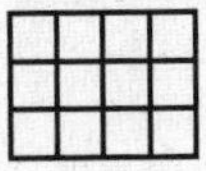

6. ______________________ 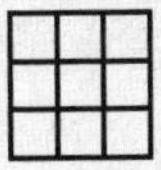

7. ______________________ 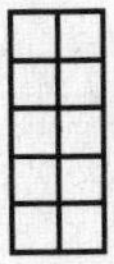

For 11–15, find the missing factor.

11. $\square \times 7 = 35$

12. $8 \times \square = 56$

13. $5 \times \square = 25$

14. $\square \times 8 = 72$

15. $8 \times \square = 16$

Name________________________________

Model Millions

One million is the counting number that comes after 999,999.
One million is written as 1,000,000.

You can use base-ten blocks to find the number of thousands there are in 1,000,000.

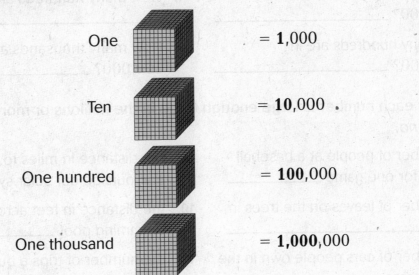

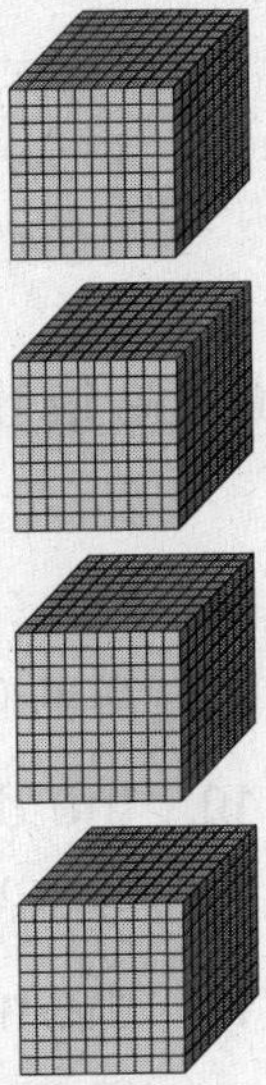

One = 1,000

Ten = 10,000

One hundred = 100,000

One thousand = 1,000,000

There are 1,000 thousands in 1,000,000.

Solve.

1. How many tens are in 1,000? ________

2. How many thousands are in 10,000? ________

3. How many hundreds are in 1,000,000? ________

4. How many tens are in 10,000? ________

5. How many ten thousands are in 1,000,000? ________

6. How many hundreds are in 1,000? ________

Tell whether each number is large enough to be in the millions or more. Write *yes* or *no*.

7. the number of people in your school ________

8. the number of hairs on your body ________

9. the distance in feet around the world ________

10. the number of pebbles on a mountain ________

11. the number of days in a year ________

12. the number of channels on TV ________

Name ____________________ **Lesson 1.2**

Model Millions

Solve.

1. How many hundreds are in 100,000? ____________

2. How many thousands are in 10,000? ____________

3. How many thousands are in 1,000,000? ____________

4. How many hundreds are in 10,000? ____________

5. How many hundreds are in 1,000,000? ____________

6. How many thousands are in 100,000? ____________

Tell whether each number is large enough to be in the millions or more. Write *yes* or *no*.

7. the number of people at a baseball stadium for one game ____________

8. the distance in miles to the nearest star outside our solar system ____________

9. the number of leaves on the trees in a forest ____________

10. the distance in feet across a swimming pool ____________

11. the number of cars people own in the United States ____________

12. the number of trips a bus might make in one day ____________

13. the number of bags of trash a family makes in one month ____________

14. the distance in miles from one city to another in your state ____________

15. the number of fourth graders in the United States ____________

16. the number of miles you might travel to reach the Moon ____________

17. the number of gallons of water in the ocean ____________

18. the number of stars in the solar system ____________

Choose the number in which the digit 5 has the greater value.

19. 435,767 or 450,767 ____________

20. 510,000 or 5,100,000 ____________

21. 125,000,000 or 521,000,000 ____________

22. 435,003 or 4,300,500 ____________

23. 1,511,672 or 115,672 ____________

24. 40,005,400 or 350,400,300 ____________

25. 135,322,000 or 9,450,322 ____________

26. 35,000,000 or 3,500,000 ____________

Name________________________________ **Lesson 1.3**

Place Value Through Millions

Look at the digit **3** in the place-value chart below. It is in the millions place.

In word form, the value of this number is three hundred million.

In standard form, the value of this number is 300,000,000.

PERIOD

MILLIONS			THOUSANDS			ONES		
Hundreds	Tens	Ones	Hundreds	Tens	Ones	Hundreds	Tens	Ones
3	6	5,	0	0	9,	0	5	8

Read the number shown in the place-value chart.

In word form, this number is written as three hundred sixty-five million, nine thousand, fifty-eight.

You can also write the number in expanded form:

300,000,000 + 60,000,000 + 5,000,000 + 9,000 + 50 + 8

Write each number in standard form.

1. 600,000,000 + 1,000,000 + 300,000 + 2,000 + 300 + 70 + 8

____

2. three million, twenty-one thousand, four hundred

____

Write each number in word form.

3. 4,000,000 + 40,000 + 300 + 70

____

____

4. 428,391,032

____

____

Write each number in expanded form.

5. 2,130,571

____

____

6. six hundred million, three hundred twenty-eight thousand, fifty-two

____

____

Use the number 46,268,921.

7. Write the value of the digit in the millions place?

____

8. Which digit is in the ten millions place?

____

Name________________________________ Lesson 1.3

Place Value Through Millions

Write each number in two other forms.

1. ninety-five million, three thousand, sixteen

____

____

2. four hundred eighty-five million, fifty-two thousand, one hundred eight

____

____

3. 507,340,015

____

____

4. 20,000,000 + 500,000 + 60,000 + 1,000 + 300 + 40

____

____

Use the number 78,024,593.

5. Write the name of the period that has the digits 24. ______________

6. Write the digit in the ten millions place. ______________

7. Write the value of the digit 8.

____

8. Write the name of the period that has the digit 5. ______________

Find the sum. Then write the answer in standard form.

9. 7 thousands 3 hundreds 4 ones + 8 ten thousands 1 thousand 5 hundreds

____

Problem Solving and Test Prep

10. The average distance from Earth to the Sun is 92,955,807 miles. What is the value of the digit 2?

____

11. The average distance from Earth to the Sun is one hundred forty-nine million, six hundred thousand kilometers. Write the number in standard form.

____

12. Which of these is the number 4,000,000 + 300,000 + 80,000 + 500 + 10?

A 4,385,100 **C** 4,380,510

B 40,308,510 **D** 4,385,010

13. Which of these is the number forty-three million, nine hundred two thousand, eleven?

A 4,392,011 **C** 43,902,011

B 43,920,011 **D** 43,902,110

Name ____________________

Compare Whole Numbers

Use **less than** (<), **greater than** (>), and **equal to** (=) to compare numbers.

Which number is greater, 631,328 or 640,009?

- Write one number under the other. Line up the digits by place value.

  631,328
  640,009

- Compare the digits, beginning with the greatest place value position.

  631,328
  ↓
  640,009

- Circle the first digits that are different.

  6(3)1,328
  ↓↓
  6(4)0,009

- 4 ten thousands > 3 ten thousands. So 640,009 > 631,328.

Compare. Write <, >, or = for each ○.

1. 4,092,332
933,793
4,092,332 ○ 933,793

2. 2,092,092
2,102,002
2,092,092 ○ 2,102,002

3. 377,493
377,943
377,493 ○ 377,943

4. 5,000,000
500,000
5,000,000 ○ 500,000

5. 7,192,322
1,797,322
7,192,322 ○ 1,797,322

6. 4,002,384
4,020,000
4,002,384 ○ 4,020,000

7. 8,344,475
8,344,475
8,344,475 ○ 8,344,475

8. 5,492,000
492,000
5,492,000 ○ 492,000

9. 4,928,399
4,923,488
4,928,399 ○ 4,923,488

10. 7,084,122
7,187,084
7,084,122 ○ 7,187,084

11. 1,203,437
1,203,445
1,203,437 ○ 1,203,445

12. 528,807,414
5,699,001
528,807,414 ○ 5,699,001

Name ______________________________

Lesson 1.4

Compare Whole Numbers

Use the number line to compare. Write the lesser number.

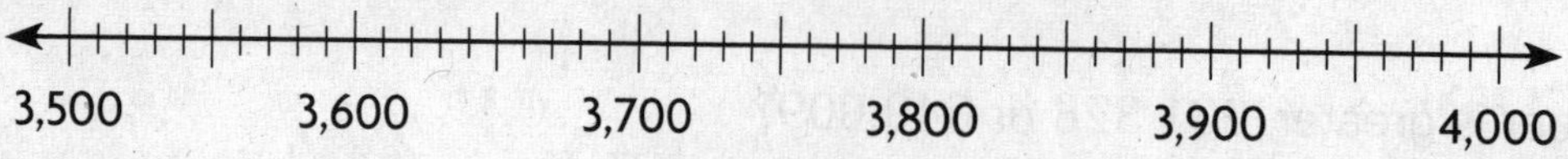

1. 3,660 or 3,590 ____________
2. 3,707 or 3,777 ____________
3. 3,950 or 3,905 ____________

Compare. Write <, >, or = for each ○.

4. 5,155 ○ 5,751
5. 6,810 ○ 6,279
6. 45,166 ○ 39,867
7. 72,942 ○ 74,288
8. 891,023 ○ 806,321
9. 673,219 ○ 73,551
10. 3,467,284 ○ 481,105
11. 613,500 ○ 1,611,311
12. 4,000,111 ○ 41,011

ALGEBRA Find all of the digits that can replace each ■.

13. 781 ≠ 78■ ____________
14. 2,4■5 ≠ 2,465 ____________
15. ■,119 ≠ 9,119 ____________

Problem Solving and Test Prep

USE DATA For 16–17 use the table.

16. Which mountain is taller: Logan or McKinley?

____

17. Which mountain is taller than 29,000 feet?

____

Tallest Mountains

Mountain	Height (in feet)
Everest	29,028
McKinley	20,320
Logan	19,551

18. Which number from the list below is the greatest?

A 34,544
B 304,544
C 43,450
D 345,144

19. Which number is less than $1,322?

A $1,521
B $1,429
C $1,319
D $1,324

Name ____________________

Order Whole Numbers

Sometimes it is important to put numbers in order. When you put whole numbers in order, you can figure out which numbers have the least and greatest values.

You can compare each number's digits to find out the order from least to greatest.

In which order would you place the numbers 55,997; 57,000; and 56,038?

- Align the numbers under each other by place value.

55,997
57,000
56,038

- Compare the digits, beginning with the greatest place value position.

55,997
↓
57,000
↓
56,038

- Start with the digit in the ten thousands place.
- The digit 5 is the same in all 3 numbers
- The next digit is in the thousands place.

- Circle the first group of digits that are different.

5(5),997
5(7),000
5(6),038

- Compare the numbers 5, 7, and 6.
- Seven is greater than 5 and 6, so 57,000 is the greatest.
- Five is less than 6, so 55,997 is smaller than 56,038.

Since $5 < 6$, and $6 < 7$, the numbers in order from least to greatest are: 55,997; 56,038; 57,000.

Write the numbers in order from least to greatest.

1. 55,997; 57,000; 56,038

_______ _______ _______

2. 787,925; 1,056,000; 789,100

_______ _______ _______

3. 94,299; 82,332; 100,554

_______ _______ _______

4. 1,354,299; 1,942,332; 1,300,554

_______ _______ _______

5. 7,234,000; 6,311,094; 2,102,444

_______ _______ _______

6. 837,570; 618,947; 609,678

_______ _______ _______

7. 65,392; 75,539; 110,628

_______ _______ _______

8. 53,294; 74,002; 80,678

_______ _______ _______

9. 754,379; 682,362; 714,954

_______ _______ _______

10. 4,299; 332; 50,599

_______ _______ _______

Name________________________________

Order Whole Numbers

Write the numbers in order from greatest to least.

1. 74,421; 57,034; 58,925

2. 2,917,033; 2,891,022; 2,805,567

3. 409,351; 419,531; 417,011

4. 25,327,077; 25,998; 2,532,707

5. 621,456; 621,045,066; 6,021,456

6. 309,423; 305,125; 309,761

7. 4,358,190; 4,349,778; 897,455

8. 5,090,115; 50,009,115; 509,155

ALGEBRA Write all of the digits that can replace each ■.

9. 389 < 3■7 < 399

10. 5,601 < 5,■01 < 5,901

11. 39,560 > 3■,570 > 34,580

12. 178,345 > 1■8,345 > 148,345

Problem Solving and Test Prep

USE DATA For 13–14, use the table.

13. Which lake has the smallest area?

14. Write the names of the lakes in order from least area to the greatest area.

Largest Lakes (area in square miles)	
Victoria	26,828
Huron	23,000
Superior	31,700
Caspian Sea	19,551

15. Which shows the numbers in order from greatest to least?

A 36,471; 36,490; 36,470

B 969,482; 979,485; 969,500

C 121,119; 121,101; 121,111

D 129,876; 129,611; 129,602

16. Which shows the numbers in order from greatest to least?

A 92,944; 92,299; 92,449

B 159,872; 159,728; 159,287

C 731,422; 731,242; 731,244

D 487,096; 487,609; 487,960

Name ______________________________ **Week 2**

Spiral Review

1. Write the following number in word form: 200,065,001

____

____

2. Write the following number in standard form: 1,000,000 + 200,000 + 5,000 + 200

____

3. What is the value of the digit 6 in the number 8,609,712?

____

4. What is the value of the digit 2 in the number 5,789,235?

____

Use the graph below.

8. Which prize was won the *most* often? Which prize was won the *least* often?

____

____

For 5–7, name each triangle by its angles.

5. 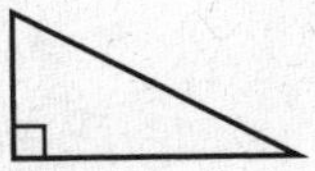

____

6. 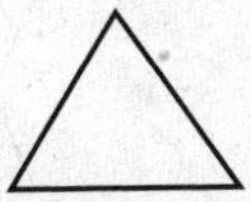

____

7. 

____

For 9–13, find the product.

9. $(3 \times 3) \times 2 =$ ________

10. $4 \times (5 \times 1) =$ ________

11. $(2 \times 3) \times 8 =$ ________

12. $5 \times (4 \times 3) =$ ________

13. $7 \times (2 \times 6) =$ ________

Name____

Problem Solving Workshop Strategy: Use Logical Reasoning

A basketball team has a score that is a 2-digit number. The sum of the digits is 8. The difference between the digits is 2. The tens digit is less than the ones digit. What is the score?

Read to Understand

1. What are you asked to find?

____

Plan

2. How can using logical reasoning help you solve the problem?

____

____

Solve

3. Solve the problem. Describe how you used logical reasoning.

____

____

____

____

4. Write your answer in a complete sentence.

____

Check

5. Look at the problem. Does the answer make sense for the problem?

____

____

Solve by using logical reasoning.

6. Mr. Lee's class sold 140 muffins and 235 cookies at the bake sale. The number of cupcakes sold was between the number of muffins and cookies sold. What is the greatest number of cupcakes the class could have sold?

____

7. Shari's soccer team won every game of the season. Her team played every Saturday and Sunday for three weeks each month. Suppose they played this schedule for three months. How many games did Shari's soccer team win?

____

Name____ **Lesson 1.6**

Problem Solving Workshop Strategy: Use Logical Reasoning

Problem Solving Strategy Practice

Use logical reasoning to solve.

1. The stadium store sells team shirts on Friday, Saturday, and Sunday. The number of shirts sold for three days were 473, 618, and 556. The least number of shirts were sold on a Friday. More than 600 shirts were sold on Saturday. How many shirts were sold each day?

____

2. Anton, Rachel, and Lamont like different baseball teams. The teams are the Yankees, the Red Sox, and the White Sox. Anton's favorite team does not have a color in its name. Lamont does not like the White Sox. Which team does each person like best?

____

Mixed Strategy Practice

3. Beth, Paulo, Lee, Maya, and Rob are standing in line to get into the movies. Beth is in front of Maya. Maya is not last in line. Rob is first. Lee is after Maya. Paulo is not last. In what order are they standing in line?

____

4. Mr. Katz bought an autographed baseball for $755. He used $50-bills, $20-bills, and $5-bills to make exactly $755. The total number of bills he used is less than 20. What combination of bills would Mr. Katz have used?

____

USE DATA For 5–6, use the information shown in the art.

5. Claire buys two items. She spends less than $100 for both of them. Which two items does she buy?

____

6. Alex wants to save money to buy the hockey stick. After 2 weeks he has $40. After 3 weeks, he has $50. After 4 weeks, he has $60. How long do you think it will take Alex to save $80?

____

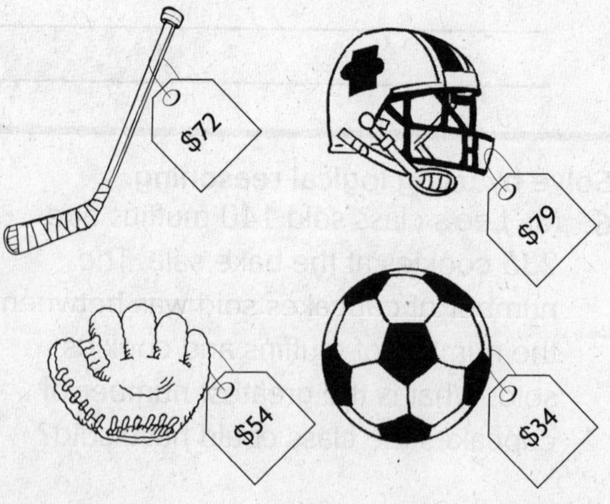

Name ____________________

Algebra: Relate Addition and Subtraction

You can solve addition and subtraction sentences by using **related facts**. For example, $7 + 8 = 15$ and $8 + 7 = 15$ are related addition facts; $15 - 8 = 7$ and $15 - 7 = 8$ are related subtraction facts. They are all in the fact family for 7, 8, and 15. Here are some ways you can find related facts.

- A **fact family** is a group of number sentences that uses the same numbers.

  Here is the fact family for 4, 7, and 11:

  $4 + 7 = 11$ $\quad 11 - 7 = 4$

  $7 + 4 = 11$ $\quad 11 - 4 = 7$

- An **inverse operation** is an opposite operation. Addition is the inverse, or opposite, operation of subtraction. Subtraction is the opposite of addition.

  $\square - 9 = 6$

  Use inverse operations to write the problem as addition.

  $9 + 6 = \boxed{15}$

  Use fact families to figure out the missing information.

  $\boxed{15} - 9 = 6$

Write a related fact. Use it to complete the number sentence.

1. $6 + \square = 13$ **2.** $9 - \square = 7$ **3.** $3 + \square = 11$ **4.** $16 - \square = 7$

______ ______ ______ ______

5. $\square + 8 = 16$ **6.** $10 - \square = 4$ **7.** $\square + 12 = 15$ **8.** $\square - 4 = 13$

______ ______ ______ ______

9. $12 + \square = 20$ **10.** $15 - \square = 8$ **11.** $9 - \square = 3$ **12.** $5 + \square = 21$

______ ______ ______ ______

Name____________________

Lesson 2.1

Algebra: Relate Addition and Subtraction

Write a related fact. Use it to complete the number sentence.

1. ■ − 7 = 8 ____________
2. 4 + ■ = 13 ____________
3. ■ + 9 = 14 ____________
4. 8 + ■ = 11 ____________
5. ■ − 4 = 8 ____________
6. 17 − ■ = 9 ____________
7. ■ − 5 = 5 ____________
8. 13 − ■ = 5 ____________
9. ■ + 7 = 16 ____________

Write the fact family for each set of numbers.

10. 6, 8, 14
11. 7, 5, 12
12. 9, 6, 15

Problem Solving and Test Prep

13. Byron can do 12 pull-ups. Malik can do 7 pull-ups. How many more pull-ups can Byron do than Malik? What related facts can you use to solve this problem?

14. Byron can do 12 pull-ups. Malik can do 7 pull-ups. Selma does more pull-ups than Malik but fewer than Byron. What are the four possible numbers of pull-ups that Selma could have done?

15. Which of the following sets of numbers cannot be used to make a fact family?

A 25,10,15
B 2,2,4
C 15,9,6
D 3,2,14

16. Which of the following sets of numbers can be used to make a fact family?

A 5,6,11
B 11,12,13
C 7,6,12
D 19,9,11

Name____

Round Whole Numbers Through Millions

When you **round** a number, you replace it with a number that is easier to work with but not as exact. You can round numbers to different place values.

Round 3,476,321 to the place of the underlined digit.

- Identify the underlined digit.
  - The underlined digit is a 4 in the hundred thousands place.
- Look at the number to the right of the underlined digit.
  - If that number is 0-4 , the underlined digit stays the same.
  - If that number is 5-9, the underlined digit is increased by 1.
  - The number to the right of the underlined digit is a 7, so the underlined digit, 4, will be increased by one; $4 + 1 = 5$.
- Change all the digits to the right of the underlined digit, hundred thousands, to zeros.

So, 3,476,321 rounded to the nearest hundred thousand is 3,500,000.

Round 8,851,342 to the nearest million.

- Identify the digit in the millions place.
  - The 8 to the left of the comma is in the millions place.
- Look at the number to the right of the the digit in the millions place.
  - If that number is 0-4 , the 8 stays the same.
  - If that number is 5-9, the 8 is increased by 1.
  - The number to the right of the millions place is an 8, so the eight in the million place will be increased by one; $8 + 1 = 9$.
- Change all the digits to the right of the millions place to zeros.

So, 8,851,342 rounded to the nearest million is 9,000,000.

Round each number to the place value of the underlined digit.

1. 3,452 ____________ **2.** 180 ____________ **3.** $72,471 ____________ **4.** 5,723,000 ____________

Round each number to the nearest thousand, hundred thousand, and million.

5. 2,472,912 ____________ ____________ ____________

6. 1,333,456 ____________ ____________ ____________

7. 7,925,248 ____________ ____________ ____________

8. 3,274,801 ____________ ____________ ____________

Name________________________________ Lesson 2.2

Round Whole Numbers Through Millions

Round each number to the place value of the underlined digit.

1. 7,803 ____________
2. 4,097 ____________
3. 23,672 ____________
4. 627,432 ____________
5. 34,809,516 ____________
6. 671,523,890 ____________

Round each number to the nearest ten, hundred, and hundred thousand.

7. 6,086,341 ____________ ____________ ____________
8. 79,014,878 ____________ ____________ ____________
9. 821,460,934 ____________ ____________ ____________

Problem Solving and Test Prep

USE DATA For 10–11, use the table.

10. Which state has a population that rounds to 5,700,000?

____

11. What is the population of Maryland, rounded to the nearest thousand?

____

Population of States in 2000 Census	
State	**Population**
Maryland	5,296,486
Tennessee	5,689,283
Wisconsin	5,363,675

12. Which number rounds to 45,000?

A 44,399 **C** 44,890

B 44,098 **D** 45,987

13. To find the rounded number that is closest to 1,234,567, to what place do you round?

____

Name____ **Lesson 2.3**

Mental Math: Addition and Subtraction Patterns

Find basic facts and use patterns to add or subtract large numbers.

Use mental math to complete the pattern.

9 + ___ = 14
___ + 50 = 140
900 + 500 = ___

- Find the basic fact.
  9 + 5 = 14
- Add the same number of zeros to each digit.
  One zero: 90 + 50 = 140
  Two zeros: 900 + 500 = 1,400
- Fill in the boxes with your answers.

9 + 5 = 14
90 + 50 = 140
900 + 500 = 1,400

130 − ___ = 50
___ − 800 = 500
13,000 − ___ = 5,000

- Find the basic fact.
  13 − 8 = 5
- Add the same number of zeros to each digit.
  One zero: 130 − 80 = 50
  Two zeros: 1,300 − 800 = 500
  Three zeros: 13,000 − 8,000 = 5,000
- Fill in the boxes with your answers.

130 − 80 = 50
1,300 − 800 = 500
13,000 − 8,000 = 5,000

Use mental math to complete the pattern.

1. 3 + 5 = ___
___ + 50 = 80
300 + 500 = ___

2. 7 − 3 = ___
70 − 30 = ___
700 − 300 = ___

3. 12 − ___ = 8
120 − 40 = ___
1,200 − ___ = 800

4. 6 + 7 = 13
___ + 70 = 130
___ + 700 = 1,300

5. 8 − ___ = 2
80 − ___ = 20
___ − 600 = 200

6. 13 + ___ = 21
130 + 80 = ___
___ + 800 = 2,100

7. 5 − 3 = ___
50 − 30 = ___
500 − 300 = ___

8. 7 − ___ = 4
70 − 30 = ___
7,000 − 3,000 = ___

9. 5 + 2 = ___
500 + 200 = ___
5,000, + 2,000 = ___

NS 3.0 Students solve problems involving addition, subtraction, multiplic ation, and division of whole numbers and understand the relationship among the operations.

Name ______________________ Lesson 2.3

Mental Math: Addition and Subtraction Patterns

Use mental math to complete the pattern.

1. ______ + 8 = 17
90 + ______ = 170
900 + 800 = ______
9,000 + 8,000 = ______

2. ______ − 4 = 8
120 − 40 = ______
1,200 − ______ = 800
12,000 − 4,000 = ______

3. ______ − 3 = 7
100 − ______ = 70
______ − 300 = 700
10,000 − 3,000 = ______

4. 7 + 9 = ______
70 + ______ = 160
700 + 900 = ______
______ + 9,000 = 16,000

5. 8 + ______ = 11
80 + ______ = 110
______ + 300 = 1,100
______ + 3,000 = 11,000

6. ______ − 5 = 9
140 − 50 = ______
1,400 − ______ = 900
______ − 5,000 = 9,000

Use mental math patterns to find the sum or difference.

7. 600 + 700 ______

8. 180 − 90 ______

9. 6,000 + 9,000 ______

10. 13,000 − 5,000 ______

11. 12,000 + 10,000 ______

12. 700 − 600 ______

13. 130,000 + 70,000 ______

14. 15,000 − 8,000 ______

Problem Solving and Test Prep

15. In 2001, there were 400 rabbits at the zoo. In 2002, there were 1,200 rabbits at the zoo. How many more rabbits were at the zoo in 2002 than 2001?

______

16. There are 600 pens in each box. How many pens are there in 2 boxes?

______

17. What number completes the sentence ■ + 3,000 = 12,000

A 90,000
B 9,000
C 8,000
D 900

18. There were 14,000 newspapers printed on Tuesday morning. By Tuesday afternoon, only 8,000 were sold. How many newspapers have not been sold yet?

______

Name ____________________ **Week 3**

Spiral Review

For 1–5, round each number to the place value of the underlined digit.

1. 1,$\underline{7}$94 ________
2. $\underline{4}$5,931 ________
3. 7$\underline{1}$3,702 ________
4. $\underline{3}$,920,703 ________
5. 9,$\underline{7}$79,999 ________

For 6–8, name each quadrilateral.

6. 

____

7. 

____

8. 

____

For 9–10, use the bag of tiles below.

9. Stephen is going to pull one tile from the bag. What are the possible outcomes?

____

____

10. Is it *likely* or *unlikely* that Stephen will pull a striped tile?

____

For 11–15, find the missing numbers.

11. $7 + \square = 9$
12. $\square + 3 = 12$
13. $4 + 2 = \square$
14. $5 + \square = 8$
15. $\square + 3 = 10$

Name ____________________

Lesson 2.4

Mental Math: Estimate Sums and Differences

When you **estimate,** you use numbers that are not as exact as the original numbers but are easier to use. To find a sum or difference that is close to the exact number, you can estimate. **Compatible numbers** are numbers that are easy to compute in your mind.

Use rounding to estimate.

Estimate the sum.

Round each number to the nearest thousands place.

Add the estimated numbers.

$$\begin{array}{r} 62{,}356 \\ +\ 37{,}922 \\ \hline \end{array} \rightarrow \begin{array}{r} 62{,}000 \\ +\ 38{,}000 \\ \hline 100{,}000 \end{array}$$

Estimate the difference.

Round each number to the nearest thousands place.

Subtract to find the estimated difference.

$$\begin{array}{r} 6{,}362 \\ -\ 1{,}714 \\ \end{array} \rightarrow \begin{array}{r} 6{,}000 \\ -\ 2{,}000 \\ \hline 4{,}000 \end{array}$$

Use compatible numbers to estimate.

Use compatible numbers to find the sum.

Find compatible numbers that would be easy to add mentally.

Add the compatible numbers.

$$\begin{array}{r} 1{,}749{,}206 \\ +\ 4{,}222{,}358 \\ \hline \end{array} \rightarrow \begin{array}{r} 1{,}750{,}000 \\ +\ 4{,}000{,}000 \\ \hline 5{,}750{,}000 \end{array}$$

Use compatible numbers to find the difference.

Find compatible numbers that would be easy to subtract mentally.

Subtract to find the difference.

$$\begin{array}{r} 88{,}796 \\ -\ 75{,}187 \\ \hline \end{array} \rightarrow \begin{array}{r} 90{,}000 \\ -\ 75{,}000 \\ \hline 15{,}000 \end{array}$$

Use rounding to estimate.

1. 78,402 + 91,113
2. 7,115 + 4,212
3. 54,823 − 36,911
4. 640 + 480
5. 7,980 + 1,341

Use compatible numbers to estimate.

6. 6,789 − 834
7. 32,116 + 71,930
8. 17,450 + 81,942
9. 22,001 − 4,663
10. 1,642 + 926

Name____

Lesson 2.4

Mental Math: Estimate Sums and Differences

Use rounding to estimate.

1. $\begin{array}{r} 6{,}356 \\ +1{,}675 \\ \hline \end{array}$

2. $\begin{array}{r} 8{,}267 \\ -2{,}761 \\ \hline \end{array}$

3. $\begin{array}{r} 38{,}707 \\ +28{,}392 \\ \hline \end{array}$

4. $\begin{array}{r} 75{,}428 \\ -19{,}577 \\ \hline \end{array}$

5. $\begin{array}{r} 187 \\ +519 \\ \hline \end{array}$

6. $\begin{array}{r} 6{,}489 \\ -1{,}807 \\ \hline \end{array}$

7. $\begin{array}{r} 24{,}655 \\ +51{,}683 \\ \hline \end{array}$

8. $\begin{array}{r} 61{,}075 \\ -29{,}732 \\ \hline \end{array}$

Use compatible numbers to estimate.

9. 5,432 − 652 ____________

10. 45,221 + 6,167 ____________

11. 392 + 47 + 89 ____________

Adjust the estimate to make it closer to the exact sum or difference.

12. 6,285 + 2,167

Estimate: 8,000

____

13. 42,819 − 11,786

Estimate: 30,000

____

14. 17,835 + 45,199

Estimate: 65,000

____

Problem Solving and Test Prep

15. In 2004, there were 398,521 visitors to the Rodeo. In 2006, there were 117,578 more visitors than in 2004. Estimate the total number of visitors to the Rodeo in 2004 and 2006.

____

16. Sara estimates the difference between 54,625 and 32,484. Her answer is 20,000. Give a closer estimate.

____

17. A plane flies 14,854 miles in one week. The next week, it flies 8,267 miles. Estimate the distance the plane flies in two weeks.

A 22,000 miles **C** 24,000 miles

B 23,000 miles **D** 25,000 miles

18. A train travels 7,824 miles the first month and travels 3,776 miles the next month. About how many more miles does the train travel in the first month than in the second month?

____

Name ____________________

Mental Math Strategies

There are strategies that can help you add or subtract mentally. The **Break Apart Strategy**, the **Friendly Number Strategy**, and the **Swapping Strategy** are three of the strategies you can use.

Break Apart Strategy	Friendly Number Strategy	Swapping Strategy
867 − 425	**43 − 27**	**145 + 213**
• Break apart the place values.	• Friendly numbers end with a zero.	• Swap digits with the same place value.
• Subtract each place value separately. 800 − 400 = 400 60 − 20 = 40 7 − 5 = 2	• Add to make one of your numbers end with zero. 27 + 3 = 30 30 is a friendly number.	• Add to make one of your numbers end with zero. 145 + 5 = 150
• Add the differences. 400 + 40 + 2 = 442	• Adjust the other number by adding the same amount. 43 + 3 = 46	• Adjust the other number by subtracting the same amount. 213 − 5 = 208
	• Subtract the new numbers. 46 − 30 = 16	• Add the new numbers. 150 + 208 = 358
So, 867 − 425 = 442.	So, 43 − 27 = 16.	So, 145 + 213 = 358.
You can use this method for both *addition* and *subtraction*.	You can only use this method for *subtraction*.	You can only use this method for *addition*.

Use a strategy to add or subtract mentally.

1. 714 + 224 ______ **2.** 322 + 138 ______ **3.** 83 − 61 ______ **4.** 150 + 350 ______ **5.** 293 − 168 ______

6. 428 − 364 ______ **7.** 69 + 81 ______ **8.** 654 − 270 ______ **9.** 36 − 22 ______ **10.** 187 + 250 ______

11. 330 − 201 ______ **12.** 92 + 63 ______ **13.** 46 − 19 ______ **14.** 96 + 81 ______ **15.** 568 − 233 ______

Name________________________________

Mental Math Strategies

Add or subtract mentally. Tell the strategy you used.

1. 73 + 15 ________
2. 87 − 48 ________
3. 57 + 91 ________
4. 152 − 68 ________
5. 542 + 148 ________
6. 515 − 151 ________
7. 799 − 231 ________
8. 387 + 73 ________
9. 945 − 425 ________
10. 452 + 339 ________
11. 396 + 265 ________
12. 594 − 496 ________

Problem Solving and Test Prep

13. Vicky has 32 baseball cards and 29 soccer cards. Use mental math to find how many cards Vicky has in all.

____

14. Kareem bowls 78 the first game and 52 the second game. Use mental math to find the difference of Kareem's scores.

____

15. Jason sells 27 tickets on Monday and 34 on Tuesday. He adds 3 to 27 to find the sum mentally. How should he adjust the sum to find the total?

A Add 3 to the sum.
B Add 4 to the sum.
C Subtract 3 from the sum.
D Subtract 4 from the sum.

16. Haley buys a baseball bat and glove that cost $25 and $42. She subtracts $2 from $42 to find the total mentally. How should Haley adjust the sum to find the total?

A Add $2 to the sum.
B Subtract $2 from the sum.
C Add $5 to the sum.
D Subtract $5 from the sum.

Name____

Lesson 2.6

Problem Solving Workshop Skill: Estimate or Exact Answer

To be an airline pilot, you must fly a total of at least 1,500 hours. Dan flew 827 hours last year and 582 hours this year. How many more hours must Dan fly to be an airline pilot?

Read to Understand

1. What are you asked to find?

____

Plan

2. Should you use an estimate or an exact answer?

____

Solve

3. Solve the problem in the space below. What operations did you use?

Solve:	Operations Used:
	____________
	____________

4. What is your answer? Write it in a complete sentence.

____

Check

5. How can you tell if your answer is correct?

____

Tell whether you should estimate or find an exact answer. Then solve.

6. Jane took a math test. She answered 134 items correctly in the first section and 113 items correctly in the second section. How many answers did Jane answer correctly in all?

____

7. Lance spent 180 days with Mr. Lee as a teacher and 176 days with Mrs. Mac as a teacher. About how many days did Lance spend in Mr. Lee's and Mrs. Mac's classes in all?

____

____

Name______________________________________ Lesson 2.6

Problem Solving Workshop Skill: Estimate or Exact Answer?

Problem Solving Skill Practice

Explain whether to estimate or find an exact answer. Then solve the problem.

1. A plane has 5 seating sections that can hold a total of 1,175 passengers. Today, the sections held 187, 210, 194, 115, and 208 passengers. Was the plane filled to capacity?

____

____

2. A small plane carries 130 gallons of fuel. It needs 120 gallons to fly a 45-mile trip. Does the pilot have enough fuel to make a 45-mile trip?

____

____

3. A movie theater has a total of 415 seats. There are 187 adults and 213 children seated in the theater. How many empty seats are there in the theater?

____

____

4. Bob drives 27 miles round trip each day for three days. Has Bob traveled more or less than 250 miles?

____

____

Mixed Applications

5. The movie theater sells 213 tickets on Monday, 187 tickets on Tuesday, and 98 tickets on Wednesday. Are there more, or less than 600 tickets sold for all three days?

____

6. The movie theater sells 209 tickets for "Canyon Trail" and 94 tickets for "A Light in the Sky". How many more tickets are sold at the theater for "Canyon Trail" than "A Light in the Sky"?

____

7. Sara sells 87 tickets for a school benefit. Josh sells 43 tickets. Marc sells 28 tickets. How many more tickets does Sara sell than Marc and Josh together?

____

8. A stamp album contains 126 stamps. Another album contains 67 stamps. Each album can hold up to 150 stamps. How many more stamps can both albums hold altogether?

____

Name________________________________ **Lesson 3.1**

Add and Subtract Through 5-Digit Numbers

There are different ways that you can add and subtract large numbers. One way is to round each number to the same place value to make an estimate. An estimate will help you know if your answer is reasonable.

Estimate. Then find the sum.

63,407 + 2,936 = ■

Estimate. Round to the nearest thousand. 63,000 + 3,000 = 66,000.

1 63,407 + 2,936 3	1 63,407 + 2,936 43	1 1 63,407 + 2,936 343	1 1 63,407 + 2,936 6,343	1 1 63,407 + 2,936 66,343
Add the ones. Regroup 10 ones.	Add the tens.	Add the hundreds. Regroup 10 hundreds.	Add the thousands.	Add the ten thousands.

Since 66,343 is close to the estimate of 66,000, the answer is reasonable.

Estimate. Then find the difference.

7,989 − 2,358 = ■

Estimate. Round to the nearest thousand. 8,000 − 2,000 = 6,000.

7,989 − 2,358 31	7,989 − 2,358 31	7,989 − 2,358 631	7,989 − 2,358 5,631
Subtract the ones.	Subtract the tens.	Subtract the hundreds.	Subtract the thousands.

Since 5,631 is close to the estimate of 6,000, the answer is reasonable.

Estimate. Then find the sum or difference.

1. 415 + 342

2. 22,314 + 13,986

3. 5,796 + 123

4. 1,427 + 427

5. 93,613 + 4,967

6. 7,435 − 251

7. 2,527 − 1,761

8. 71,376 − 33,149

9. 4,931 − 3,187

10. 65,168 − 44,291

Name ____________________

Add and Subtract Through 5-Digit Numbers

Estimate. Then find the sum or difference.

1. 414 + 727	**2.** 784 − 149	**3.** 5,305 + 848	**4.** 7,322 − 616
5. 2,673 + 4,548	**6.** 3,357 + 1,219	**7.** 8,452 − 2,621	**8.** 9,344 − 5,667
9. 4,955 + 978	**10.** 9,999 − 901	**11.** 7,593 + 1,475	**12.** 8,891 − 1,490
13. 13,069 + 1,956	**14.** 16,560 − 15,699	**15.** 71,948 − 51,052	**16.** 37,326 + 42,673

ALGEBRA Find the missing digit.

17. 9■8 + 247 = 1,175	**18.** 77,895 − 21,23■ = 56,661	**19.** ■,689 − 726 = 3,963	**20.** 61,357 + 29,7■6 = 91,113

Problem Solving and Test Prep

21. Jan drove 324 miles on Monday, then 483 miles on Tuesday. How many miles did Jan drive in all?

____

22. A baseball team scores 759 runs in a season. The next season the team scores 823 runs. How many runs are scored in all?

____

23. An airplane will fly a total of 4,080 miles this trip. The plane has flown 1,576 miles so far. How many more miles will the plane need to travel?

A 2,504 miles **C** 2,594 miles
B 2,514 miles **D** 5,656 miles

24. There are 35,873 soccer fans at the first game. There are 23,985 fans at the second game. How many more fans are at the first game?

____

Name ______________________________ **Week 4**

Spiral Review

For questions 1–4, compare using <, >, or =.

1. 5,327 ◯ 5,341

2.

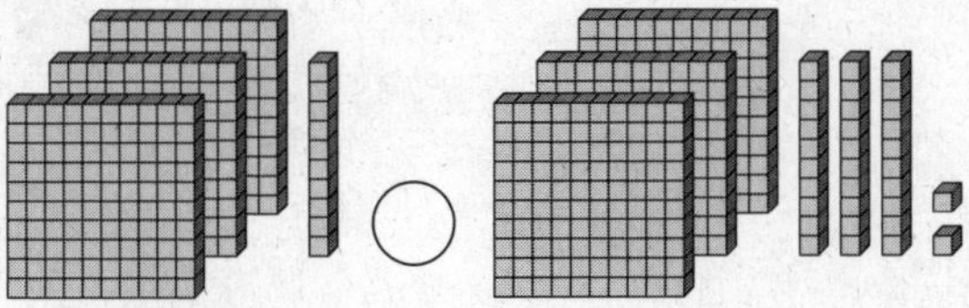

3.

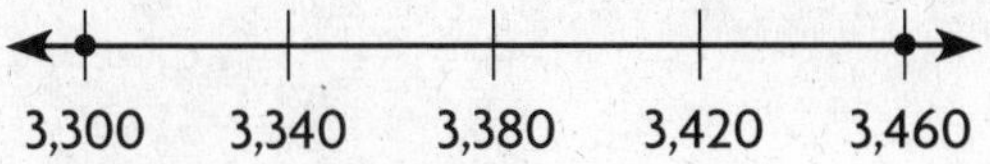

3,300 ◯ 3,460

4. 9,304 ◯ 4,039

For 5–7, name the solid figure that each object is shaped like.

5.

____

6.

____

7.

____

For 8–10, use the table below to tell whether each event is *likely, unlikely,* or *impossible*.

Warren's Bag of Marbles	
Solid	
Spotted	
Striped	

8. Warren will pull out a striped marble.

____

9. Warren will pull out a yellow marble.

____

10. Warren will pull out a solid marble.

____

For 11–12, write an expression. Then write an equation to solve.

11. Eliza and 8 of her friends went to the movies. Each paid $7 for a movie ticket. How much did Eliza and her friends pay in all?

____

____

____

12. There were 22 children and 34 adults at the community swimming pool. How many people were at the pool in all?

____

____

____

Name ____________________

Subtract Across Zeros

When the bottom digit is greater than the top digit, you need to regroup to subtract. Regroup from the next greater place value to the left with a digit that is not zero.

Estimate. Then find the difference.
3,000 − 1,076

Estimate: 3,000 − 1,000 = 2,000.

First, line up the digits in the ones place, tens place, hundreds place, and thousands place of the numbers you are subtracting.

Start with the ones place.
Regroup from the left.

The greater digit to the left that is not zero is the 3 in the thousands place.

Regroup 3 thousand as 2 thousand and 10 hundreds.	Regroup 10 hundreds as 9 hundreds and 10 tens	Regroup 10 tens as 9 tens and 10 ones.	Subtract.
2 10 ~~3~~,~~0~~00 − 1,076	9 2 ~~10~~ 10 ~~3~~,~~0~~~~0~~0 − 1,076	9 9 2 ~~10~~ ~~10~~ 10 ~~3~~,~~0~~~~0~~~~0~~ − 1,076	9 9 2 ~~10~~ ~~10~~ 10 ~~3~~,~~0~~~~0~~~~0~~ − 1,076 1,924

So, 3,000 − 1,076 = 1,924.
1,924 is close to the estimate of 2,000, so it is reasonable.

Estimate. Then find the difference.

1. 6,000 − 4,321

2. 2,300 − 1,482

3. 9,000 − 6,814

4. 2,400 − 1,256

5. 7,000 − 5,988

6. 5,500 − 4,432

7. 3,000 − 2,322

8. 8,200 − 5,756

9. 6,000 − 5,987

10. 2,000 − 1,776

Name ______________________

Subtract Across Zeros

Estimate. Then find the difference.

1. 3,078 −678

2. 760 −194

3. 6,004 −452

4. 7,030 −4,265

5. 8,056 −2,109

6. 9,000 −2,708

7. 4,890 −1,405

8. 6,902 −3,440

9. 670 – 413

10. 4,700 – 876

11. 5,030 – 2,125

____

Choose two numbers from the box to make each difference.

4,200	4,000	3,020
3,402	424	

13. 3,776 ________

14. 1,180 ________

15. 2,596 ________

15. 598 ________

Problem Solving and Test Prep

17. One of the largest volcanic eruptions occurred in 1883 on the Indonesian Island of Krakatoa. How many years before 2006 had this eruption occured?

____

18. Jessie estimates the distance from New York to San Diego to be 3,000 miles. The actual distance is 2,755 miles. What is the difference between Jessie's estimate and the actual distance?

____

19. Helena starts a trip with 4,345 miles on her car. She finishes the trip with 8,050 miles on her car. How many miles did Helena travel on her trip?

A 12,395

B 4,705

C 3,805

D 3,705

20. A mountain peak reaches 3,400 feet in elevation. A mountain climber has climbed 1,987 feet so far. How many more feet does the climber need to go before reaching the top of the peak?

____

Name____ **Lesson 3.3**

Add and Subtract Greater Numbers

When you add or subract large numbers, choose the method that works best for the problem you are going to solve. Think about the kind of numbers in the problem, then choose a method.

Pencil and Paper works best when the numbers need to be regrouped and the answer has to be exact.

- Subtract each place value starting with the ones.
- Regroup when needed.

```
                 7 13  9 9 10
  84,000   →     8 4, 0 0 0
- 39,075   →   - 3 9, 0 7 5
               -------------
                 4 4, 9 2 5
```

Mental Math is a good method for working with numbers that end in zero or numbers that don't need to be regrouped.

- Add using mental math.

```
  280,000   →     280,000
+ 300,200   →   + 300,200
                ---------
                  580,200
```

Find the sum or difference. Write the method you used.

1. 378,452 − 193,511

2. 1,420,300 + 1,678,500

3. 620,000 − 419,500

4. 3,721,682 + 6,161,248

5. 7,573,150 + 1,200,000

6. 621,899 + 213,655

7. 3,470,360 + 2,681,000

8. 701,000 + 107,000

9. 1,321 + 4,982

10. 17,672 − 13,827

11. 932,887 + 214,203

12. 2,300 − 2,200

Name______________________________ Lesson 3.3

Add and Subtract Greater Numbers

Find the sum or difference. Write the method you used.

1. 56,684 + 37,925 ________

2. 45,002 − 8,000 ________

3. 369,021 + 488,627 ________

4. 90,451 − 89,693 ________

5. 4,500 + 1,001 ________

6. 56,634 + 9,378 ________

7. 359,000 − 109,000 ________

8. 411,800 − 288,236 ________

9. 30,550 − 10,220 ________

10. 621,100 + 123,300 ________

11. 41,067 − 13,968 ________

ALGEBRA Find the missing digit.

12. 45,■22 + 12,534 = 57,856

13. 3■2,400 − 141,200 = 181,200

14. 17,648 + ■4,537 = 72,185

15. 630,489 − 241,■25 = 388,764

Problem Solving and Test Prep

16. Fast Fact Jupiter's radius at its equator is 71,492 km. Earth's radius at its equator is 6,378 km. How many more km is the radius of Jupiter than the radius of Earth?

____

17. Miguel scores 208,700 points in a video game. Sera scores 290,550 points. How many more points does Sera score than Miguel?

____

18. A plane travels 89,102 miles in a month. The next month it travels 106,448 miles. How many miles does the plane travel in these two months?

A 17,346 miles **C** 195,550 miles

B 185,540 miles **D** 295,550 miles

19. In one season, 187,197 fans attend a minor league's baseball games. The next season, 216,044 fans attend. How many total fans attend the games for these two seasons?

____

Name ____________________

Problem Solving Workshop Skill: Too Much/Too Little Information

Alameda County has 3,628 miles of roads. There are 2,945 miles of city roads. There are also county roads and state highways. How many miles of county roads are there?

1. What are you asked to find?

2. What information do you know?

3. Is there too much, or too little, information to solve the problem? Why?

4. What is the answer to the question? Write your answer in a complete sentence.

Tell if you have *too much* or *too little* information. Solve the problem, if possible.

5. Jake has 325 baseball trading cards, 468 football trading cards, and 229 soccer trading cards. What is the total number of soccer and baseball trading cards that Jake has?

6. **Challenge** Jenny drove 1,255 miles last week. She drove 187 miles on Monday. On Tuesday, she drove 53 more miles than on Monday. On Wednesday, she drove 26 more miles than on Tuesday. How many more miles did Jenny drive on Monday and Tuesday together than on Wednesday and Thursday together?

NS 3.0 Students solve problems involving addition, subtraction, multiplication, and division of whole numbers and understand the relationships among the operations.

Name______________________________________ **Lesson 3.4**

Problem Solving Workshop Skill: Too Much/Too Little Information

Problem Solving Skill Practice

Tell if you have *too much* or *too little* information. Identify the extra or missing information. Then solve the problem, if possible.

1. Juan takes a road trip for three days. On Day 1, he drives 278 miles On day 2, he drives 367 miles. On day 3, he drives 316 miles. Each day he takes one hour out of driving time to eat lunch. How many miles in all does Juan drive?

____

2. There are a total of 720 tickets available for a school concert. Dom sells 93 tickets and Oscar sells 123 tickets. How many tickets do Dom and Oscar sell in all?

____

3. Ms. Jackson buys two packages of grass seed for $14.95 each and a new hose for $16.79. How much does Ms. Jackson spend on the grass seed?

____

4. Betty's class sells a total of 516 red and blue sports caps. The caps cost $4.79 each. How many more red caps were sold than blue caps?

____

Mixed Applications

USE DATA For 5–6, use the table.

5. Jeff wants to drive round trip from San Francisco to Los Angeles. How many miles will he travel in all?

____

6. Amelia drives 200 miles each day. Can she make the trip to Seattle from San Francisco in 4 days?

____

One Way Distances from San Francisco

City	Number of Miles
to Boise, ID	655
to Los Angeles, CA	390
to Salt Lake, UT	750
to Seattle, WA	825

7. A city park has 2,574 visitors in one day. The zoo has 3,078 visitors on the same day. Write a number sentence to show the total number of visitors to both the city park and the zoo.

____

8. Hannah bowls 3 games. Her total score is 188. If she bowled a 73 the first game and a 47 the second game, what was Hannah's score in the last game?

____

Name

Addition Properties

Commutative Property

The Commutative Property states that you can add numbers in any order and the sum will be the same. For example, $2 + 3 = 5$, and $3 + 2 = 5$. The order in which you add numbers does not change the sum.

Find the missing number.

47 + ■ = 56 + 47

56 + 47 uses the same numbers as 47 + ■, but in a different order.

47 + 56 = 56 + 47

So, ■ = 56.

Associative Property

The Associative Property states that the way addends are grouped does not change the sum. For example, $(4 + 2) + 5$ and $4 + (2 + 5)$ both equal 11. The way you group numbers does not change the sum.

Find the missing number

■ + (31 + 18) = (24 + 31) + 18

(24 + 31) + 18 uses the same numbers as ■ + (31 + 18), except grouped together differently.

24 + (31 + 18) = (24 + 31) + 18

So, ■ = 24.

Identity Property

The Identity Property states that when you add zero to any number, the sum is that number. For example, $45 + 0 = 45$.

Find the missing number

62 + ■ = 62

Adding zero to 62 does not change 62.
So, 62 + 0 = 62.

Find the missing number. Tell which property you used.

1. 39 + ■ = 39 ______
2. 22 + ■ = 13 + 22 ______
3. (7 + ■) + 8 = 7 + (2 + 8) ______
4. ■ + 40 = 40 + 22 ______
5. ■ + 88 = 88 ______
6. 3 + (14 + 6) = (3 + ■) + 6 ______

Name____________________

Addition Properties

Find the missing number. Tell which property you used.

1. ■ + 0 = 0 + 23
2. 15 + 5 = ■ + 15
3. 12 + (2 + 7) = (■ + 2) + 7
4. ■ + 7 = 7 + 36
5. ■ + 45 = 45 + 0
6. (22 + ■) + 11= 22 + (44 + 11)

Change the order or group the addends so that you can add mentally. Find the sum. Tell which property you used.

7. 120 + 37 + 280 ____________
8. 25 + 25 + 30 ____________
9. 60 + 82 + 40 ____________
10. 28 + 21 + 32 + 19 ____________
11. 66 + 27 + 44 ____________
12. 133 + 25 + 247 ____________
13. 45 + 22 + 25 ____________
14. 61 + 57 + 39 + 23 ____________

Problem Solving and Test Prep

USE DATA For 15–16, use the table.

Sam's Marble Collection	
Type	**Number**
Blue Chinese Checkers	32
Cat's Eyes	81
Speckled Spots	18
Shooter Stripes	59

15. Use the Associative Property to find the total number in Sam's collection.

____

16. Sam buys another 15 Shooter stripes. How many marbles are in Sam's collection now?

____

17. Which shows the Identity Property of Addition?

A 16 + 0 = 16

B 12 + 1 = 13

C 29 + 29 = 28

D 1 + 1 = 2

18. Which shows the Commutative Property of Addition?

A 11 + 9 = 10 + 10

B 0 + 7 = 0

C 20 + 20 = 40

D 5 + 7 = 7 + 5

Name ______________________ **Week 5**

Spiral Review

For 1–4, estimate. Then the find the sum or difference.

1. 2,345 + 1,179

____

2. 4,845 − 2,954

____

3. 9,678 − 928

____

4. 6,429 + 3,218

____

Erika asked her friends how many times they had been to the zoo. Her results are shown in the line plot below.

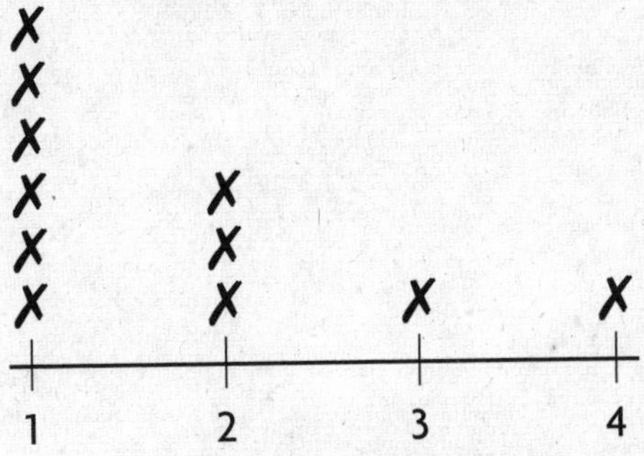

Visits to the Zoo

9. What is the range of the data shown in the line plot? ________

10. How many times have most of Erika's friends been to the zoo? How can you tell?

____

____

For 5–8, describe the lines. Write *intersecting* or *parallel*.

5. 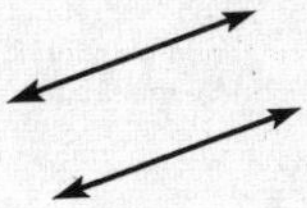

6. 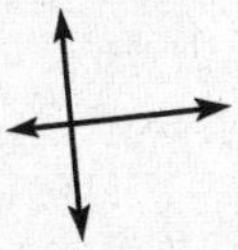

______________ ______________

7. 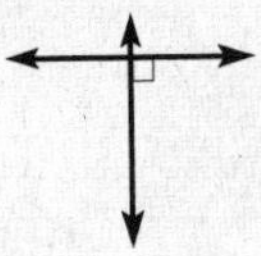

8. 

______________ ______________

For 11–14, identify each property of addition. Write *Commutative, Associative,* or *Identity*.

11. 13 + 4 = 4 + 13

____

12. 79 + 0 = 79

____

13. 7 + (9 + 8) = (7 + 9) + 8

____

14. 23 + 19 = 19 + 23

____

Name________________________

Write and Evaluate Expressions

An expression is a part of a number sentence that has numbers and signs such as "+", "−", and "×," but does not have an "=" sign. You can find the value of an expression by adding, subtracting, multiplying, and/or dividing. When there are parentheses in an expression, you need to do the operation in parentheses first.

Tell what you do first. Then find the value of the expression.

15 − (5 + 2)

Always do the operation in parentheses first. Add 5 + 2. — 15 − (5 + 2)

Subtract this number from 15 to find the value of the expression. — 15 − 7

So, 15 − (5 + 2) is 8. — 8

Tell what you do first. Then find the value of the expression.

12 − (8 − 3)

Always do the operation in parentheses first. Subtract 8 − 3. — 12 − (8 − 3)

Subtract this answer from 12 to find the value of the expression. — 12 − 5

So, 12 − (8 − 3) is 7. — 7

Tell what you do first. Then find the value of the expression.

1. 9 + (6 + 7) ________

2. 8 + (9 − 2) ________

3. 12 − (7 + 1) ________

4. 13 − (2 + 4) ________

5. (10 − 3) + 7 ________

6. (9 + 3) + 4 ________

7. 11 − (8 − 4) ________

8. 6 + (7 − 3) ________

9. (4 − 2) − 1 ________

Write and Evaluate Expressions

Tell what you do first. Then find the value of each expression.

1. 12 − (4 + 3) ________ **2.** 5 + (15 − 3) ________ **3.** (17 − 3) + 5 ________ **4.** 5 + (18 − 2) ________

5. (18 + 22) − 15 ________ **6.** (31 − 16) − 8 ________ **7.** 9 + (25 − 9) ________ **8.** (31 + 5) − 21 ________

Place the parentheses so the expression has a value of 7.

9. 12 − 10 + 5 ________ **10.** 5 + 9 − 7 ________ **11.** 16 − 10 + 1 ________ **12.** 40 − 36 + 3 ________

13. 10 + 6 − 9 ________ **14.** 4 + 4 − 1 ________ **15.** 12 − 6 + 1 ________ **16.** 13 − 9 + 3 ________

Problem Solving and Test Prep

17. Will buys 17 toy racing cars. He gives 7 to Paul and 6 to Bill. How many cars does Will have left? Write the expression and solve.

____

18. There are 12 fish in the class tank. Asa and Troy each took 3. Mrs. Hampton then buys 9 more fish to put in the tank. How many fish are in the tank now?

____

19. What is the value of the following expression?
(17 − 12) + 4

A 1
B 2
C 9
D 11

20. What is the value of the following expression?
4 + (15 − 9)

A 10
B 5
C 2
D 6

Name ____________________

Expressions with Variables

You can use a variable, or a letter or symbol to represent any number that you do not know. Then you can write an expression.

Choose a variable. Write an expression. Tell what the variable represents.

Pete hangs 3 posters on the wall. Then he hangs some more posters on the wall.

Write a variable to represent a number in the problem. — Let p represent the posters Peter hangs.

Pete started with 3 posters and added some more posters to his wall.

Write the expression. — $3 + p$

So, the expression is $3 + p$, and the variable p represents the other posters Pete hangs on the wall.

Jenna found some shells on the beach. She gave 4 of them to Lance.

Choose a variable to represent the unknown number in the problem. — Let s represent the shells Jenna found.

Jenna gave away 4 shells, so 4 were subtracted from her total.

Write the expression. — $s - 4$

So, the expression is $s - 4$, and the variable s represents the shells Jenna found.

Choose a variable. Write an expression. Tell what the variable represents.

1. Mara has some pens. She gives 6 to Sue.

   ____

2. Pat has some rolls. He finds 5 more.

   ____

3. Barry has some envelopes. He uses 4 of them.

   ____

4. Dave drew 3 pictures. Then he drew some more.

   ____

5. Pa had some coins. He gave 7 away.

   ____

6. Greg has some hats. He gets 3 more.

   ____

AF 1.0 Students use and interpret variables, mathematical symbols, and properties to write and simplify expressions and sentences.

Name____________________

Expressions with Variables

Choose a variable. Write an expression. Tell what the variable represents.

1. Sara had some cards. She gave away 5 of them.

____

2. Raymondo had 9 stickers and bought some more.

____

3. Tan added $15 to his bank account.

____

4. Gee gave away some of her 20 pins.

____

Find the value of each expression if $a = 3$ and $b = 8$.

5. $a + 7$ ________

6. $17 - b$ ________

7. $(b - 3) + 18$ ________

8. $(a + 9) - 5$ ________

9. $a + (b - 1)$ ________

10. $b + (a + 15)$ ________

11. $25 - (8 - a)$ ________

12. $(b - 6) + 14$ ________

Problem Solving and Test Prep

USE DATA For 13–14, use the table.

13. Write an expression that tells how many dolls Lisa will have in all if she gets some miniature dolls.

____

Lisa's Doll Collection	
Type Doll	**Number**
Baby	15
Fashion	10
Foreign	8
Rag	17

14. Lisa gave some of her fashion dolls to a charity. Write an expression that tells how many dolls total Lisa has left.

____

15. What is the value of the expression below if $x = 9$?

$(6 + x) - 4$

A 14
B 13
C 12
D 11

16. What is the value of the expression below if $p = 7$?

$(p - 3) + 4$

A 0
B 9
C 10
D 8

ame______________________________________ Lesson 4.4

Addition and Subtraction Equations

hoose a variable for the unknown. Write an equation for each situation. ell what the variable represents.

1. Rickie has 15 model cars. Some are red and 8 are blue.

____

2. Wendy had \$12. Her mother gave her some more so she now has \$17.

____

olve the equation.

3. $19 - 4 = n$ 4. $6 + \blacksquare = 19$ 5. $r - 12 = 21$ 6. $t + 14 = 31$

$n =$ ______ $\blacksquare =$ ______ $r =$ ______ $t =$ ______

rite words to match the equation.

. $b + 5 = 12$

____

____

____

8. $a - 9 = 2$

____

____

____

. $16 - w = 4$

____

____

____

10. $y + 7 = 29$

____

____

____

oblem Solving and Test Prep

8 hearing dogs graduated in February, 5 in May, and 9 in November. Write and solve an equation that tells how many hearing dogs graduated in all.

____

12. 13 dogs graduated in May. There were 5 hearing dogs, 4 service dogs and some tracking dogs. Write an equation that shows the total number of dogs that graduated in May.

____

Jed watched 10 minutes of previews and a 50-minute dog movie. Which equation tells the total time Jed was in the theater?

A $10 + 50 = t$ **C** $t - 10 = 50$

B $50 - t = 10$ **D** $t + 10 = 50$

14. Haley's favorite picture book is 27 pages. 11 of the pages have pictures of dogs. The rest have pictures of birds. Which equation can be used to find how many pages have birds?

A $27 + 11 = b$ **C** $b - 11 = 27$

B $11 + 27 = b$ **D** $b + 11 = 27$

Name____ Lesso

Addition and Subtraction Equations

An equation is a number sentence stating that two amounts are equal.

Choose a variable for the unknown. Write an equation. Tell what the variable represents.

Emil has 18 stamps. After he uses some stamps, he has 12 stamps left.

Choose a variable for the missing amount. The missing amount is the number of stamps Emil used. — Let s represent the stamps Emil used.

Write a subtraction equation using the variable. — $18 - s = 12$

So, the subtraction equation is $18 - s = 12$, and s represents the number of stamps Emil used.

Solve the equation.

$13 - n = 4$

Use mental math to find which number n represents. Ask yourself, "13 minus what number equals 4?" — $13 - n = 4$

Replace n with 9 in the equation. — $13 - 9 = 4$

The answer is correct. — $4 = 4$

So, $n = 9$.

Choose a variable for the unknown. Write an equation for each. Tell what the variable represents.

1. Tim has 16 socks. After he loses some of his socks, he has 10 socks left.

   ____

2. There are 30 bowls on the sl There are some white bowls 12 green bowls.

   ____

Solve the equation.

3. $8 + \blacksquare = 20$ ________
4. $n - 9 = 8$ ________
5. $4 + k = 12$ ________
6. $y + 5 = 13$ ________
7. $6 + \blacksquare = 12$ ________
8. $11 - \blacksquare = 8$ ________

Name____

Adds Equals to Equals

Both sides of an equation have the same value. For example, in the equation $3 + 5 = 2 + 6$, both sides of the equation have a value of 8. When you add the same amount to both sides, the equation will still be equal.

Tell whether the values on both sides of the equation are equal. Write *yes* or *no*. Explain your answer.

Find the value of each side.

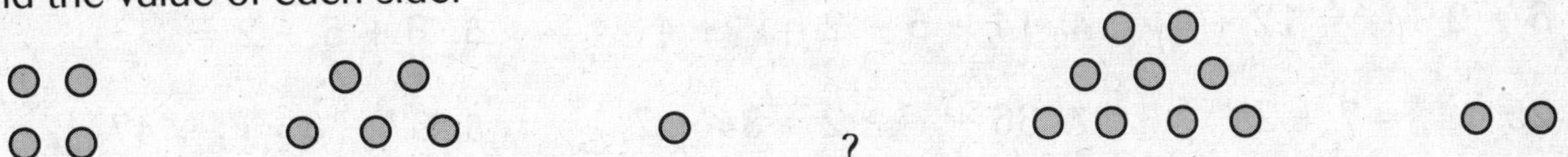

$4 + 5 + 1 \stackrel{?}{=} 9 + 2$

Left side: $4 + 5 + 1 = 10$ Right side: $9 + 2 = 11$

The values on both sides are not equal.

So, *no*, the sides of the equation are not equal.

Complete to make the equation true.

(9 – ■) + 1 = 6 + 1

Find the value of the right side of the equation.

Think: What number should go in the box so the left side of the equation also equals 7.

So, the number in the box should be 3.

Tell whether the values on both sides of the equation are equal. Write *yes* or *no*. Explain your answer.

1. 
$2 + 3 + 1 \stackrel{?}{=} 3 + 3$

____

2. 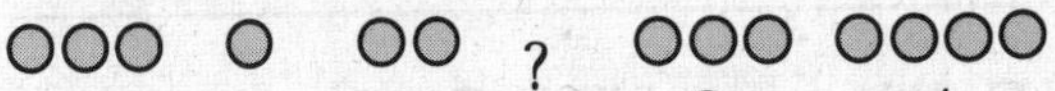
$3 + 1 + 2 \stackrel{?}{=} 3 + 4$

____

Complete to make the equation true.

3. 14 + 7 = ■ + 7

4. 6 + 8 – ■ = 19 – 8

5. 13 – 3 = ■ – 2

6. 8 + 2 = ■ + 7

7. 6 + ■ = 12 + 0

8. 3 + 4 – ■ = 2 + 3

Name______________________________ **Lesson 4.5**

Add Equals to Equals

Tell whether the values on both sides of the equation are equal. Write *yes* or *no*.

1. $4 + 6 + 1 \stackrel{?}{=} 9 + 2$

2. $2 + 4 + 1 \stackrel{?}{=} 5 + 2 + 1$

Complete to make the equation true.

3. 6 + 2 + ☐ = 12 + 4

4. 14 − 5 − 2 = ☐ + 4

5. 8 + 5 − 2 = ☐

6. 25 + ☐ = 7 + 25

7. 36 + ☐ − 2 = 34 + 7

8. 67 − 8 = ☐ + 47

9. 13 + 5 + 9 = 45 − ☐

10. 42 − 24 − ☐ = 12 + 4 − 9

11. 10 + 5 + 15 = 46 − ☐

Add to or subtract from both sides of the equation. Find the new value.

12. Add 17.
23 − 5 = 18

13. Subtract 11.
32 + 12 = 44

14. Add 9.
16 − 12 = 4

Problem Solving and Test Prep

15. Mary has 7 roses. Sae has 9 roses. If each girl adds 3 roses to her bunch, how many more roses does Mary need to add to have the same number of roses as Sae?

16. Mike has 15 packets of seeds and Jamal has 8. Mike gives 3 packets to Jamal. How many packets of seeds must Jamal buy if he wants to have as many packets as Mike?

17. Deb has 3 goldfish. Dan has 2 goldfish and one beta fish. They each get one zebra fish. Write an equation that shows the number of fish each person has. Is the equation true? Explain.

18. The letters *x* and *y* stand for numbers. If $x = y + 1$, which statement is true?

A $x + 4 = y + 4$

B $x - 1 = y$

C $x - 1 = y + 2$

D $x - 1 = y + 3$

Spiral Review

For 1–2, write a fraction in numbers and in words that names the shaded part.

1. 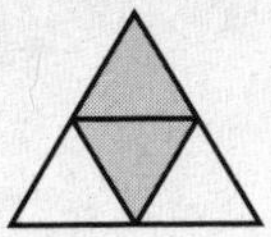

____

____

2. 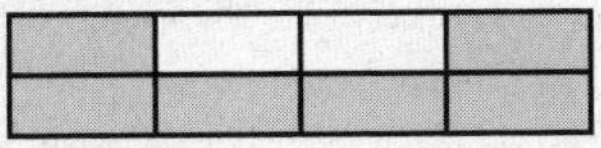

____

____

For 3–5, name the solid figure. Then tell how many faces.

3. 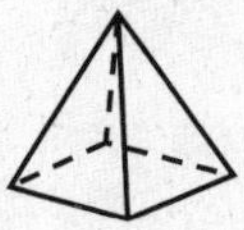

____

____

4. 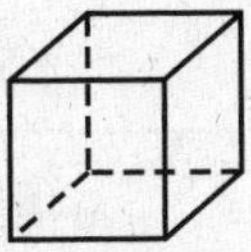

____

____

5. 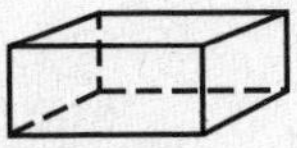

____

____

For 6–8, tell whether each event is *certain*, *likely*, *unlikely* or *impossible*.

Mandy has a bag filled with 25 tiles. There are 12 blue, 2 green, 4 yellow, and 7 red tiles.

6. pulling a blue tile ______________

7. pulling a green tile ______________

8. pulling an orange tile ______________

For 9–12, tell which operation to do first. Then find the value of each expression.

9. $(9 - 8) + 19$

____

____

10. $3 + (23 + 13)$

____

____

11. $(34 - 6) - 12$

____

____

12. $49 - (37 + 12)$

____

____

Name____ **Lesson 4.6**

Problem Solving Workshop Strategy: Work Backward

Many volunteer teams must patrol and clean the lion preserve. Twelve teams leave the preserve on patrol. Seven teams arrive to clean. There are 23 teams at the preserve now. How many volunteer teams were there originally?

Read to Understand

1. What are you asked to find?

____

Plan

2. How can you use the work backward strategy to solve this problem?

____

Solve

3. Solve the problem. Show your work below.

4. How many volunteer teams were there originally? ____________

Check

5. Is there another strategy you could use to solve the problem?

____

Work backwards to solve.

6. Pete got home at 4:30 P.M. It took him 20 minutes to walk from school to the library. He was at the library for 1 hour. It took Pete 5 minutes to walk from the library to his home. At what time did Pete leave school?

____

7. There were some cars in a parking lot in the morning. Later, 15 more cars arrived. Then 9 cars left. There are now 60 cars in the lot. How many cars were originally in the lot?

____

Name____

Lesson 4.6

Problem Solving Workshop Strategy: Work Backward

Problem Solving Strategy Practice

Work backward to solve.

1. Leon arrived at the preserve at 11:00 A.M. He began the morning by taking 45 minutes to feed his pets at home and driving 2 hours to get to the preserve. What time did Leon begin?

____

2. Kit read a 25-page book about lions. Seven pages were about hunts, 15 pages about habitat, and the rest were about prides. How many pages were about prides?

____

3. Twelve lions in the pride did not go on a hunt. When more lions returned from the hunt, there were 21. How many lions were on the hunt?

____

4. Polly ate lunch and then took 15 minutes to walk to Cher's house. They rode bikes for 35 minutes and then studied for 20 minutes. If they finished at 2:30, when did Polly finish lunch?

____

Mixed Strategy Practice

5. Five prides were sent from the zoo to a preserve. Two prides were returned. Now there are 17 prides at the zoo. How many prides were at the zoo before the 5 were sent away?

____

6. Red, blue, green, and brown teams lined up for their assignments. The brown team was ahead of the red team. The blue team was not last. The green team was first. Which team was last?

____

7. **USE DATA** Use the information in the table below to draw a bar graph.

Preserve Lion Population

Age	Number
Cubs	18
Adolescents	14
Mature	2
Older	7

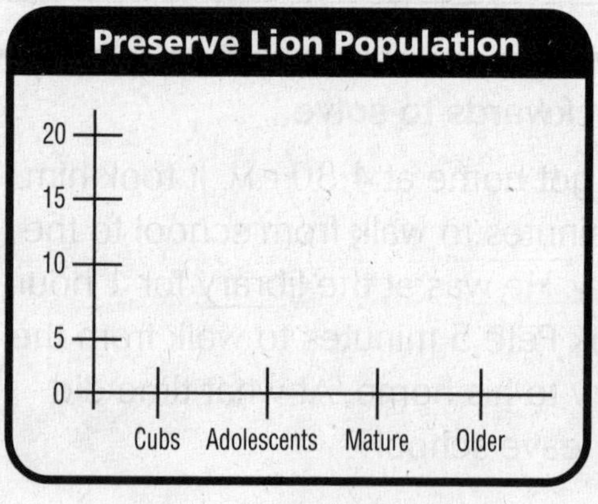

Name____

Patterns: Find a Rule

An **ordered pair** is two numbers written in parentheses. For example, (3,7) is an ordered pair. The two numbers are related by a rule. In this example, 3 and 7 are related by the rule "add 4," because when you add 4 to 3, you get 7. Another ordered pair related by "add 4" is (1,5).

Find a rule. Write your rule as an equation.
Use the equation to extend your pattern.

Input	a	12	25	31	43	59
Output	b	20	33	39	■	■

- Find a relationship between each input number and output number.

(12,20) $12 + 8 = 20$ (25,33) $25 + 8 = 33$ (31,39) $31 + 8 = 39$

- The rule is add 8 to each input number to get the output number. Write the rule as an equation.

$a + 8 = b$

- Use the rule to complete the table.

Input	a	12	25	31	43	59
Output	b	20	33	39	51	67

(43,■) $43 + 8 = 51$ (43,51)

(59,■) $59 + 8 = 67$ (59,67)

So, the rule is add 8 to a, or $a + 8 = b$. The ordered pairs are (43,51) and (59,67).

Find a rule. Write your rule as an equation.
Use the equation to extend your pattern.

1.

Input	a	62	58	47	■	■
Output	b	57	53	42	31	24

____

2.

Input	a	31	46	59	73	90
Output	b	42	57	70	■	■

____

3.

Input	a	18	24	57	■	■
Output	b	9	15	48	62	84

____

4.

Input	a	14	17	22	34	56
Output	b	4	7	12	■	■

____

Name____

Patterns: Find a Rule

Find a rule. Write your rule as an equation. Use the equation to extend your pattern.

1.

Input	*f*	10	15	20	25	30
Output	*g*	5	10	15		

2.

Input	*c*	88	86	84	82	80
Output	*d*	66	64	62		

3.

Input	*s*	17	13	9		
Output	*t*	70	66	62	65	68

4.

Input	*x*	15	14	13	12	11
Output	*y*	30	29	28		

Use the rule and the equation to make an input/output table.

5. Add 7 to *m*.
$m + 7 = n$

Input	*m*				
Output	*n*				

6. Subtract 14 from *a*.
$a - 14 = b$

Input	*a*				
Output	*b*				

Problem Solving and Test Prep

USE DATA For 7–8, use the input/output table.

7. A figure is made of a row of squares. One square has a perimeter of 4. Two squares has a perimeter of 6, and so on. Finish the input-output table to show the pattern.

Input	*s*	1	2	3	4	5
Output	*p*	4	6			

8. What will be the perimeter of 10 squares in a row?

9. Which equation describes the data in the table?

Input	*c*	0	2	3	4
Output	*d*	13	15	16	17

A $d + 13 = c$

B $c + 13 = d$

C $c - 13 = d$

D $d - 13 = c$

10. What is the rule for the table?

Input	*g*	1	3	5	7
Output	*h*	5	7	9	11

A add 5 to *g*

B subtract 5 from *g*

C add 4 to *g*

D subtract 4 from *g*

Name____ **Lesson 5.1**

Algebra: Relate Operations

When you add or **multiply**, you join equal-size groups.
When you subtract or **divide**, you separate into equal-size groups or find out how many are in each group.

Write the related multiplication sentence.
Draw a picture that shows the sentence.

$3 + 3 + 3 + 3 + 3$

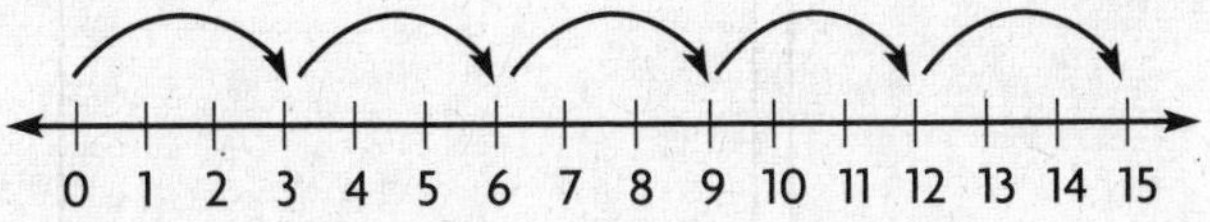

Start with 0 on the number line. Skip count by 3s five times. Stop at 15.

5 groups of 3 is the same as 5×3.

So, the related multiplication sentence is $5 \times 3 = 15$.

Write the related division sentence.
Draw a picture that shows the sentence.

$10 - 2 - 2 - 2 - 2 - 2 = 0$

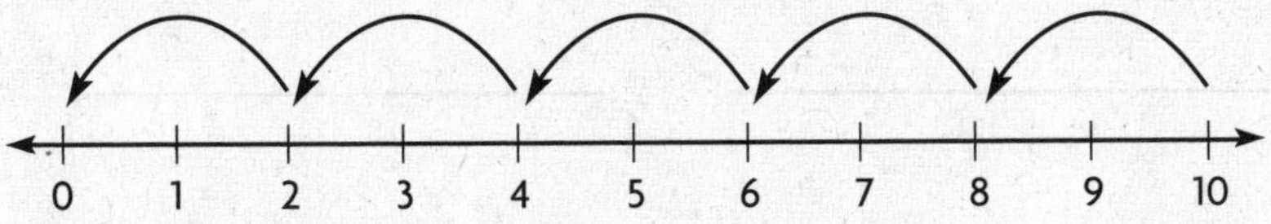

Start with 10 on the number line. Subtract 2 five times. Stop at 0.

A group of 10 divided into equal groups of 2 is the same as $10 \div 2$.

So, the related division sentence is $10 \div 2 = 5$.

Write the related multiplication or division sentence.
Draw a picture that shows the sentence.

1. $4 + 4 + 4 + 4 = 16$

____

2. $24 - 3 - 3 - 3 - 3 - 3 - 3 - 3 - 3$

____

3. $5 + 5 + 5 + 5 = 20$

____

4. $18 - 6 - 6 - 6 = 0$

____

Name____________________

Algebra: Relate Operations

Write the related multiplication or division sentence. Draw a picture that shows the sentence.

1. $20 - 4 - 4 - 4 - 4 - 4$

2. $5 + 5 + 5 = 15$

3. $6 - 2 - 2 - 2$

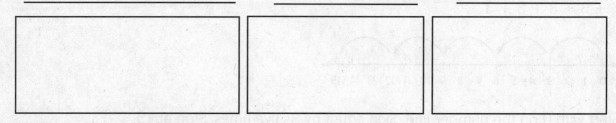

For 4–6, tell whether the number sentence is *true* or *false*. If false, explain how you know.

4. $5 + 5 + 5 + 5 \stackrel{?}{=} 4 \times 5$

5. $3 \times 4 \stackrel{?}{=} 4 + 4 + 4 + 4$

6. $2 \times 7 \stackrel{?}{=} 7 + 7 + 7$

Problem Solving and Test Prep

7. A class of 21 students will go on 3 different rides at the fair. The same number of students will go on each ride. How many students will go on each ride?

8. Jake plays 7 different games at the fair. He plays each game 2 times. How many games does Jake play in all?

9. Which of these is another way to write $36 - 9 - 9 - 9 - 9 = 0$?

A $4 \times 9 = 36$
B $36 \div 4 = 9$
C $9 \times 4 = 36$
D $36 \div 9 = 4$

10. Which of these is another way to write $8 + 8 + 8 + 8 + 8 = 40$?

A $5 \times 8 = 40$
B $8 \times 5 = 40$
C $40 \div 8 = 5$
D $40 \div 5 = 8$

Name____ **Lesson 5.2**

Algebra: Relate Multiplication and Division

Multiplication and division are opposite operations, or **inverse operations**. A **fact family** is a set of multiplication and division sentences that use the same numbers.

Find the fact family for the set of 2, 3, 6.

$2 \times 3 = 6$

2 rows of 3 triangles each.
So, $2 \times 3 = 6$.

$6 \div 3 = 2$

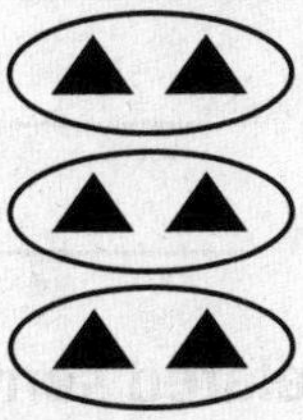

Divide 6 triangles into 3 equal groups.

There are 2 triangles in each group.
So, $6 \div 3 = 2$.

$3 \times 2 = 6$

3 rows of 2 triangles each.
So, $3 \times 2 = 6$.

$6 \div 2 = 3$

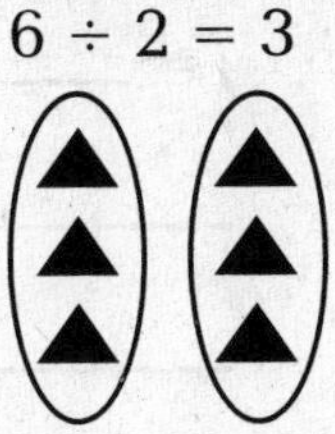

Divide 6 triangles into 2 equal groups.

There are 3 triangles in each group.

So, $6 \div 2 = 3$.

The fact family for 2, 3, and 6 is $2 \times 3 = 6$; $3 \times 2 = 6$; $6 \div 3 = 2$; $6 \div 2 = 3$.

Write the fact family for the set of numbers.

1. 2, 9, 18

____

2. 3, 5, 15

____

3. 1, 8, 8

____

4. 2, 8, 16

____

5. 5, 8, 40

____

6. 3, 8, 24

____

Algebra: Relate Multiplication and Division

Write the fact family for the set of numbers.

1. 4, 2, 8

2. 7, 2, 14

3. 8, 9, 72

4. 6, 1, 6

Find the value of the variable. Then write a related sentence.

5. $4 \times 7 = c$
$c =$ ____

6. $81 \div m = 9$
$m =$ ____

7. $16 \div j = 4$
$j =$ ____

8. $8 \times n = 16$
$n =$ ____

9. $64 \div 8 = r$
$r =$ ____

10. $7 \times 8 = w$
$w =$ ____

11. $9 \times 5 = p$
$p =$ ____

12. $10 \times 3 = a$
$a =$ ____

Problem Solving and Test Prep

13. Laura colors every picture in each of her 5 coloring books. There are 9 pictures in each book. How many pictures does Laura color in all?

14. Carlos has 63 crayons. He puts them into 7 equal groups for his classmates to use. How many crayons are in each group?

15. Which fact belongs to the same family as $6 \times 7 = 42$?

A $6 + 7 = 13$ **C** $42 - 6 = 36$
B $42 \div 7 = 6$ **D** $42 + 7 = 49$

16. Which fact belongs to the same family as $36 \div 9 = 4$?

A $4 \times 9 = 36$ **C** $36 + 9 = 45$
B $36 - 4 = 32$ **D** $9 + 4 = 13$

Name ____________________ **Week 7**

Spiral Review

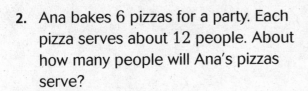

For 1–2, tell whether you need an exact answer or an estimate. Then solve.

1. There are 800 students going on the class picnic trip by bus. One bus can carry 64 people. How many buses are needed?

____

2. Ana bakes 6 pizzas for a party. Each pizza serves about 12 people. About how many people will Ana's pizzas serve?

____

____

For 3–5, find the surface area of each solid figure.

3. 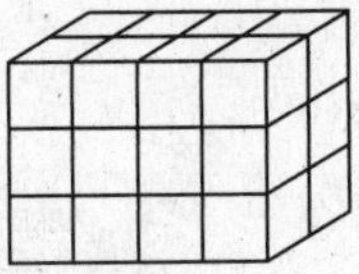

____

4. 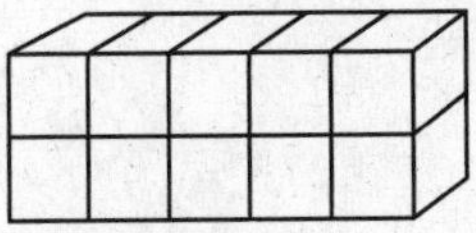

____

5. 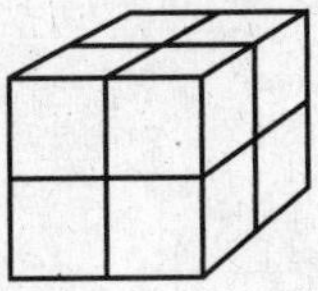

____

For 6–7, use the graph below.

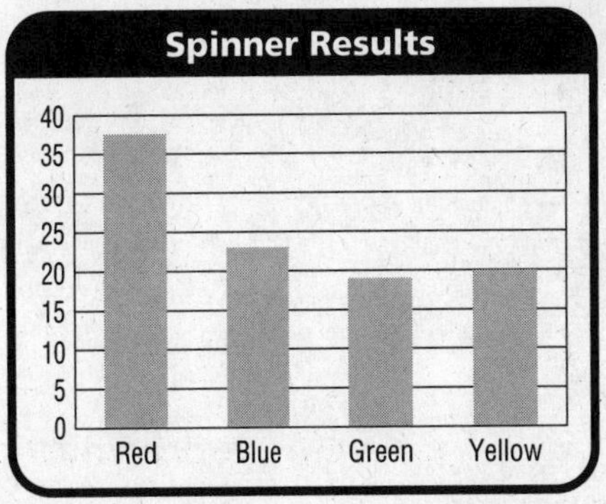

6. Which color did the spinner stop on most often? _______________

7. Kim is going to spin a spinner. Predict which color it will probably stop on.

____

For 8–11, multiply both sides of the equation by the given number. Find the new values.

8. $(15 - 9) = (3 \times 2)$; multiply by 5

____

9. $(4 + 4) = (56 \div 7)$; multiply by 3

____

10. $(36 - 24) = (4 \times 3)$; multiply by 7

____

11. $(12 - 2) = (2 \times 5)$; multiply by 9

____

Name____ **Lesson 5.3**

Multiply and Divide Facts Through 5

You can use a multiplication table to help you find **products** and **quotients**.

Find the product.

3×3

- The first factor is 3. Find the row labeled 3.
- The second factor is 3. Find the column labeled 3.
- Trace with your fingers across row 3 and down column 3.
- Where row 3 and column 3 meet is the product, 9.
- So, $3 \times 3 = 9$.

×	0	1	2	3	4	5
0	0	0	0	0	0	0
1	0	1	2	3	4	5
2	0	2	4	6	8	10
3	0	3	6	9	12	15
4	0	4	8	12	16	20
5	0	5	10	15	20	25
6	0	6	12	18	24	30
7	0	7	14	21	28	35
8	0	8	16	24	32	40
9	0	9	18	27	36	45

Find the quotient.

$15 \div 5$

- The divisor is 5. Find the row labeled 5.
- The dividend is 15. Follow across row 5 until you find the number 15.
- From 15, follow your finger up to the top of the column.
- The number you find is the quotient, 3.
- So, $15 \div 5 = 3$.

×	0	1	2	3	4	5
0	0	0	0	0	0	0
1	0	1	2	3	4	5
2	0	2	4	6	8	10
3	0	3	6	9	12	15
4	0	4	8	12	16	20
5	0	5	10	15	20	25
6	0	6	12	18	24	30
7	0	7	14	21	28	35
8	0	8	16	24	32	40
9	0	9	18	27	36	45

Find the product or quotient.

1. $8 \div 2$ ______
2. 5×3 ______
3. 4×2 ______
4. $25 \div 5$ ______
5. $24 \div 3$ ______
6. 4×6 ______
7. $20 \div 5$ ______
8. 3×6 ______
9. $24 \div 4$ ______
10. 7×4 ______

Name____________________

Multiply and Divide Facts Through 5

Find the product or quotient.

1. 4×3 ______
2. $5 \div 1$ ______
3. 4×8 ______
4. 3×5 ______
5. 2×7 ______
6. $8 \div 2$ ______
7. $35 \div 5$ ______
8. $32 \div 4$ ______
9. $16 \div 4$ ______
10. 3×7 ______
11. 4×10 ______
12. $14 \div 2$ ______
13. 1×7 ______
14. 3×8 ______
15. $20 \div 4$ ______
16. $9 \div 3$ ______

Algebra Find the value of $a \times 3$ for each value of a.

17. $a = 2$ ______
18. $a = 5$ ______
19. $a = 1$ ______
20. $a = 4$ ______

Problem Solving and Test Prep

21. Sue has 32 marbles and wants to put them into 4 equal groups. How many marbles will be in each group?

______

22. Joe eats 5 apples every week. How many apples will Joe eat in 6 weeks?

______

23. Laurie makes a quilt pattern that is 10 squares wide and 3 squares long. How many squares does the quilt have in all?

A 3
B 10
C 13
D 30

24. Mrs. Long delivers 30 quilts to a hospital. She delivers the same number of quilts on each of the 6 floors. How many quilts does Mrs. Long hand out on each floor?

A 5
B 6
C 7
D 8

Name______________________________

Multiply and Divide Facts Through 10

Find the product or quotient. Show the strategy you used.

$50 \div 5$

- Solve using **inverse operations**, so use multiplication facts you know.
  Think: $5 \times 10 = 50$
- So, $50 \div 5 = 10$.

9×3

Draw a rectangular **array** that is 9 units wide and 3 units long.

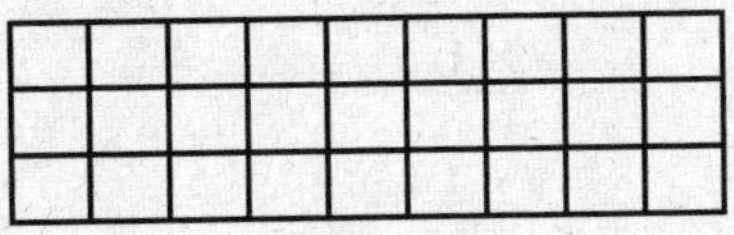

Count the squares in the array.

There are a total of 27 squares.

So, $9 \times 3 = 27$.

Find the product or quotient. Show the strategy you used.

1. $5 \times 4 =$ ____ **2.** $30 \div 5 =$ ____ **3.** $7 \times 8 =$ ____ **4.** $24 \div 6 =$ ____

5. $5 \times 9 =$ ____ **6.** $6 \times 3 =$ ____ **7.** $44 \div 4 =$ ____ **8.** $32 \div 8 =$ ____

9. $12 \times 2 =$ ____ **10.** $81 \div 9 =$ ____ **11.** $7 \times 7 =$ ____ **12.** $6 \times 9 =$ ____

13. $9 \times 7 =$ ____ **14.** $28 \div 4 =$ ____ **15.** $4 \times 0 =$ ____ **16.** $64 \div 8 =$ ____

Name___________________________________

Multiply and Divide Facts Through 10

Find the product or quotient. Show the strategy you used.

1. 8×8 ______
2. 7×9 ______
3. 8×5 ______
4. 9×6 ______

5. $56 \div 8$ ______
6. $81 \div 9$ ______
7. $100 \div 10$ ______
8. $72 \div 9$ ______

9. $\begin{array}{r} 10 \\ \times 9 \\ \hline \end{array}$
10. $\begin{array}{r} 7 \\ \times 8 \\ \hline \end{array}$
11. $\begin{array}{r} 9 \\ \times 8 \\ \hline \end{array}$
12. $\begin{array}{r} 6 \\ \times 6 \\ \hline \end{array}$

13. $8\overline{)64}$
14. $9\overline{)36}$
15. $7\overline{)49}$
16. $6\overline{)54}$

Problem Solving and Test Prep

17. Jeff had 10 checkers left at the end of each of the 7 games he played. How many checkers did Jeff have at the end of 7 games?

____

18. Kim played checkers for 6 days and won a total of 24 games. She won the same number of games each day. How many games did Kim win each day?

____

19. There are 6 rows of chairs with 7 chairs in each row. How many chairs are there? Describe the strategy you used to find the answer.

____

____

20. There are 8 rows of checkers in one box. Each row has 9 checkers. How many checkers are in one box?

A 17

B 32

C 56

D 72

Name____ **Lesson 5.5**

Multiplication Table Through 12

You can use a multiplication table to multiply and divide numbers through 12.

Use the multiplication table to find the product of 7 × 11.

- The first factor is 7. Find the row labeled 7.
- The second factor is 11. Find the column labeled 11.
- Trace your finger across row 7 and down column 11.
- Where row 7 and column 11 meet is the product, 77.
- So, 7 × 11 = 77.

columns (across the top); rows (down the side)

×	0	1	2	3	4	5	6	7	8	9	10	11	12
0	0	0	0	0	0	0	0	0	0	0	0	0	0
1	0	1	2	3	4	5	6	7	8	9	10	11	12
2	0	2	4	6	8	10	12	14	16	18	20	22	24
3	0	3	6	9	12	15	18	21	24	27	30	33	36
4	0	4	8	12	16	20	24	28	32	36	40	44	48
5	0	5	10	15	20	25	30	35	40	45	50	55	60
6	0	6	12	18	24	30	36	42	48	54	60	66	72
7	0	7	14	21	28	35	42	49	56	63	70	77	84
8	0	8	16	24	32	40	48	56	64	72	80	88	96
9	0	9	18	27	36	45	54	63	72	81	90	99	108
10	0	10	20	30	40	50	60	70	80	90	100	110	120
11	0	11	22	33	44	55	66	77	88	99	110	121	132
12	0	12	24	36	48	60	72	84	96	108	120	132	144

Use the rule to find the missing number.

Find the row for the divisor, 12.

Trace with your finger across row 12 to find the dividend, 36.

Since 36 is in the column for 3, 3 is the output.

So, 36 ÷ 12 = 3.

Divide the input by 12.

Input	Output
36	
72	6
108	9
144	12

Find the product or quotient. Show the strategy you used.

1. 2 × 12 ________

2. 12 ÷ 3 ________

3. 4 × 11 ________

4. 66 ÷ 6 ________

Algebra Use the rule.

5. **Multiply by 12.**

Input	Output
4	☐
☐	60
7	☐
☐	108

6. **Divide by 11.**

Input	Output
33	☐
77	☐
☐	8
☐	10

Name____

Multiplication Table Through 12

Find the product or quotient. Show the strategy you used.

1. $110 \div 11$ ________

2. 8×11 ________

3. 12×9 ________

4. $99 \div 11$ ________

5. 7×12 ________

6. 6×11 ________

7. $84 \div 12$ ________

8. $48 \div 12$ ________

9. 11×11 ________

10. $132 \div 11$ ________

11. $108 \div 12$ ________

12. 12×12 ________

13. $60 \div 12$ ________

14. $63 \div 7$ ________

15. 11×9 ________

16. 11×12 ________

ALGEBRA **Use the rule to find the missing numbers.**

17. Multiply by 11.

Input	Output
2	■
4	■
6	■

18. Multiply by 12.

Input	Output
3	■
■	48
5	■

19. Divide by 11.

Input	Output
99	■
■	10
121	■

20. **WRITE Math** What could the missing factors be in ■ × ■ = 36? Find as many factor pairs as you can. Explain how you found them.

____

____

____

Name ______________________________

Patterns on the Multiplication Table

When a whole number is multiplied by another whole number, the product is called a **multiple.** For example, $5 \times 3 = 15$, so 15 is a multiple of 5 and 3.

When a number is multiplied by itself, the product is called a **square number.** For example, $4 \times 4 = 16$, so 16 is a square number.

Which multiples have only even numbers?

Look at the rows in the multiplication table.

Look for a row where all the multiples are even numbers.

All the multiples for the numbers 2, 4, 6, 8, 10, and 12 are even.

column

row

×	0	1	2	3	4	5	6	7	8	9	10	11	12
0	0	0	0	0	0	0	0	0	0	0	0	0	0
1	0	1	2	3	4	5	6	7	8	9	10	11	12
2	0	2	4	6	8	10	12	14	16	18	20	22	24
3	0	3	6	9	12	15	18	21	24	27	30	33	36
4	0	4	8	12	16	20	24	28	32	36	40	44	48
5	0	5	10	15	20	25	30	35	40	45	50	55	60
6	0	6	12	18	24	30	36	42	48	54	60	66	72
7	0	7	14	21	28	35	42	49	56	63	70	77	84
8	0	8	16	24	32	40	48	56	64	72	80	88	96
9	0	9	18	27	36	45	54	63	72	81	90	99	108
10	0	10	20	30	40	50	60	70	80	90	100	110	120
11	0	11	22	33	44	55	66	77	88	99	110	121	132
12	0	12	24	36	48	60	72	84	96	108	120	132	144

Use the multiplication table to find the square number for 11 × 11.

Find the column and row for the number 11.

Use your finger to move along the row and down the column.

Your fingers meet at 121.

$11 \times 11 = 121$, so 121 is the square number.

column

row

×	0	1	2	3	4	5	6	7	8	9	10	11	12
0	0	0	0	0	0	0	0	0	0	0	0	0	0
1	0	1	2	3	4	5	6	7	8	9	10	11	12
2	0	2	4	6	8	10	12	14	16	18	20	22	24
3	0	3	6	9	12	15	18	21	24	27	30	33	36
4	0	4	8	12	16	20	24	28	32	36	40	44	48
5	0	5	10	15	20	25	30	35	40	45	50	55	60
6	0	6	12	18	24	30	36	42	48	54	60	66	72
7	0	7	14	21	28	35	42	49	56	63	70	77	84
8	0	8	16	24	32	40	48	56	64	72	80	88	96
9	0	9	18	27	36	45	54	63	72	81	90	99	108
10	0	10	20	30	40	50	60	70	80	90	100	110	120
11	0	11	22	33	44	55	66	77	88	99	110	121	132
12	0	12	24	36	48	60	72	84	96	108	120	132	144

Find the square number.

1. 5×5 ________
2. 8×8 ________
3. 2×2 ________
4. 4×4 ________
5. 7×7 ________
6. 10×10 ________
7. 9×9 ________
8. 12×12 ________

Use the multiplication table.

9. Do any numbers have only odd-numbered multiples?

   ____

10. What pattern do you see in the multiples of 5?

   ____

NS 3.0 Students solve problems involving addition, subtraction, multiplication, and division of whole numbers and understand relationships among the operations.

Name____

Patterns on the Multiplication Table

Find the square number.

1. 9×9 ________

2. 5×5 ________

3. 10×10 ________

4. 4×4 ________

5. 2×2 ________

For 6–7, use the multiplication table.

6. What pattern do you see in the first 9 multiples of 11?

____

____

____

7. What pattern do you see in the first 9 multiples of 9?

____

____

____

×	0	1	2	3	4	5	6	7	8	9	10	11	12
0	0	0	0	0	0	0	0	0	0	0	0	0	0
1	0	1	2	3	4	5	6	7	8	9	10	11	12
2	0	2	4	6	8	10	12	14	16	18	20	22	24
3	0	3	6	9	12	15	18	21	24	27	30	33	36
4	0	4	8	12	16	20	24	28	32	36	40	44	48
5	0	5	10	15	20	25	30	35	40	45	50	55	60
6	0	6	12	18	24	30	36	42	48	54	60	66	72
7	0	7	14	21	28	35	42	49	56	63	70	77	84
8	0	8	16	24	32	40	48	56	64	72	80	88	96
9	0	9	18	27	36	45	54	63	72	81	90	99	108
10	0	10	20	30	40	50	60	70	80	90	100	110	120
11	0	11	22	33	44	55	66	77	88	99	110	121	132
12	0	12	24	36	48	60	72	84	96	108	120	132	144

Problem Solving and Test Prep

8. Niko has a square number that is less than 50. The digits add up to 9. What is Niko's number?

____

9. Use the rule *1 less than 3 times the number* to make a pattern. *Start with* 5. What is the 4th number in the pattern?

____

10. Which number has multiples with a repeating pattern of fives and zeros in the ones place?

A 1

B 5

C 10

D 20

11. The multiples of which number are three times the multiples of 4?

A 8

B 12

C 40

D 84

Spiral Review

For 1–4, write the fraction and decimal for each shaded part.

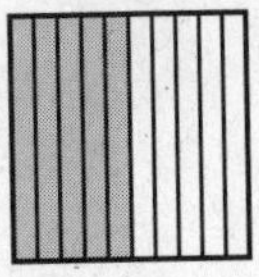

1. ____________ 2. ____________

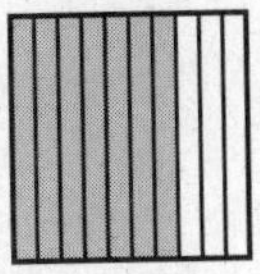

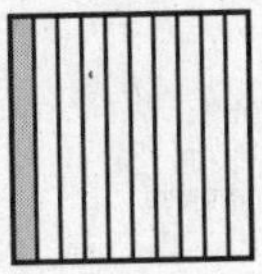

3. ____________ 4. ____________

For 8–9, use an organized list to solve.

At a sandwich bar, you can choose white or wheat bread, cheddar or Swiss cheese, and ham or turkey.

8. Find all the possible sandwich combinations. How many are there?

____

9. How did making an organized list help you solve this problem?

____

For 5–7, name each triangle by the length of its sides.

5.

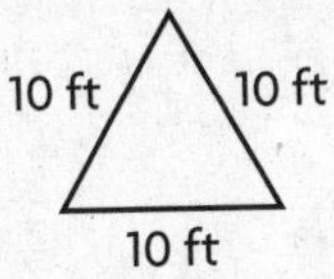

____

6.

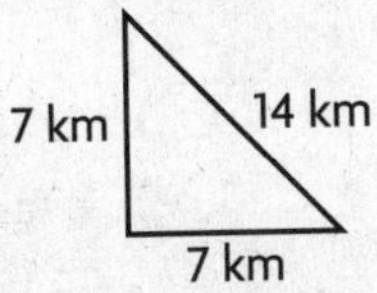

____

7.

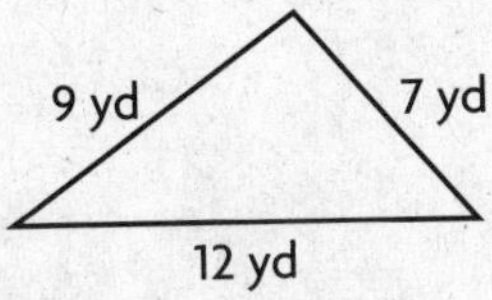

____

For 10–13, find the missing factor.

10. $6 \times t = 42$

____

11. $m \times 10 = 80$

____

12. $h \times 12 = 60$

____

13. $7 \times y = 49$

____

Name ____

Problem Solving Workshop Skill: Choose the Operation

The hiking trip is on Friday. On Thursday, the high temperature was 67°F. Russell hopes that it will be 8° warmer on Friday. What temperature is Russell hoping for?

1. What are you asked to find?

____

2. Will you separate a group into smaller groups to solve? Why or why not?

____

3. Which operation will you use to answer the question?

____

4. What was the high temperature on Thursday? How many degrees will you add to Thursday's temperature to get Friday's temperature?

____

5. What temperature is Russell hoping for on Friday? Write a number sentence to solve.

____

6. How can you check your answer?

____

Tell which operation you would use to solve the problem. Then solve the problem.

7. Maria has 24 apples. She wants to give an equal number of apples to each of her 4 friends. How many apples should Maria give each friend?

____

8. Gina has 5 boxes. She places 6 books in each box. How many books does Gina have in all?

____

Name______________________________ Lesson 5.7

Problem Solving Workshop Skill: Choose the Operation

Problem Solving Skill Practice

Tell which operation you would use to solve the problem. Then solve the problem.

1. Sally takes 24 gallons of juice to the school picnic. The students at the picnic drink 2 gallons of juice every hour. How many hours will it take the students to drink all the juice?

____

2. Each student in Lori's class brings 12 cookies for the bake sale. There are 12 students in Lori's class. How many cookies does the class have for the bake sale?

____

Mixed Applications

3. Greg sells 108 mini muffins at the bake sale. He sold the mini muffins in bags of 12. How many bags of mini muffins does Greg sell? Which fact family did you use?

____

____

4. Julie wants to know how many workbooks she will use for the school year. The subjects she is studying are math, science, and reading. Each subject has 2 workbooks. Write a number sentence to show how many workbooks Julie will be using this year.

____

USE DATA For 5 -6, use the information in the table.

5. At the bake sale, 9 people buy slices of pie. Each person buys the same number of slices for $2 each. How many slices of pie does each person buy?

____

Bake Sale Final Sales	
cupcakes	147
cookies	211
slices of pie	54
slices of cake	39
brownies	97

6. How many cookies, brownies, and cupcakes were sold in all? _______________

Name ____________________

Algebra: Find Missing Factors

You can draw models and use fact families to find missing factors.

Find the missing factor.

$10 \times g = 70$

Use a model to find the value of g.

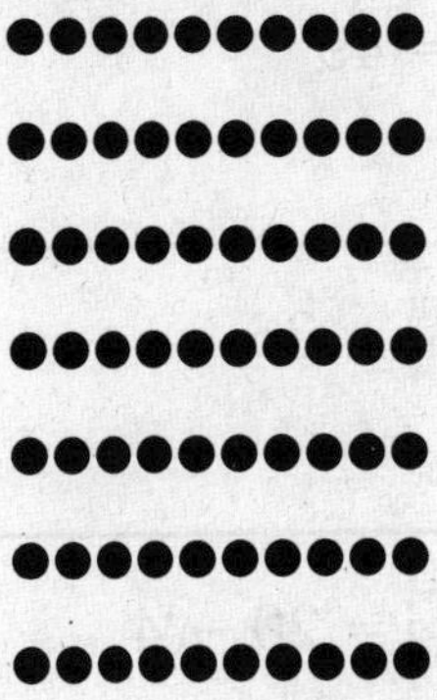

Use the product, 70, and the factor, 10, to draw 70 counters in rows of 10.

Count the number of rows to find the missing factor.

There are 7 rows of 10, so the missing factor, g, is 7.

So, $10 \times 7 = 70$.

Find the missing factor.

$■ \times 11 = 132$

Use a related division sentence to find the missing factor, ■.

Think of the fact family and write the related division sentence. $132 \div 11 = ■$

Solve for ■. $132 \div 11 = 12$, or $■ = 12$

The missing factor is 12.

So, $12 \times 11 = 132$, or $■ = 12$.

Find the missing factor.

1. $7 \times n = 28$ ______

2. $5 \times p = 30$ ______

3. $m \times 5 = 45$ ______

4. $6 \times n = 48$ ______

5. $8 \times g = 72$ ______

6. $w \times 11 = 55$ ______

7. $4 \times j = 32$ ______

8. $9 \times v = 27$ ______

9. $4 \times g = 12$ ______

10. $9 \times v = 81$ ______

11. $3 \times n = 36$ ______

12. $10 \times z = 80$ ______

AF 1.1 Use letters, boxes, or other symbols to stand for any number in simple expressions or equations (e.g., demonstrate an understanding and the use of the concept of a variable).

Name ____________________

Lesson 5.8

Algebra: Find Missing Factors

Find the missing factor.

1. $4 \times g = 20$

$g =$ ____________

2. $y \times 3 = 27$

$y =$ ____________

3. $8 \times w = 48$

$w =$ ____________

4. $7 \times a = 49$

$a =$ ____________

5. $■ \times 2 = 24$

$■ =$ ____________

6. $9 \times r = 81$

$r =$ ____________

7. $4 \times ■ = 36$

$■ =$ ____________

8. $7 \times s = 77$

$s =$ ____________

9. $5 \times ■ = 23 + 2$

$■ =$ ____________

10. $8 \times ■ = 20 - 4$

$■ =$ ____________

11. $6 \times ■ = 11 + 7$

$■ =$ ____________

12. $10 \times ■ = 15 + 5$

$■ =$ ____________

13. $7 \times ■ = 12 + 2$

$■ =$ ____________

14. $3 \times ■ = 16 + 5$

$■ =$ ____________

15. $4 \times ■ = 13 + 3$

$■ =$ ____________

Problem Solving and Test Prep

16. Each season, a total of 32 tickets are given away. Each chosen family is given 4 free tickets. Write a number sentence that can be used to find the number of families that will receive tickets.

____

17. The manager of the Antelopes orders 4 uniforms for each new player. This year, the manager orders 16 uniforms. Write a number sentence that can be used to find the number of new players.

____

18. What is the missing factor in $11 \times ■ = 121$?

A 10
B 11
C 12
D 13

19. What is the missing factor in $■ \times 12 = 120$?

A 0
B 11
C 12
D 10

Name ______________________________________

Multiplication Properties

Use properties and mental math to find the product.

Associative Property

The Associative Property states that the way factors are grouped does not change the product.

$8 \times 3 \times 3$

- $(8 \times 3) \times 3$ uses the same numbers as $8 \times (3 \times 3)$, except they are grouped together differently.
- Choose which set of factors you can easily use mental math to solve.

$8 \times (3 \times 3)$ or $(8 \times 3) \times 3$

$8 \times 9 = 72$ $24 \times 3 = 72$

So, $8 \times 3 \times 3 = 72$.

Distributive Property

The Distributive Property states that you can break apart one of the factors and multiply its parts by the other factor.

7×12

- Break 12 apart so you can easily use mental math.
  $12 = 10 + 2$
- Multiply both parts by 7, and then add the products.
  $7 \times 10 = 70$
  $7 \times 2 = 14$
  $70 + 14 = 84$

So, $7 \times 12 = 84$.

Commutative Property

The Commutative Property states that you can multiply numbers in any order and the product will be the same. The order in which you multiply numbers does not change the product. For example, 10×3 and 3×10 both have a product of 30.

Zero Property

The Zero Property states that the product of any number and zero is zero. For example, $12 \times 0 = 0$, $56 \times 0 = 0$, and $864 \times 0 = 0$.

Identity Property

The Identity Property states that the product of any number and 1 is that number. For example, $9 \times 1 = 9$, $56 \times 1 = 56$, and $864 \times 1 = 864$.

Use the properties and mental math strategies to find the product.

1. $6 \times 2 \times 4$ ________

2. 8×12 ________

3. $6 \times 0 \times 4$ ________

4. $2 \times 13 \times 1$ ________

5. 5×14 ________

6. $5 \times 5 \times 3$ ________

7. $4 \times 6 \times 3$ ________

8. $2 \times 0 \times 3$ ________

9. $8 \times 1 \times 5$ ________

Name________________________________

Multiplication Properties

Use the properties and mental math to find the product.

1. $3 \times 4 \times 2$ ________
2. $4 \times 5 \times 5$ ________
3. $7 \times 4 \times 0$ ________
4. $7 \times 12 \times 1$ ________

Find the missing number. Name the property you used.

5. $(5 \times 3) \times 4 = 5 \times (■ \times 4)$ ________
6. $3 \times 5 = 5 \times ■$ ________
7. $8 \times ■ = (2 \times 10) + (6 \times 2)$ ________
8. $3 \times (7 - ■) = 3$ ________
9. $8 \times (5 - 3 - 2) = ■$ ________
10. $3 \times (2 \times 4) = ■ \times (2 \times 3)$ ________

Make a model and use the Distributive Property to find the product.

11. 14×6 ________
12. 5×15 ________
13. 9×17 ________

Show two ways to group by using parentheses. Find the product.

14. $12 \times 5 \times 6$ ________ ________
15. $4 \times 3 \times 2$ ________ ________
16. $9 \times 3 \times 8$ ________ ________

Problem Solving and Test Prep

17. The pet store window has 5 kennels with 4 puppies each and 6 kennels with 6 kittens each. How many animals are in the window?

____

18. Jake takes his border collie on a walk for exercise. They walk four blocks that are 20 yards each. How many yards do Jake and his border collie walk?

____

19. Each packet of catnip toys has 7 toys. Each box of packets has 20 packets. How many toys are there in 5 boxes of catnip toys?

A 500
B 600
C 700
D 800

20. Is the number sentence true? Explain.
$5 \times (4 - 3) = 5$

____

____

Spiral Review

For 1–5, estimate. Then find the product.

1. $70 \times 590 =$ ____________

2. $63 \times 801 =$ ____________

3. $1{,}234 \times 50 =$ ____________

4. $\$44.19 \times 21 =$ ____________

5. $\$90 \times 7932 =$ ____________

For 10–12, use the bag of marbles below.

10. Which outcome is most likely? ____

11. Which outcome is least likely? ____

12. Which outcomes are equally likely?

____

For 6–9, tell if each angle is *right, obtuse* or *acute*.

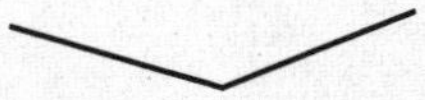

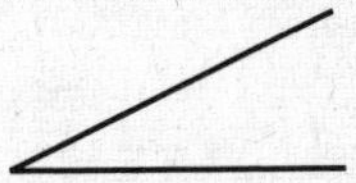

6. ____________ **7.** ____________

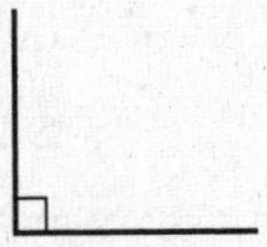

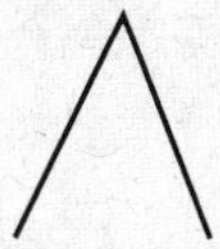

8. ____________ **9.** ____________

For 13–16, write an expression with a variable. Tell what the variable represents.

13. 2 times a number of marbles

____

14. some books separated into 5 equal piles

____

15. some pennies divided equally among 7 students

____

16. a number of socks times 30

____

Name____ Lesson 6.3

Expressions with Parentheses

When there are **parentheses** in an expression, do what is in the parentheses first. Then, follow the order of operations and divide and multiply from left to right. Finally, add and subtract from left to right.

Follow the order of operations to find the value of the expression.

$3 \times (6 - 2 + 4) \div 2$

$6 - 2 + 4$ is in parentheses, so perform this operation first.

$3 \times (6 - 2 + 4) \div 2$
$3 \times (4 + 4) \div 2$
$3 \times 8 \div 2$

Multiply and divide from left to right.

$24 \div 2 = 12$

So, $3 \times (6 - 2 + 4) \div 2 = 12$.

Choose the expression that matches the words.
Savon has 4 pages with 5 stamps on each one. He used 3 stamps.

a. $(4 \times 5) - 3$ **b. $4 \times (5 - 3)$**

There are 4 pages and each page has 5 stamps.
Multiply 4 groups of 5 stamps or 4×5.
There are 20 stamps.

Savon uses 3 of the 20 stamps. He now has 3 stamps less than he did before.
This is subtraction or $- 3$.

So, choice **a.** $(4 \times 5) - 3$ matches the words because Savon had 20 stamps or 4×5 stamps and used 3 or subtracted 3.

Choose the expression that matches the words.

1. There are 6 shelves with 10 books on each shelf. Dan took 5 of the books.
   a. $(6 \times 10) - 5$ **b.** $6 \times (10 - 5)$
   ____

2. Sue had \$9. Then she worked for 8 hours for \$5 an hour.
   a. $\$9 \times (8 \times \$5)$ **b.** $\$9 + (8 \times \$5)$
   ____

Follow the order of operations to find the value of expression.

3. $(12 + 8) \div (4 - 2)$ ____________
4. $(8 \times 3) - 6$ ____________
5. $4 \times (3 \times 2)$ ____________
6. $14 - (5 \times 2)$ ____________
7. $(2 \times 4) \times (5 \times 1)$ ____________
8. $5 + (10 \div 2)$ ____________

Name________________________________ **Lesson 6.3**

Expressions with Parentheses

Follow the order of operations to find the value of each expression.

1. $2 - 3 \times 8 \div 12$ _______
2. $(5 + 28) \div 3 - 5$ _______
3. $(15 + 9) \div 2 - 1$ _______
4. $(2 + 7) \times 6 - 3$ _______

Choose the expression that matches the words.

5. Gene divided 12 toy soldiers into 2 equal groups. Then he bought 6 more.

   A $12 \div 2 + 6$ **B** $12 \div (2 + 6)$

6. Sabrina brought 6 bunches of 5 flowers each. Then she threw out 4 bunches that had wilted.

   A $6 \times (5 - 4)$ **B** $6 \times 5 - 4$

Write words to match the expression.

7. $49 \div 7 + 2$ _______
8. $6 \times 7 + 28$ _______
9. $(4 \times 9) \div (16 - 14)$ _______

Use parenthesis to make the number sentence true.

10. $44 \div 2 + 2 = 11$
11. $81 \div 7 + 2 + 4 = 13$
12. $3 \times 21 + 2 - 3 = 66$

Problem Solving and Test Prep

13. There were 5 birds nesting in each of 7 trees. Jim fed all but 2. How many birds did Jim feed?

    ____

14. Grace went on a bird watch for 7 days. Each day she saw 3 quail, 5 wrens, and a lark. How many birds did Grace see in all?

    ____

15. Which expression has a value of 14?

    A $10 + (4 \times 2) - 6$

    B $44 \div 11 + 12$

    C $27 \div 9 + 11$

    D $18 \times 2 - 14$

16. Find the value of the expression.

    $(12 \times 6) \div (3 + 3)$

    ____

Name____

Write and Evaluate Expressions

Some multiplication and division expressions have variables that represent numbers. A **variable** is a letter or symbol that stands for any number. Examples of multiplication and division expressions are $9 \times k$ and $h \div 3$.

Write an expression that matches the words.
A handful of keys divided equally and put on 4 key chains.

Write a variable to represent a number in the problem. — Let k represent the keys.

The keys are being separated into four groups, so the operation is division. — $k \div 4$

So, the expression is $k \div 4$.

Find the value of the expression.

$s \div 3$ if $s = 27$.

The variable s represents 27. Replace s with 27.

$s \div 3$
$27 \div 3$

Solve using division. — $27 \div 3 = 9$

So, the value of $s \div 3$ is 9, if $s = 27$.

Write an expression that matches the words.

1. 7 times more books than before ____________

2. the price of some pens, at $1 each ____________

3. 40 plates divided equally and put on some tables ____________

4. some shoes divided equally and put into 8 boxes ____________

Find the value of the expression.

5. $b \times 5$ if $b = 4$ ____________

6. $60 \div k$ if $k = 10$ ____________

7. $8 \times w$ if $w = 6$ ____________

8. $m \div 4$ if $m = 36$ ____________

9. $s \times 9$ if $s = 5$ ____________

10. $h \div 7$ if $h = 49$ ____________

Name________________________________

Write and Evaluate Expressions

Write an expression that matches the words.

1. Stamps s divided equally in 6 rows

____

2. Some peas p in each of 10 pods

____

3. Some marbles m on sale at 15¢ each

____

4. 42 cookies divided among several students s

____

Find the value of the expression.

5. $y \times 5$ if $y = 6$ ______

6. $63 \div b$ if $b = 7$ ______

7. $9 \times a$ if $a = 2$ ______

8. $r \div 6$ if $r = 54$ ______

Match the expression with the words.

9. $4 \times t + 8$ ______

10. $t \times 12 \div 4$ ______

11. $t \div 2 - 8$ ______

a. a number, t, divided by 2 minus 8

b. 4 times a number, t, plus 8

c. a number, t, times 12 and separated into 4 pieces

Problem Solving and Test Prep

12. Ella has some pages with 15 stickers to a page. Write an expression for the number of stickers she has.

____

13. Look at Exercise 12. Suppose Ella has 5 pages. How many stickers does she have in all?

____

14. Robert has 7 times as many soap box racers as Xavier. Let r represent the number of soap box racers Robert has. Which expression tells the number of racers Xavier has?

A $7 + r$

B $r - 7$

C $7 \times r$

D $r \div 7$

15. Fran spent 350 cents on stamps. Write an expression for the number of stamps that Fran bought. How many stamps did she buy if each stamp cost 35 cents? Explain.

____

____

Multiplication and Division Equations

Write an equation for each. Choose a variable for the unknown. Tell what the variable represents.

A number of rings divided equally among 4 friends is 2 rings for each friend.

- Decide on a variable to represent the unknown amount. — The variable r will represent the total number of rings.
- Choose an operation to solve the problem. — The rings are being separated into four equal groups, so the operation is division.
- The rings are divided into equal groups of 2 rings. — $r \div 4 = 2$

So, the equation is $r \div 4 = 2$, and the variable r represents the total number of rings.

Solve the equation.

$a \times 7 = 63$

- Use mental math to figure out what number is being represented by a. — What number times 7 equals 63?
- You know that $9 \times 7 = 63$. — $a = 9$
- Replace a with 9. — $9 \times 7 = 63$

  $63 = 63$

So, the value of a is 9.

Write an equation for each. Choose a variable for the unknown. Tell what the variable represents.

1. 5 bowls with a number of grapes in each is 30 grapes.

____

____

2. $40 divided equally among a number of friends is $10.

____

____

Solve the equation.

3. $b \times 7 = 35$ ______________

4. $60 \div k = 10$ ______________

5. $9 \div j = 45$ ______________

6. $36 \div m = 6$ ______________

7. $s \times 3 = 12$ ______________

8. $h \div 6 = 7$ ______________

Name______________________________

Multiplication and Division Equations

Write an equation for each. Choose the variable for the unknown. Tell what the variable represents.

1. Three students divide 27 bracelets equally among them.

____

2. Two pounds of beads put equally in bags makes a total of 50 pounds.

____

3. Maddie plants 3 seeds each in 15 pots.

____

4. Jesse divides 36 ornaments equally and puts them into 9 bags.

____

Solve the equation.

5. $a \times 6 = 48$ $a =$ ______
6. $d \div 4 = 7$ $d =$ ______
7. $3 \times w = 27$ $w =$ ______
8. $63 \div n = 9$ $n =$ ______
9. $b \div 5 = 5$ $b =$ ______
10. $22 \div t = 11$ $t =$ ______
11. $4 \times k \times 3 = 24$ $k =$ ______
12. $5 \times h \times 3 = 45$ $h =$ ______

Problem Solving and Test Prep

13. Phyllis is making rings. Each ring has 3 beads. If she can make 7 rings, how many beads does Phyllis have?

____

14. Ted divided 56 colored blocks into 8 bags. How many blocks were in each bag?

____

15. If $t = 3$, which equation below can be used to find the value of t?

A $t \div 12 = 4$ **C** $t \times 5 = 30$
B $36 \div t = 12$ **D** $15 \times t = 60$

16. What is the value of p?
$21 \div p = 7$

A 21 **C** 4
B 7 **D** 3

Name____ Lesson 6.6

Multiply Equals by Equals

When you multiply both sides of an equation by the same number, the sides stay equal. For example, if you multiply both sides of the equation "$8 = 2 \times 4$" by 5, each side will have a value of 40. The left and right sides of the equation will still be equal.

Tell whether the equation is true. If not, explain why.
$(3 \times 3) \times 6 = (36 \div 4) \times 7$

Find the value for each side.

$(3 \times 3) \times 6 \quad = \quad (36 \div 4) \times 7$

$(3 \times 3) \times 6 \qquad (36 \div 4) \times 7$

$9 \times 6 \qquad 9 \times 7$

$54 \qquad 63$

Compare the two amounts. $54 \neq 63$

So, the equation is not true, because the sides of the equation are not equal.

Multiply both sides of the equation by the given number. Find the new values.

$12 - 4 = 40 \div 5$; multiply by 3.

First find the value of each side. $12 - 4 = 40 \div 5$

$8 = 8$

Then multiply both sides of the equation by 3. $8 \times 3 = 8 \times 3$

$24 = 24$

So, the new values for the equation are $24 = 24$.

Tell whether each equation is true. If not, explain why.

1. $(3 \times 4) \times 2 = 4 \times (12 - 5)$

____

2. $(8 + 2) \times 5 = (45 \div 9) \times 2 \times 5$

____

Multiply both sides of the equation by the given number. Find the new values.

3. $8 = 4 \times 2$; multiply by 9.

____

4. $(3 + 7) = (2 \times 5)$; multiply by 6.

____

5. $4 \times 3 = (6 + 6)$; multiply by 2.

____

6. $(15 - 6) = (8 + 1)$; multiply by 5.

____

7. $9 = 3 \times 3$; multiply by 8.

____

8. $(24 \div 6) = (4 \times 1)$; multiply by 7.

____

Name____________________________________

Multiply Equals by Equals

Tell whether each equation is true. If not, explain why.

1. $(5 + 2) \times 3 \stackrel{?}{=} 3 \times 7$ ____________

2. $(16 - 7) \times 8 \stackrel{?}{=} (4 + 5) \times (6 + 2)$ ____________

3. $54 \div 6 + 7 \stackrel{?}{=} 4 \times 3$ ____________

4. $2 \times 12 - 3 \stackrel{?}{=} 3 \times 11 + 1$ ____________

5. $(64 \div 8) + 14 \stackrel{?}{=} (22 - 11) \times 2$ ____________

6. $4 + 63 \div 9 \stackrel{?}{=} 5 \times 11 - 33$ ____________

Multiply both sides of the equation by the given number. Find the new values.

7. $11 - 2 = 5 + 4$; multiply by 4. ____________

8. $3 \times 3 = 36 \div 4$; multiply by 6. ____________

9. $15 - 11 = 28 \div 7$; multiply by 8. ____________

10. $144 \div 12 = 3 + 9$; multiply by 5. ____________

What number makes the equation true?

11. $(5 + 3) \times 4 = \square \times 16$

12. $(14 - \square) \times 5 = 3 \times 10 + 5$

13. $66 \div 6 + 13 = 2 \times \square + 6$

Problem Solving and Test Prep

14. Peg picked 4 times plus 6 as many apples as Joe. If Joe picked 5 apples from each of 2 trees, how many apples did Peg pick? ____________

15. Peg and Joe put their apples together. Then they divided them equally among 8 children. How many apples did each child get? ____________

16. The letters *A* and *B* stand for numbers. If $A \times 4 = B + 2$, which statement is true?

    A $A = B$
    B $A < B$
    C $A > B$
    D $A + 2 = B \times 4$

17. What number goes into the box to make this number sentence true? Explain how you know.

    $(15 - 8) \times 2 = (\square \div 12) \times 7$

    ____

    ____

Spiral Review

For 1–4, find the sum or difference. Write the method you used.

1. $\begin{array}{r} 4{,}986 \\ +3{,}578 \\ \hline \end{array}$

2. $\begin{array}{r} 78{,}005 \\ -62{,}000 \\ \hline \end{array}$

______________ ______________

3. $\begin{array}{r} 57{,}692 \\ -42{,}128 \\ \hline \end{array}$

4. $\begin{array}{r} 87{,}002 \\ +12{,}000 \\ \hline \end{array}$

______________ ______________

For 5–7, name each quadrilateral.

5. 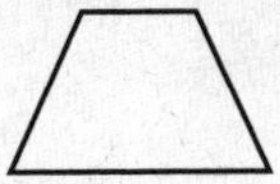

____

6. 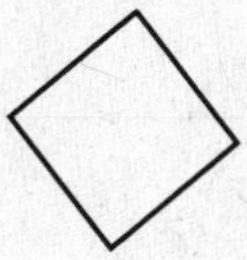

____

7. 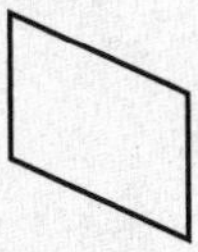

____

For 8–10, use the bar graph below.

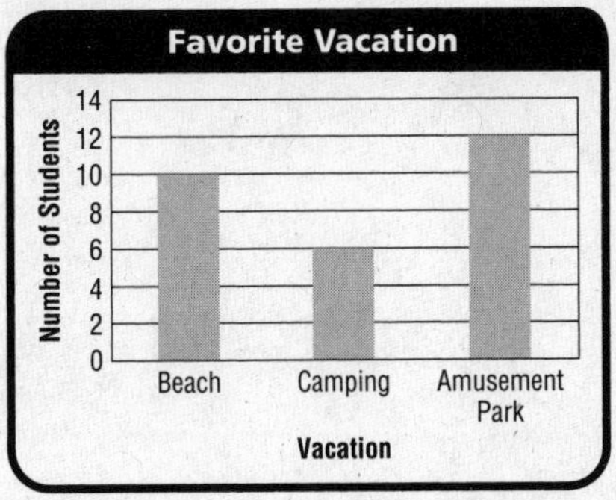

8. Which vacation got the most votes?

____

9. Which vacation got the least votes?

____

10. How many more votes did the beach get than camping?

____

For 11–15, follow the order of operations to find the value of each expression.

11. $3 + 5 \times 2$

____

12. $4 \times 3 + 2$

____

13. $36 + 7 \div 7$

____

14. $18 - 12 + 3$

____

15. $12 - 3 \div 3$

____

Name______________________________

Problem Solving Workshop Strategy: Predict and Test

Marc likes to solve word scrambles and maze puzzles. Yesterday he solved 10 word scrambles and maze puzzles in all. He solved 2 more word scrambles than maze puzzles. How many word scrambles did Marc solve yesterday?

Read to Understand

1. What are you asked? ______________________________

____

Plan

2. How can you use the predict and test strategy to help you solve the problem?

____

____

Solve

3. Finish solving the problem.

Predict	Test		Does it Check?
	Difference	Total	
8 scrambles 6 puzzles			
5 scrambles 5 puzzles			

Number of word scrambles Marc solved: ______________

Check

4. Does the answer make sense for the problem? Explain. ______________

____

Predict and test to solve.

5. The product of two numbers is 42. Their sum is 13. What are the numbers?

____

6. Sue has three times as much money as Dan. Together they have $40. How much money does each person have?

____

____

Name________________________________

Problem Solving Workshop Strategy: Predict and Test

Problem Solving Strategy Practice

Predict and test to solve.

1. Betty likes to solve number puzzles. Here is the most recent one she found. The product of two numbers is 48. Their sum is 14. What are the two numbers?

____

2. Kim is thinking of two numbers. The quotient is 4 and the difference is 27. What are the two numbers?

____

3. Kyle and Ellie played basketball. Kyle scored half as many points as Ellie. Together they scored 27 points. How many did each player score?

____

4. Don bought two puzzle books. Together, they cost $19. One cost $5 more than the other. How much did each book cost?

____

Mixed Strategy Practice

USE DATA For 5-10, complete the table.

	Sum	Product	Difference	Two Numbers
5.	10	21	4	☐, ☐
6.	10	25	0	☐, ☐
7.	10	16	6	☐, ☐
8.	9	18	3	☐, ☐
9.	15	54	3	☐, ☐
10.	8	16	0	☐, ☐

Name____

Lesson 6.8

Patterns: Find a Rule

When you are asked to to find a rule, look for a pattern in the numbers. See how different pairs of numbers are related.

Find a rule. Write your rule as an equation.
Use your rule to find the missing numbers.

Input, b	90	70	60	50	30
Output, c	9	7	6	☐	☐

- Look for a pattern in the numbers to find a rule.
  Study how the input numbers relate to the output numbers.
- The first ordered pair is 90 and 9. $90 \div 10 = 9$.
  The second ordered pair is 70 and 7. $70 \div 10 = 7$.
  The rule is "divide by 10."
- Write the rule as an equation.
  You are dividing b by 10 to get c, so the equation is $b \div 10 = c$.
- Write in the rest of the numbers in the table using the equation.
  $50 \div 10 = 5$; $30 \div 10 = 3$.

So, the rule is "divide by 10," the equation for the rule is $b \div 10 = c$, and the remaining numbers in the table are 5 and 3.

Find a rule. Write your rule as an equation.
Use your rule to find the missing numbers.

1.

Input, r	2	3	5	6	8
Output, s	18	27	45	☐	☐

____

2.

Input, a	3	6	7	9	10
Output, b	12	24	28	☐	☐

____

3.

Input, r	6	18	21	33	9
Output, s	2	6	7	☐	☐

____

4.

Input, p	2	4	6	8	9
Output, m	12	24	36	☐	☐

____

5.

Input, e	50	30	25	10	☐
Output, f	10	6	5	☐	1

____

6.

Input, w	80	70	40	☐	☐
Output, z	40	35	20	10	5

____

AF 1.5 Understand that an equation such as $y = 3x + 5$ is a prescription for determining a second number when a first number is given.

Name____ **Lesson 6.8**

Patterns: Find a Rule

Find a rule. Write your rule as an equation. Use your rule to find the missing numbers.

1.

Input, *c*	4	8	32	128	512
Output, *d*	1	2	8	■	■

____

2.

Input, *r*	4	5	6	7	8
Output, *s*	8	10	12	■	■

____

3.

Input, *a*	10	20	30	40	50
Output, *b*	1	2	3	■	■

____

4.

Input, *m*	85	80	75	70	65
Output, *n*	17	16	15	■	■

____

Use the rule and the equation to fill in the input/output table.

5. Multiply *a* by 3, subtract 1.
$a \times 3 - 1 = b$

Input, *a*	1	2	3	4	5
Output, *b*	2	■	■	■	■

____

6. Divide *c* by 2, add 1.
$c \div 2 + 1 = d$

Input, *c*	2	4	6	8	10
Output, *d*	2	■	■	■	■

____

Problem Solving and Test Prep

7. Use Data Use the label. Hal has 3 servings of milk a day. How many grams of protein will he get in 5, 6, and 7 days? Write an equation.

____

____

____

____

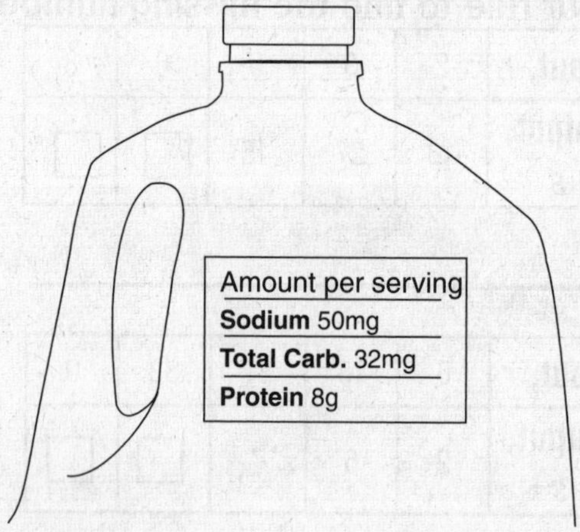

8. What equation shows a rule for the table?

Input, *p* (pints)	1	2	3	4	5
Output, *c* (cups)	2	4	6	8	10

____

9. What equation shows a rule for the table?

Input, *p*	2	4	6	8	10
Output, *g*	6	12	18	24	30

____

Name_______________________________________

Mental Math: Multiplication Patterns

You can use basic facts and mental math to multiply whole numbers by multiples of 10, 100, 1,000.

Use mental math to complete the pattern.

- 9×6 is a basic math fact. — $9 \times 6 = 54$
- Use the basic fact to find 9×60. Since 60 has one zero, add one zero to the product. — $9 \times \mathbf{60} = \mathbf{540}$
- Use the basic fact to find 9×600. Since 600 has two zeros, add two zeros to the product. — $9 \times \mathbf{600} = \mathbf{5{,}400}$

Use mental math to complete the pattern.

- 4×5 is a basic math fact. — $4 \times 5 = 20$
- Use the basic fact to find 4×50. Since 50 has one zero, add one zero to the product. — $4 \times \mathbf{50} = \mathbf{200}$
- Use the basic fact to find 4×500. Since 500 has two zeros, add two zeros to the product. — $4 \times \mathbf{500} = \mathbf{2{,}000}$

Use mental math to complete the pattern.

1. $7 \times 5 = 35$

$7 \times 50 =$ ________

$7 \times 500 =$ ________

2. $3 \times 4 = 12$

$3 \times 40 =$ ________

$3 \times 400 =$ ________

3. $9 \times 8 = 72$

$9 \times 80 =$ ________

$9 \times 800 =$ ________

4. $8 \times 6 = 48$

$8 \times 60 =$ ________

$8 \times 600 =$ ________

5. $2 \times 7 = 14$

$2 \times 70 =$ ________

$2 \times 700 =$ ________

6. $4 \times 4 = 16$

$4 \times 40 =$ ________

$4 \times 400 =$ ________

7. $6 \times 3 =$ ________

$6 \times 30 =$ ________

$6 \times 300 =$ ________

8. $7 \times 7 =$ ________

$7 \times 70 =$ ________

$7 \times 700 =$ ________

9. $4 \times 6 =$ ________

$4 \times 60 =$ ________

$4 \times 600 =$ ________

Name____

Mental Math: Multiplication Patterns

Use mental math to complete the pattern.

1. $7 \times 6 = 42$
$7 \times 60 =$ ______
$7 \times 600 =$ ______
$7 \times 6{,}000 =$ ______

2. $3 \times 8 = 24$
$3 \times 80 =$ ______
$3 \times 800 =$ ______
$3 \times 8{,}000 =$ ______

3. $9 \times 7 = 63$
$9 \times 70 =$ ______
$9 \times 700 =$ ______
$9 \times 7{,}000 =$ ______

Use patterns and mental math to find the product.

4. 2×30 ______

5. 3×700 ______

6. $9 \times 4{,}000$ ______

7. 7×800 ______

ALGEBRA **Find the value of** ***n*****.**

8. $2 \times n = 42{,}000$
$n =$ ______

9. $7 \times 400 = n$
$n =$ ______

10. $8 \times n = 16{,}000$
$n =$ ______

11. $n \times 500 = 4{,}500$
$n =$ ______

Problem Solving and Test Prep

12. Windsurfing costs $20 a day at New State Park. Jen windsurfed for 5 days. Paul windsurfed for 7 days. How much more did Paul pay than Jen?

______

13. Every carload of people entering the state park pays $7. In January, there were 200 cars that entered the park. In July, there were 2,000 cars that entered the park. How much more money did the park collect in July than in January?

______

14. Which number is missing from this equation?

$\blacksquare \times 7 = 3{,}500$

______

15. Which number is missing from this equation?

$8 \times \blacksquare = 32{,}000$

______

Name________________________________ Lesson 7.2

Mental Math: Estimate Products

You can use rounding and compatible numbers to estimate products.

Round the money amount. Then use mental math to estimate the product.

6 × $3.25

- Round $3.25 to the nearest dollar. — $3.25 rounds to $3.00
- Use mental math. — $6 \times 3 = 18$; $6 \times 30 = 180$; $6 \times 300 = 1{,}800$
- Add a decimal point and dollar sign. — 6 × $3.00 = $18.00

So, 6 × $3.25 is about $18.00.

Estimate the product. Write the method.

7 × $7.59

- Use a compatible number that is easy to compute mentally. — 7 × $7.59 ↓ 7 × $8.00
- Use mental math. — $7 \times 8 = 56$; $7 \times 80 = 560$; $7 \times 800 = 5{,}600$
- Add a decimal point and dollar sign. — 7 × $8.00 = $56.00

So, 7 × $7.59 is about $56.00.

Round the greater factor. Then use mental math to estimate the product.

1. 6 × 316 ____________

2. 5 × 29 ____________

3. 4 × 703 ____________

Estimate the product. Write the method you used.

4. 3 × 508 ____________

5. 7 × 22 ____________

6. 8 × 3,061 ____________

Name________________________________

Mental Math: Estimate Products

Estimate the product. Write the method.

1. 2×49 **2.** 7×31 **3.** 5×58 **4.** 4×73

5. 3×27 **6.** 8×26 **7.** 4×25 **8.** 5×82

9. 6×53 **10.** 9×47 **11.** 6×71 **12.** 5×31

13. 88×2 **14.** 29×8 **15.** 65×4 **16.** 39×7

Problem Solving and Test Prep

USE DATA For 17–18, use the table.

17. About how many pencils will Haley use in 8 months?

18. About how many more pencils will Haley use in ten months than Abby will use in ten months?

Pencils Used Each Month

Name	Number of Pencils
Haley	18
Abby	12
Bridget	17
Kelsey	21

19. Which number sentence gives the best estimate of 6×17?

A 6×20

B 6×25

C 6×10

D 6×5

20. Which number sentence would give the best estimate of 6×51?

A 6×5

B 6×45

C 6×50

D 6×55

Name ______________________

Spiral Review

For 1–2, compare. Write <, >, or = for each ◯.

1. $\frac{1}{2}$ ◯ $\frac{3}{4}$

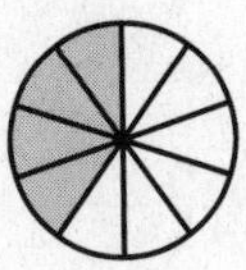

2. $\frac{4}{10}$ ◯ $\frac{6}{10}$

For 6–7, use the table.

Samantha's Sock Drawer	
Color	**Number of Socks**
Black	6
Blue	2
White	10

6. Is it *likely* or *unlikely* that Samantha will pull one white sock from her drawer?

____

7. Is it *likely* or *unlikely* that Samantha will pull one blue sock from her drawer?

____

For 3–5, use the map.

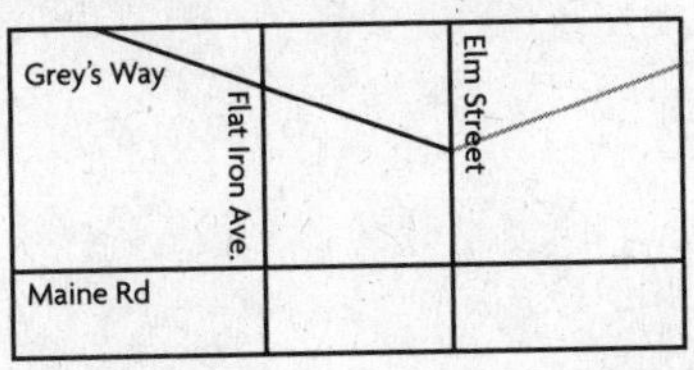

3. Which street appears to be parallel to Elm street?

____

4. Which streets appears to intersect Maine Rd?

____

____

5. Which streets appears to intersect Grey's Way?

____

For 8–13, complete to make the equation true.

8. $2 + \square + 7 = 5 + 7$

9. $1 + 9 - 5 = \square - 5$

10. $25 - 15 - \square = 10 - 6$

11. $4 + 3 + 2 = \square + 2$

12. $12 - 4 + 2 = \square + 2$

13. $\square - 8 + 2 = 13 + 2$

Name ____________________

Problem Solving Workshop Strategy: Draw a Diagram

Carol, Juan, Tami, and Brad are the first four people in line to see the Open Ocean exhibit. Carol is not first in line. Tami has at least two people ahead of her in line. Juan is third. Give the order of the four people in line.

Read to Understand

1. What are you asked to find out in this problem?

____

Plan

2. What kind of diagram can you draw to help you solve this problem?

____

Solve

3. What is the order of the four people in line? How did you solve this problem?

____

Check

4. What other strategy could you use to solve this problem?

____

Draw a diagram to solve.

5. Ann, Tom, and Jed live in the same neighborhood. Ann's yard is 80 feet longer than Jed's yard. Tom's yard is 20 feet longer than twice the length of Jed's yard. Jed's yard is 40 feet long. Who has the longest yard?

____

6. Sue, Adam, Kat, and Paul are in line. Sue is third in line. Adam is not last in line. Paul is two places ahead of Sue in line. Give the order of these four people in line.

____

Name________________________________

Problem Solving Workshop Strategy: Draw a Diagram

Problem Solving Strategy Practice

Draw a diagram to solve.

1. Jan walks 5 blocks north, 1 block east, and 3 more blocks north. Then she walks 1 block west and 1 block south. How far is Jan from where she started?

____

2. Nick's toy boat is 24 inches long. Ben has 10 toy boats, but they are each only 6 inches long. How many of Ben's boats, laid end to end, would it take to match the length of Nick's boat?

____

Mixed Strategy Practice

USE DATA For 3–6, use the information in the table.

3. How many times greater is the maximum lifespan of 6 Bowhead Whales than that of 1 Fin Whale?

____

4. List the types of whales shown in order from shortest lifespan to longest lifespan.

____

____

Whales' Maximum Life Span

Whale Type	Years
Pilot	60
Orca	90
Fin	60
Blue	80
Bowhead	130

5. Look at Exercise 3. Write a similar problem using two different types of whales.

____

____

____

6. Write three different expressions that equal the life span of the Bowhead whale, using one or more operations.

____

____

____

Name____ **Lesson 7.4**

Model 3-Digit by 1-Digit Multiplication

The number sentence 2×152 means "152 + 152." It also means "show 152 two times." You can use base-ten blocks to model this problem in order to solve it.

Use base-ten blocks to model the product. Record your answer.

2×432

- The smaller number, 2, tells how many times to model the larger number, 432.
- So, 2×432 means to model 432 two times, or make 2 models of 432.
- Add the hundreds, the tens, and the ones to find the product.
- There are 8 hundreds which is equal to 800. There are 6 tens which is equal to 60. There are 4 ones, which is equal to 4.
- $800 + 60 + 4 = 864$
- So, $2 \times 432 = 864$.

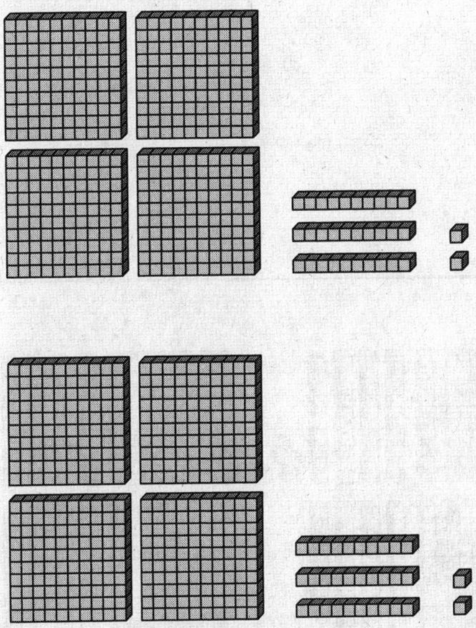

Use base-ten blocks to model the product. Record your answer.

1. 3×214 ______________

2. 4×121 ______________

3. 3×326 ______________

4. 2×224 ______________

5. 4×133 ______________

6. 5×116 ______________

7. 3×421 ______________

8. 4×331 ______________

9. 3×123 ______________

10. 3×144 ______________

11. 5×161 ______________

12. 7×112 ______________

Name ____________________

Model 3-Digit by 1-Digit Multiplication

Find the product.

1.

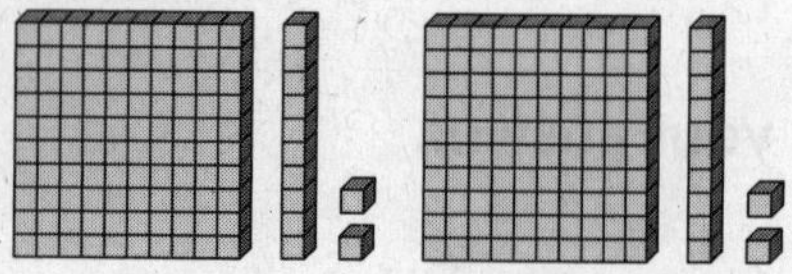

____

2.

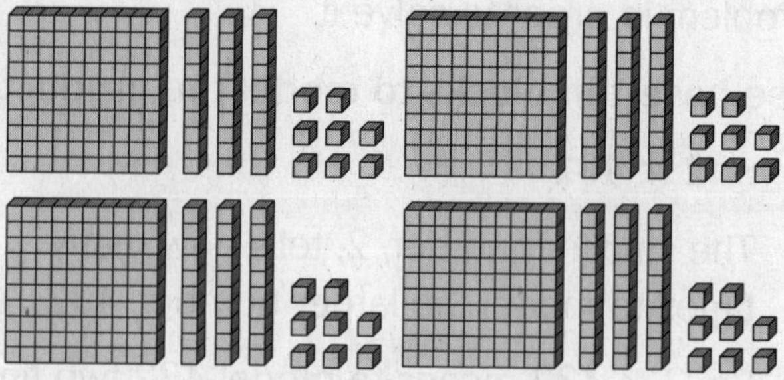

____

3.

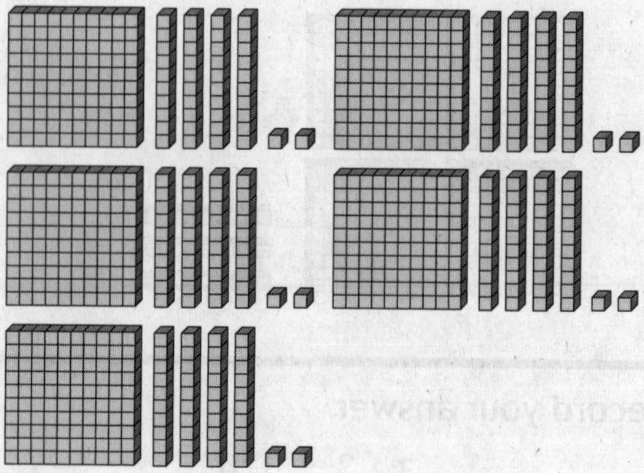

____

4.

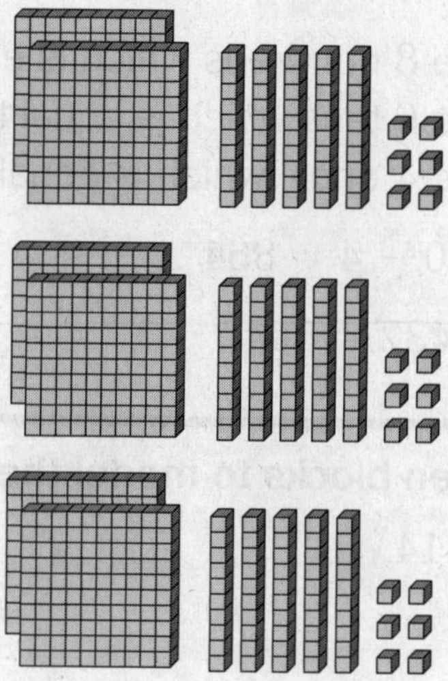

____

Use base-ten blocks to model the product. Record your answer.

5. 2 × 101 ______

6. 3 × 310 ______

7. 5 × 192 ______

8. 4 × 257 ______

9. 3 × 436 ______

10. 6 × 288 ______

11. 7 × 285 ______

12. 5 × 437 ______

Name____________________________________ Lesson 7.5

Record 3-Digit by 1-Digit Multiplication

Use partial products and place value with regrouping to multiply.

Use partial products.

7 × 332

Estimate the product. 332 rounds to 300; 7 × 300 = 2,100

Multiply the 2 ones, or 2, by 7.

$$\begin{array}{r}332\\ \times\ 7\\ \hline\end{array} \quad \text{or} \quad \begin{array}{r}2\\ \times\ 7\\ \hline 14\end{array}$$

Multiply the 3 tens, or 30, by 7.

$$\begin{array}{r}332\\ \times\ 7\\ \hline\end{array} \quad \text{or} \quad \begin{array}{r}30\\ \times\ 7\\ \hline 210\end{array}$$

Multiply the 3 hundreds, or 300, by 7.

$$\begin{array}{r}332\\ \times\ 7\\ \hline\end{array} \quad \text{or} \quad \begin{array}{r}300\\ \times\ 7\\ \hline 2{,}100\end{array}$$

Add the products. 14 + 210 + 2,100 = 2,324

So, 7 × 332 = 2,324. Since 2,324 is close to the estimate of 2,100, it is reasonable.

Estimate. Then record the product.

1. 2 × 181 ______

2. 3 × 314 ______

3. 4 × 156 ______

4. 5 × 212 ______

5. 6 × 313 ______

6. 2 × 427 ______

7. 5 × 367 ______

8. 6 × 148 ______

9. 7 × 134 ______

10. 4 × 510 ______

11. 2 × 615 ______

12. 4 × 311 ______

Name______________________________ Lesson 7.5

Record 3-Digit by 1-Digit Multiplication

Estimate. Then record the product.

1. 3×518 ______
2. 7×336 ______
3. 5×731 ______
4. 6×492 ______
5. 8×254 ______
6. 4×836 ______
7. 8×633 ______
8. 9×126 ______

ALGEBRA Find the missing digit.

9. $55\square \times 4 = 2{,}224$
10. $52\square \times 6 = 3{,}138$
11. $815 \times \square = 2{,}445$
12. $\square 76 \times 5 = 1{,}880$
13. $157 \times 8 = 1{,}2\square 6$
14. $676 \times 2 = 1{,}\square 52$
15. $424 \times 7 = \square{,}968$
16. $128 \times 9 = 1{,}15\square$

Problem Solving and Test Prep

17. Sergio's media player contains 135 classical selections. It contains 5 times as many country selections as classical selections. How many selections does Sergio have in all?

______

18. Marie has 6 different boxes of jazz sheet music. Each box holds 112 pages. Write an equation to show how many pages of sheet music Marie has in all. Solve the equation.

______

19. Which expression shows how to multiply 4×657 using place value and expanded form?
    - A $4 \times 600 + 4 \times 50 + 4 \times 7$
    - B $4 \times 6 + 4 \times 5 + 4 \times 7$
    - C $4 + 6 + 4 + 5 + 4 + 7$
    - D $4 \times 600 + 4 \times 500 + 4 \times 7$

20. What expression shows how to multiply 4×367 using place value and expanded form?
    - A $3 \times 400 + 3 \times 70 + 3 \times 60$
    - B $7 \times 600 + 7 \times 40 + 7 \times 30$
    - C $6 \times 400 + 6 \times 30 + 6 \times 7$
    - D $4 \times 300 + 4 \times 60 + 4 \times 7$

Name____

Multiply 4-Digit Numbers and Money

When you multiply 3- and 4-digit numbers, you will often need to regroup.

When you multiply money, you ignore the dollar sign and the decimal, and add them to your final product.

Estimate. Then find the product.

$$\begin{array}{r} \$13.24 \\ \times \quad 7 \\ \hline \end{array}$$

Estimate the product.	$13.24 rounds to $10; $10 × 7 = $70	
Multiply the 4 ones by 7. Regroup the 2 tens.	$\begin{array}{r} {\scriptstyle 2} \\ 132\mathbf{4} \\ \times \ \mathbf{7} \\ \hline 8 \end{array}$	$4 \times 7 = 28$
Multiply the 2 tens by 7. Add the regrouped tens. Regroup 1 hundred.	$\begin{array}{r} {\scriptstyle 1\,2} \\ 13\mathbf{2}4 \\ \times \ \mathbf{7} \\ \hline 68 \end{array}$	$2 \times 7 = 14$ $14 + 2 = 16$
Multiply 3 hundreds by 7. Add the regrouped hundreds. Regroup 2 thousands.	$\begin{array}{r} {\scriptstyle 2\,1\,2} \\ 1\mathbf{3}24 \\ \times \ \mathbf{7} \\ \hline 268 \end{array}$	$3 \times 7 = 21$ $21 + 1 = 22$
Multiply 1 thousand by 7. Add the regrouped thousands.	$\begin{array}{r} {\scriptstyle 2\,1\,2} \\ \mathbf{1}324 \\ \times \ \mathbf{7} \\ \hline 9268 \end{array}$	$1 \times 7 = 7$ $7 + 2 = 9$
Add a dollar sign and a decimal point to your answer.	$92.68	

So, $13.24 × 7 = $92.68.
Since $92.68 is close to the estimate of $70, it is reasonable.

Estimate. Then find the product.

1. $\begin{array}{r} \$3,184 \\ \times \quad 2 \\ \hline \end{array}$

2. $\begin{array}{r} 1,828 \\ \times \quad 4 \\ \hline \end{array}$

3. $\begin{array}{r} \$26.37 \\ \times \quad 5 \\ \hline \end{array}$

4. $\begin{array}{r} 6,916 \\ \times \quad 7 \\ \hline \end{array}$

Name____________________

Multiply 4-Digit Numbers and Money

Estimate. Then find the product.

1. 1,379 × 4

2. $64.11 × 3

3. $4,279 × 8

4. 1,563 × 9

5. $5,218 × 3

6. 4,156 × 7

7. $81.27 × 5

8. 2,453 × 6

Compare. Write <, >, or = for each ◯.

9. 2 × 9,736 ◯ 3 × 3,299

10. 6 × $17.50 ◯ 7 × $15.00

11. 9 × 3,998 ◯ 6 × 4,557

12. 5 × $6,115 ◯ 4 × $7,676

13. 7 × 2,115 ◯ 2 × 7,449

14. 4 × 3,441 ◯ 6 × 2,113

Problem Solving and Test Prep

15. What number is 630 less than 4 times 4,721?

____

16. Charlie buys 3 bear statues for $21.45 each. He gives the cashier a $100 bill. How much change will Charlie receive?

____

17. It is 3,014 miles one way from Rob's house in Florida to Lynn's house in California. What is the round–trip distance?

A 6,028 miles
B 6,000 miles
C 3,014 miles
D 3,000 miles

18. It is 1,260 miles from San Diego to Seattle. What is the round–trip distance?

A 2,420
B 1,262
C 2,520
D 1,462

Spiral Review

For 1–4, divide.
Write the method you used.

1. 268 ÷ 4 ________

2. 6,442 ÷ 2 ________

3. $9\overline{)810}$ ________

4. $7\overline{)3,976}$ ________

For 5–8, tell if each figure is a polygon. Write *yes* or *no*.

5. ________

6. ________

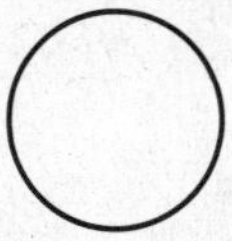

7. ________

8. ________

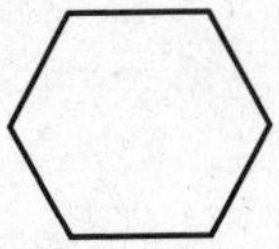

For 9–11, list the possible outcomes for each object.

9. 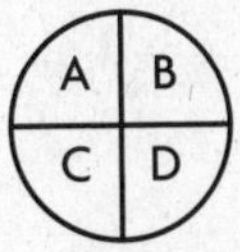

____

10.

1	2	3
4	5	6

____

11. 

____

For 12–15, use the rule and equation to fill in the input/output table.

12. add three, $l + 3 = k$

Input	l	1	2	3	4
Output	k				

13. subtract 3, $g - 3 = h$

Input	g	3	4	5	6
Output	h				

14. add 15, $r + 15 = t$

Input	r	5	6	7	8
Output	t				

15. subtract 10, $w - 10 = z$

Input	w	10	15	20	25
Output	z				

Name____ **Lesson 7.7**

Multiply With Zeros

Estimate. Then record the product.

$$\begin{array}{r} \$10.20 \\ \times \quad 7 \\ \hline \end{array}$$

Estimate the product. $10.20 rounds to $10; $10 × 7 = $70

Multiply 0 ones, or 0, by 7.

$$\begin{array}{r} 1{,}02\mathbf{0} \\ \times \quad \mathbf{7} \\ \hline \end{array} \qquad \begin{array}{r} 0 \\ \times\ 7 \\ \hline 0 \end{array}$$

Multiply 2 tens, or 20, by 7.

$$\begin{array}{r} 1{,}020 \\ \times \quad 7 \\ \hline 0 \end{array} \qquad \begin{array}{r} 2 \\ \times\ 7 \\ \hline 14 \end{array}$$

Place the 4 in the tens column.
Place the 1 above the hundreds column.

$$\begin{array}{r} \scriptstyle 1 \quad\ \ \\ 1{,}020 \\ \times \quad 7 \\ \hline 40 \end{array}$$

Multiply 0 hundreds, or 0, by 7.
Add the regrouped hundred to the product: 0 + 1 = 1.

$$\begin{array}{r} \scriptstyle 1 \quad\ \ \\ 1{,}\mathbf{0}20 \\ \times \quad 7 \\ \hline 40 \end{array} \qquad \begin{array}{r} 0 \\ \times\ 7 \\ \hline 0 \end{array}$$

Place the 1 in the hundreds column.

$$\begin{array}{r} \scriptstyle 1 \quad\ \ \\ 1{,}020 \\ \times \quad 7 \\ \hline 140 \end{array}$$

Multiply 1 thousand, or 1,000, by 7.

$$\begin{array}{r} \scriptstyle 1 \quad\ \ \\ \mathbf{1}{,}020 \\ \times \quad 7 \\ \hline 140 \end{array} \qquad \begin{array}{r} 1 \\ \times\ 7 \\ \hline 7 \end{array}$$

Place the 7 in the thousands place.
Add a decimal point and dollar sign to the product.

$$\begin{array}{r} \scriptstyle 1 \quad\ \ \\ 1{,}020 \\ \times \quad 7 \\ \hline 7{,}140 \end{array}$$

So, $10.20 × 7 = $71.40. Since $71.40 is close to the estimate of $70, it is reasonable.

Estimate. Then record the product.

1. $\begin{array}{r} 63 \\ \times\ 3 \\ \hline \end{array}$ **2.** $\begin{array}{r} 83 \\ \times\ 5 \\ \hline \end{array}$ **3.** $\begin{array}{r} 36 \\ \times\ 8 \\ \hline \end{array}$ **4.** $\begin{array}{r} 67 \\ \times\ 6 \\ \hline \end{array}$

Name________________________________

Multiply with Zeros

Estimate. Then find the product.

1. 3,044 × 3

2. 4,700 × 5

3. $75.05 × 6

4. $43.05 × 4

5. 8,077 × 2

6. 1,130 × 7

7. $30.45 × 6

8. $51.03 × 8

9. 4 × 2,340

10. 6 × $30.55

11. 7 × $1,023

12. 5 × 3,405

13. 3,240 × 3

14. 4,860 × 5

15. 2,106 × 8

16. 6,004 × 9

Problem Solving and Test Prep

17. Saya pays $35.90 for one ticket to the circus. How much will 8 tickets cost?

18. Raul buys 3 packs of sports stickers. Each pack has 105 stickers. How many stickers does Raul buy in all?

19. Mr. Bench buys 4 pairs of pajamas for $20.98 each. How much does Mr. Bench spend?

A $80.92

B $81.92

C $82.92

D $83.92

20. Carl buys 6 books for summer reading. Each book has 203 pages. How many pages will Carl read over the summer?

A 1,209

B 818

C 1,218

D 809

Name________________________________ Lesson 8.1

Mental Math: Multiplication Patterns

You can use basic math facts and patterns to help you multiply multiples of 10, 100, and 1,000.

Use patterns and mental math to find the product.

12 × 6,000

- 12 × 6 is a basic math fact. — 12 × 6 = 72
- Use the basic math fact to find 12 × 6,000.
- Since 6,000 has three zeros, add three zeros to the product. — 12 × 6,**000** = 72,**000**

So, 12 × 6,000 = 72,000.

11 × 1,000

- 11 × 1 is a basic math fact. — 11 × 1 = 11
- Use the basic math fact to find 11 × 1,000.
- Since 1,000 has three zeros, add three zeros to the product. — 11 × 1,**000** = 11,**000**

So, 11 × 1,000 = 11,000.

Use mental math and patterns to find the product.

1. 22 × 10 ______	**2.** 35 × 100 ______	**3.** 67 × 100 ______	**4.** 48 × 1,000 ______
5. 40 × 100 ______	**6.** 31 × 1,000 ______	**7.** 49 × 10 ______	**8.** 50 × 10 ______
9. 80 × 20 ______	**10.** 10 × 300 ______	**11.** 42 × 1,000 ______	**12.** 90 × 900 ______
13. 19 × 10 ______	**14.** 93 × 100 ______	**15.** 60 × 100 ______	**16.** 17 × 1,000 ______

Name ____________________

Mental Math: Multiplication Patterns

Use mental math and patterns to find the product.

1. $50 \times 3,000$ ______

2. 7×40 ______

3. $8 \times 1,000$ ______

4. 50×700 ______

5. $12 \times 2,000$ ______

6. 70×200 ______

7. 11×120 ______

8. 90×80 ______

ALGEBRA Copy and complete the tables using mental math.

9. 1 roll = 20 nickels

Number of rolls	20	30	40	50	600
Number of Nickles	400	■	■	■	■

10. 1 roll = 60 dimes

Number of rolls	20	30	40	50	600
Number of Dimes	1,200	■	■	■	■

	x	7	60	700	8,000
11.	40	280	■	■	■
12.	60	■	■	■	480,000

	x	8	40	500	9,000
13.	50	400	■	■	■
14.	90	■	■	■	810,000

Problem Solving and Test Prep

USE DATA For 15–16, use the table.

Insect Lengths	
Insect	**Length (in mm)**
Carpenter Bee	19
Drywood Termite	12
Fire Ant	4
Termite	12
Wasp	15

15. How long would a drywood termite magnified by 6,000 appear to be?

______

16. Which would appear longer, a drywood termite magnified 1,200 times or a wasp magnified 900 times?

______

17. How many zeros are in the product of 400×500?

A 4 **C** 6

B 5 **D** 7

18. How many zeros must be in the product of 1,000 and any factor?

______

Name____ Lesson 8.2

Multiply by Tens

Whenever you multiply a number by a multiple of 10, the answer will have a zero in the ones place. For example, $24 \times 60 = 1{,}440$.

Choose a method. Then find a product.

78 × 60

Step 1: Find the number in the problem that is a multiple of 10. Take the zero away from that number.

60 is the multiple of 10. Change 60 to 6.

$$\begin{array}{r} {}^{4} \\ 78 \\ \times\ 6 \\ \hline 468 \end{array}$$

Step 2: Multiply 78 × 6. Regroup as needed.

Step 3: Replace the zero in the ones place.

Think: The product of 78 × 6 is 468. Placing a zero in the ones place makes it 4,680.

So, 78 × 60 = 4,680.

Choose a method. Then find a product.

1. 90 × 18 ______
2. 46 × 50 ______
3. 50 × 32 ______
4. 22 × 40 ______
5. 60 × 28 ______
6. 36 × 30 ______
7. 12 × 20 ______
8. 47 × 40 ______
9. 66 × 60 ______

Name______________________________ **Lesson 8.2**

Multiply by Tens

Choose a method. Then find the product.

1. 20×17 ________
2. 15×60 ________
3. 66×50 ________
4. 78×30 ________
5. 96×40 ________
6. 90×46 ________
7. 52×80 ________
8. 70×29 ________

ALGEBRA Find the missing digit.

9. $22 \times 3■ = 660$ ________
10. $60 \times 37 = 2,■20$ ________
11. $5■ \times 80 = 4,480$ ________
12. $■0 \times 77 = 3,080$ ________
13. $40 \times 44 = ■,760$ ________
14. $90 \times 83 = 7,4■0$ ________

Problem Solving and Test Prep

USE DATA For 15–17, use the table.

15. How many frames does it take to produce 60 seconds of *Snow White*?

____

16. Are there more frames in 30 seconds of *Pinocchio* or 45 seconds of *The Enchanted Drawing*?

____

Animated Productions

Title	Date Released	Frames Per Second
The Enchanted Drawing©	1900	20
Little Nemo©	1911	16
Snow White and the Seven Dwarfs©	1937	24
Pinocchio©	1940	19
The Flintstones™	1960–1966	24

17. Sadie runs 26 miles each week. How many miles will Sadie run in 30 weeks?

A 780
B 720
C 690
D 700

18. If gourmet cookies cost $12 a pound, how much does it cost to purchase 30 pounds of cookies?

A $360
B $3,600
C $36
D $36,000

Name________________________________ Lesson 8.3

Mental Math: Estimate Products

You can use compatible numbers and rounding to estimate products.

Use rounding and mental math to estimate the product.

$23 × 62

- Round both numbers to the nearest ten.

$23 rounds to $20
62 rounds to 60

- Rewrite the problem using the rounded numbers.
- Use mental math.

$20 × 60
$2 × 6 = $12
$20 × 6 = $120
$20 × 60 = $1,200

So, $23 × 62 is about $1,200.

Use compatible numbers and mental math to estimate products.

41 × 178

41 × 178

- Use compatible numbers that are easy to compute mentally.

40 × 200

- Use mental math.

4 × 2 = 8
4 × 20 = 80
4 × 200 = 800
40 × 200 = 8,000

So, 41 × 178 is about 8,000.

Estimate the product. Choose the method.

1. 78 × 21 ________ **2.** $46 × 59 ________ **3.** 81 × 33 ________ **4.** 67 × 102 ________

5. $42 × 88 ________ **6.** 51 × 36 ________ **7.** 73 × 73 ________ **8.** $44 × 99 ________

9. 92 × 19 ________ **10.** 26 × 37 ________ **11.** 193 × 18 ________ **12.** 58 × 59 ________

Name______________________________

Mental Math: Estimate Products

Choose the method. Estimate the product.

1. 34 × 34 ________

2. 27 × 42 ________

3. 41 × 55 ________

4. 17 × 39 ________

5. 72 × 21 ________

6. 54 × 67 ________

7. 58 × 49 ________

8. 64 × 122 ________

9. 93 × 93 ________

10. 19 × 938 ________

11. 42 × 666 ________

12. 71 × 488 ________

Problem Solving and Test Prep

13. **Fast Fact** A serving of watermelon has 27 grams of carbohydrate. About how many grams of carbohydrate do 33 servings contain?

____

14. There are 52 homes in Ku's neighborhood. If the door on each refrigerator in each home is opened 266 times a week, and each home has one refrigerator, about how many times are the doors opened in all?

____

15. Choose the best estimate for the product of 48 × 637.

A 20,000
B 24,000
C 30,000
D 34,000

16. An assembly line produces enough cotton for 1,500 T-shirts a day. How could you estimate the number of T-shirts 45 assembly lines produce?

A 1,500 × 50
B 30 × 1,200
C 2,000 × 100
D 150 × 4,500

Spiral Review

For 1–2, solve a simpler problem.

1. Mrs. Torino's class sold wrapping paper to raise money for a field trip to the marine park. For each roll of paper that they sold, they earned $1.50. The class sold 52 rolls of paper. How much money did they raise?

____

2. For a snack, Jill is going to serve cookies. Each person will get 3 cookies. If there are 63 people, how many cookies will Jill need?

____

5. The bar graph below shows the temperatures recorded at the same time each day for seven days. Based on the data, what is the temperature on Day 4?

____

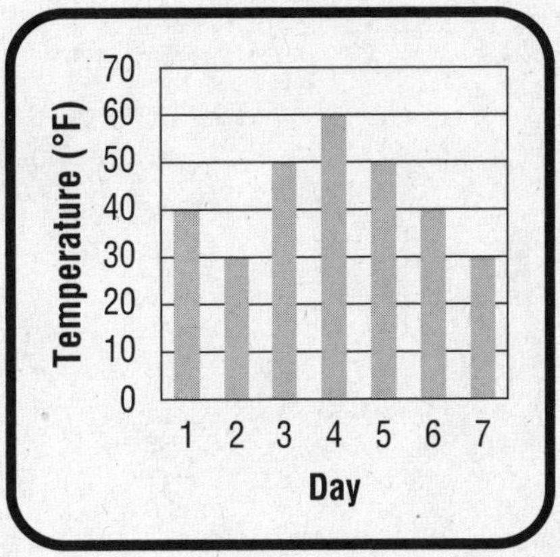

For 3–4, find the missing length.

3. 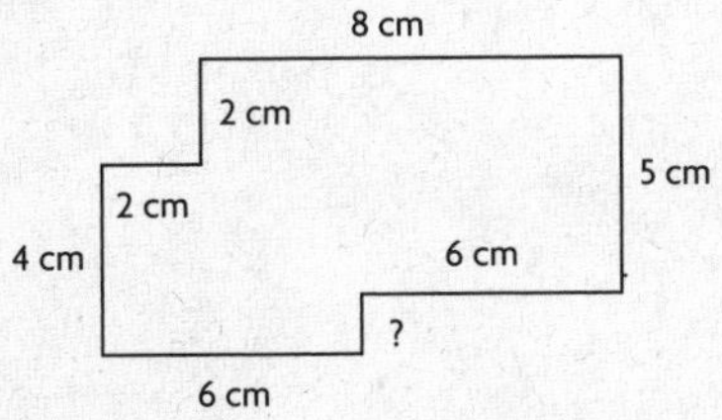

Perimeter = 34 cm

____

4. 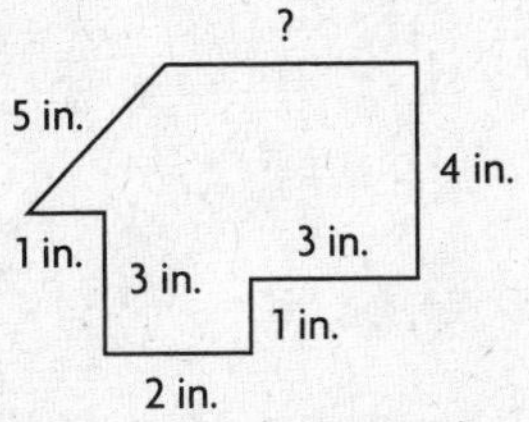

Perimeter = 24 in.

____

For 6–10, find the value of each expression if $x = 2$ and $y = 7$.

6. $7 + x$

____

7. $12 - y$

____

8. $36 - (x + y)$

____

9. $(y + 2) - x$

____

10. $14 + x$

____

Name____________________

Problem Solving Workshop Skill: Multistep Problems

Rides at a carnival cost $2 each. There were 20 rides. A group of 22 students went on every ride once. How much money was spent to ride all of the rides?

1. What are you asked to find out?

2. Can you solve the problem using just one step? Why or why not?

3. Which operation or operations do you need to use to solve the problem?

4. How much money was spent on all the rides?

5. What can you do to check your answer?

Solve the problem.

6. Gina owns a music store. On Saturday, her store sold 56 CDs for $8 each. On Sunday, her store sold 32 CDs for $7 each. How much money did Gina's store receive that weekend?

7. Dan and Peter both enjoy hiking through the woods. Dan hiked 6 miles a day for 5 days. Peter hiked 9 miles a day for 4 days. How many more miles did Peter hike than Dan?

NS 3.2 Demonstrate an understanding of, and the ability to use, standard algorithms for multiplying a multidigit number by a two-digit number and for dividing a multidigit number by a one-digit number; use relationships between them to simplify computations and to check results.

Name______________________________

Problem Solving Workshop Skill: Multistep Problems

Problem Solving Skill Practice

1. The Pacific Wheel is a ferris wheel that can carry 6 passengers in each of 20 cars in one ride. How many passengers can it carry on a total of 45 rides?

____

2. Bus A travels 532 miles one way. Bus B travels 1,268 miles round trip. Which bus travels the most round-trip miles if Bus A makes 6 trips and Bus B makes 5 trips?

____

3. There are 62 students in all. Twenty-five take only band class. Thirty-four take only art class. The rest take both band and art class. How many students take both band and art?

____

4. Trin bought 6 T-shirts at $17 each. Ron bought 7 shirts at the same price. How much did Trin and Ron spend altogether?

____

Mixed Applications

USE DATA For 5–6, use the table.

5. The Smiths are a family of 7. How much will they spend for admission to the carnival if they go on Saturday night?

____

6. How much will the Smiths save if they go on Monday instead of Saturday?

____

Carnival One Night Admission Tickets

Night	Cost
Monday through Wednesday	$12
Thursday through Friday	$15
Saturday	$20

7. A local carnival has a Ferris wheel with 20 cars that seat 4 people each. Each ride is 10 minutes with 5 minutes to unload and reload. How many people can the Ferris wheel carry in 3 hours?

____

8. Rosa rode the Ferris wheel, the go-carts for 10 minutes, the merry-go-round for 25 minutes, and the roller coaster for 35 minutes. She was on rides for 1 hour and 30 minutes. How long did she ride the Ferris wheel?

____

Name____ **Lesson 9.1**

Model 2-Digit by 2-Digit Multiplication

You can use arrays to multiply 2-digit numbers by 2-digit numbers.

Use the model and partial products to solve.
18 × 19

The array is 18 units wide and 19 units long.

19

18

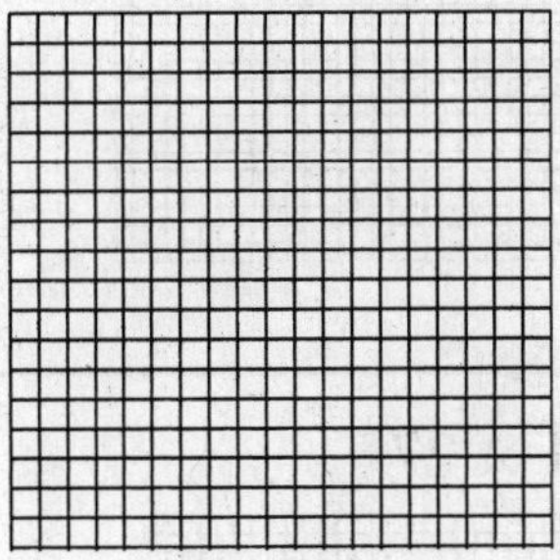

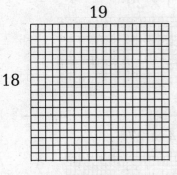

Divide the array into 4 smaller arrays.
Use products you know.

10 9

10

8

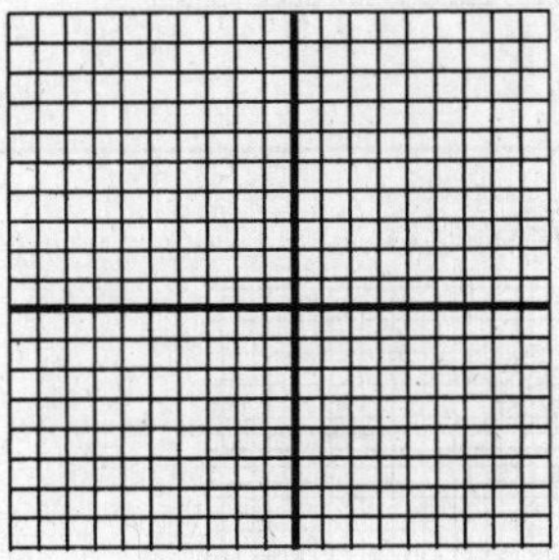

Find the products of smaller arrays.

$10 \times 10 = 100$
$10 \times 8 = 80$
$10 \times 9 = 90$
$9 \times 8 = 72$

Find the sum of the products. $100 + 80 + 90 + 72 = 342$

So, $18 \times 19 = 342$.

Use the model and partial products to solve.

1.

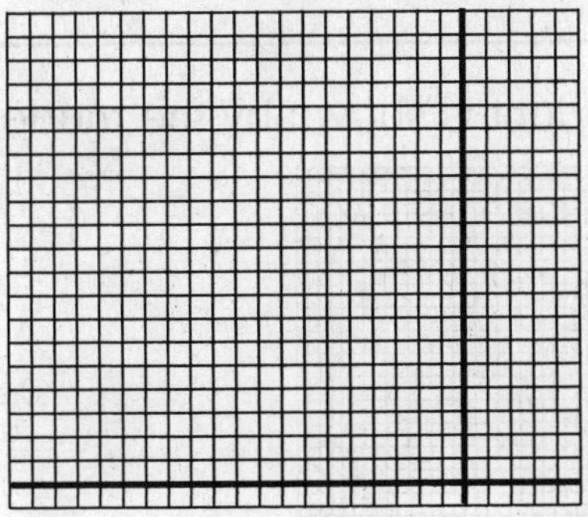

21×25

____

2.

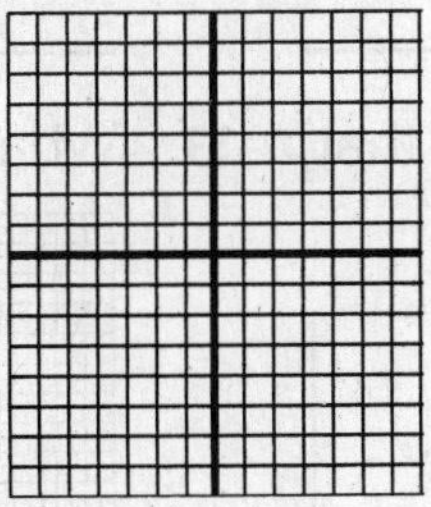

16×14

____

3.

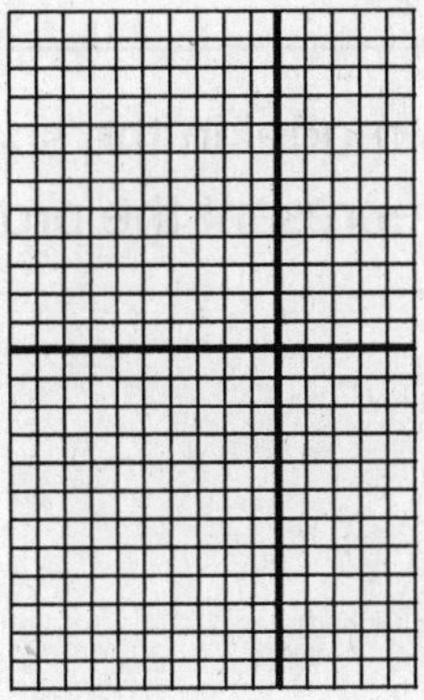

24×15

____

Name_______________________________________

Lesson 9.1

Model 2-Digit by 2-Digit Multiplication

Use the model and partial products to solve.

1. 15 × 29

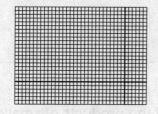

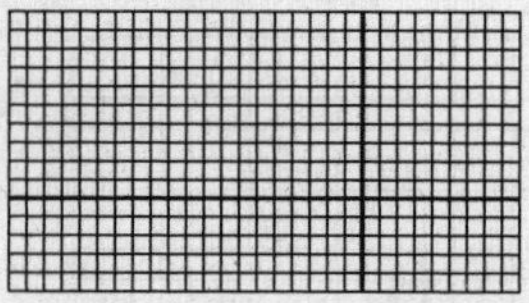

____

2. 17 × 32

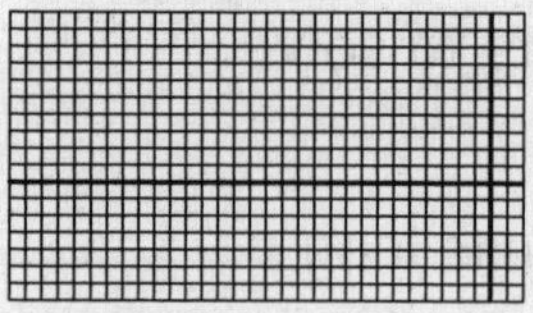

____

3. 19 × 25

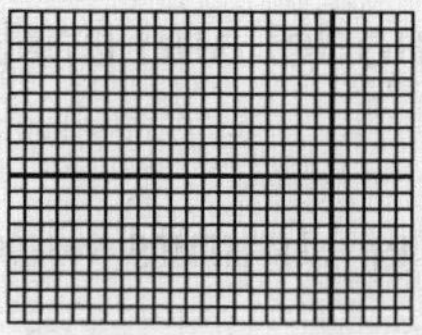

____

4. 14 × 27

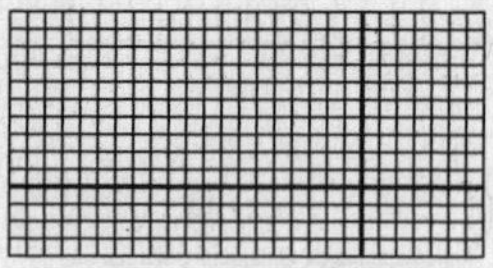

____

5. 16 × 28

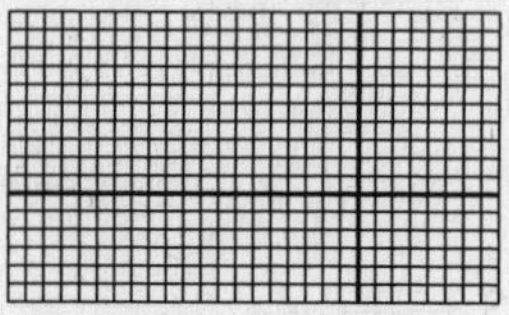

____

6. 19 × 24

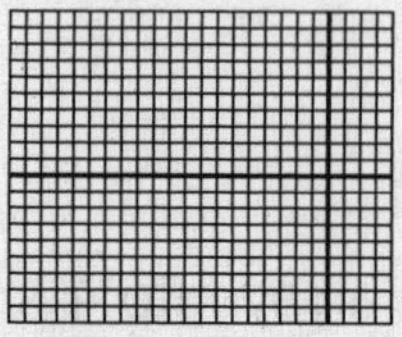

____

7. 17 × 26

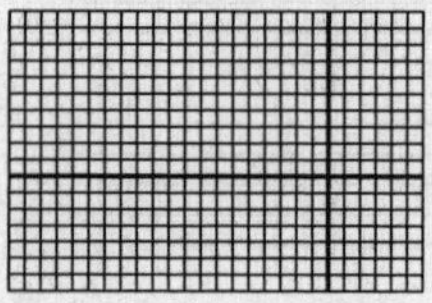

____

8. 18 × 21

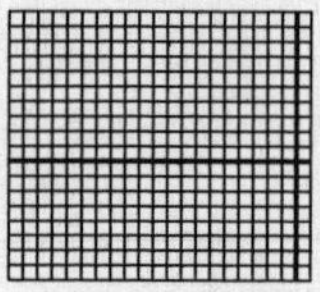

____

9. 26 × 36

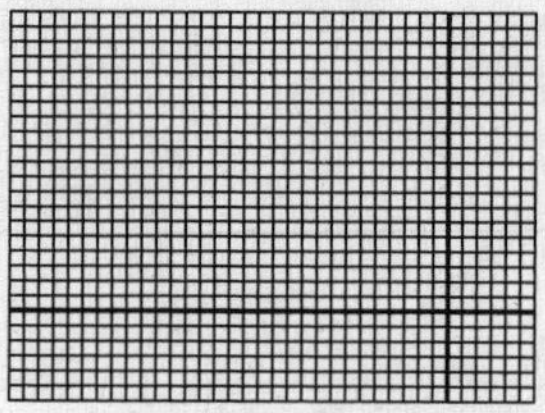

____

Problem Solving and Test Prep

10. The apples from an average tree will fill 20 bushel-sized baskets. If an orchard has 17 average trees, how many baskets of apples can it produce?

____

11. If each student eats about 65 apples a year, how many apples will the 27 students in Mrs. Jacob's class eat in all?

____

12. Draw a model in the space below that could represent the product 64.

____

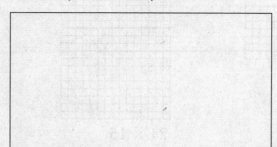

13. What product is shown by the model?

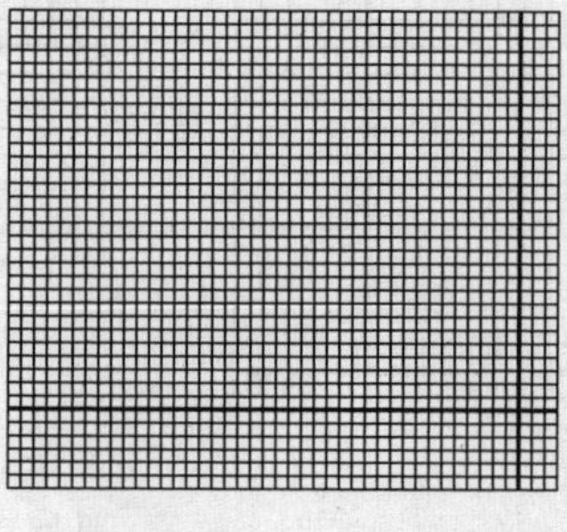

____

Name____________________

Record 2-Digit by 2-Digit Multiplication

Estimate. Then find the product.

89
× 47

Estimate the product.		89 rounds to 90; 47 rounds to 50 90 × 50 = 4,500
Multiply the 9 ones by 7 ones. Regroup 6 tens.	6 89 × 47 3	7 ones × 9 ones = 63 ones
Multiply the 8 tens by 7 ones. Add the regrouped tens.	6 89 × 47 623	7 ones × 8 tens = 56 tens 56 tens + 6 tens = 62 tens
Multiply the 9 ones by 4 tens, or 40. Regroup the 3 tens.	3 89 × 47 623 60	9 ones × 40 = 360
Multiply the 8 tens by 4 tens, or 40. Add the regrouped tens.	3 89 × 47 623 3560	8 tens × 4 tens = 32 tens 32 tens + 3 tens = 35 tens
Add the partial products.	89 × 47 623 + 3560 4183	

So, 89 × 47 = 4,183. Since 4,183 is close to the estimate of 4,500, it is reasonable.

Estimate. Then find the product.

1. 76 × 31

2. 24 × 35

3. 14 × 28

4. 64 × 56

NS 3.2 Demonstrate an understanding of, and the ability to use, standard algorithms for multiplying a multidigit number by a two-digit number and for dividing a multidigit number by a one-digit number; use relationships between them to simplify computations and to check results.

Name______________________________ **Lesson 9.2**

Record 2-Digit by 2-Digit Multiplication

Estimate. Then choose either method to find the product.

1. $\begin{array}{r} 28 \\ \times 19 \\ \hline \end{array}$ ________

2. $\begin{array}{r} 36 \\ \times 53 \\ \hline \end{array}$ ________

3. $\begin{array}{r} \$76 \\ \times 25 \\ \hline \end{array}$ ________

4. $\begin{array}{r} 64 \\ \times 31 \\ \hline \end{array}$ ________

5. 76×83 ________

6. 41×69 ________

7. 57×65 ________

8. $82 \times \$48$ ________

Problem Solving and Test Prep

USE DATA For 9–10, use the bar graph.

9. Sun Beach Parasail had 19 riders each windy day. How many riders in all parasailed last year on windy days?

____

10. On each of 75 sunny days, Sun Beach Parasail had 62 riders. How many riders in all parasailed on those 75 days?

____

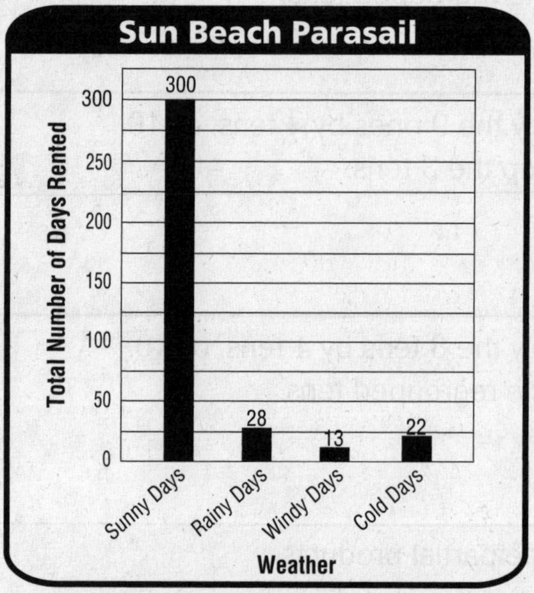

11. Willa bought 16 arborvitae trees for her backyard. Each tree cost $33. How much did the trees cost in all?

A $300
B $480
C $528
D $600

12. There are 47 members in the Fun in the Sun Parasail Club. Each member spent 88 hours last year parasailing. How many hours did the club members spend parasailing last year in all?

A 6,413
B 4,136
C 4,230
D 7,236

Name____ Lesson 9.3

Multiply 2- and 3-Digit Numbers and Money

When you multiply money, it is the same as multiplying whole numbers. Just add the decimal (.) and dollar sign ($) to your answer.

Estimate. Then find the product.

$3.99
× 30

Estimate the product.	$3.99 rounds to $4; 30 stays at 30 $4 × 30 = $120
Multiply 399 by 0 ones.	$3.99 × 30 000 399 × 0 ones = 000
Multiply 399 by 3 tens, or 30. Add the regouped tens.	2 2 $3.99 × 30 000 11970 or 399 × 30 = 11,970
Add the partial products.	$3.99 × 30 000 + 11970 11970
Add the decimal. There will always be 2 digits to the right of the decimal in money. Add a dollar sign to your answer.	$119.70

So, $3.99 × 30 = $119.70.
Since $119.70 is close to the estimate of $120, it is reasonable.

Estimate. Then find the product.

1. 126 × 21

2. 241 × 33

3. $4.19 × 42

4. 386 × 37

5. $2.45 × 16

6. 406 × 24

7. $6.20 × 44

8. 187 × 29

NS 3.2 Demonstrate an understanding of, and the ability to use, standard algorithms for multiplying a multidigit number by a two-digit number and for dividing a multidigit number by a one-digit number; use relationships between them to simplify computations and to check results.

Name________________________ **Lesson 9.3**

Multiply 2- and 3-Digit Numbers and Money

Estimate. Then find the product.

1. 58×39 ______

2. $\$4.28 \times 45$ ______

3. 622×76 ______

4. 199×37 ______

5. $\$3.97 \times 36$ ______

6. 544×47 ______

7. $37 \times \$638$ ______

8. 747×23 ______

9. $\$9.32 \times 42$ ______

10. 81×422 ______

11. $\$1.23 \times 71$ ______

12. 15×602 ______

Problem Solving and Test Prep

13. **Reasoning** Sally found 9×15 using the break apart strategy. Show how Sally found the product.

______

______

______

14. Each of 41 electric cars can drive 50 miles in one hour. How many miles total can all the cars travel in one hour?

______

15. Kip likes a multigrain bread that costs \$3.89 per loaf. If his family consumes one loaf every week, how much will they spend on this bread in one year?

A \$202.28
B \$206.17
C \$206.89
D \$208.21

16. How many minutes are there in 24 hours?

A 1,380
B 1,500
C 1,440
D 1,540

Name ____________________ **Week 14**

Spiral Review

For 1–4, write the numbers in order from greatest to least.

1. 632,296; 69,999; 620,955

____

2. 787,529; 1,000,056; 700,189

____

3. 56,977; 59,000; 55,036

____

4. 8,325; 8,835; 8,915

____

For 8–11, tell whether the data is numerical or categorical data.

8. colors of bikes

____

9. number of *A*s on a test

____

10. votes for class president

____

11. favorite animals

____

For 5–7, name the figure that each object is shaped like.

5. ____________________

6. ____________________

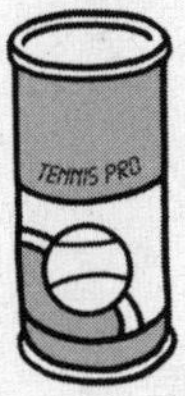

7. ____________________

For 12–17, find the product.

12. $(5 \times 3) \times 6 =$ ____________________

13. $9 \times (8 \times 0) =$ ____________________

14. $7 \times (9 \times 1) =$ ____________________

15. $(4 \times 2) \times 4 =$ ____________________

16. $10 \times (1 \times 2) =$ ____________________

17. $0 \times (2 \times 2) =$ ____________________

Name____

Multiply Greater Numbers

You can multiply larger numbers using mental math if the numbers are easy to work with, such as multiples of 100. If the numbers are more dificult to work with, use pencil and paper.

$$\begin{array}{r} 3{,}500 \\ \times\ 21 \\ \hline \end{array}$$

3,500 is an easy number to work with. Use mental math.

Estimate: 4,000 × 20 = 80,000.

Think:

35 × 1 = 35, so 3,500 × 1 = 3,500.

35 × 2 = 70, so 3,500 × 20 = 70,000

70,000 + 3,500 = 73,500.

So, 3,500 × 21 = 73,500.

Since 73,500 is close to the estimate of 80,000, it is reasonable.

$$\begin{array}{r} 5{,}464 \\ \times\ 46 \\ \hline \end{array}$$

These numbers are more difficult to work with. Use pencil and paper.

Estimate: 5,000 × 50 = 250,000

Think:

$$\begin{array}{r} 5{,}464 \\ \times\ 46 \\ \hline 32{,}784 \\ 218{,}560 \\ \hline 251{,}344 \end{array}$$

So, 5,464 × 46 = 251,344.

Since 251,344 is close to the estimate of 250,000, it is reasonable.

Estimate. Then find the product. Write the method used.

1. $\begin{array}{r} 250 \\ \times\ 30 \\ \hline \end{array}$

2. $\begin{array}{r} 297 \\ \times\ 52 \\ \hline \end{array}$

3. $\begin{array}{r} 528 \\ \times\ 44 \\ \hline \end{array}$

4. $\begin{array}{r} 4{,}000 \\ \times\ 45 \\ \hline \end{array}$

5. $\begin{array}{r} 6{,}482 \\ \times\ 26 \\ \hline \end{array}$

6. $\begin{array}{r} 812 \\ \times\ 48 \\ \hline \end{array}$

7. $\begin{array}{r} 7{,}000 \\ \times\ 28 \\ \hline \end{array}$

8. $\begin{array}{r} 2{,}500 \\ \times\ 32 \\ \hline \end{array}$

NS 3.2 Demonstrate an understanding of, and the ability to use, standard algorithms for multiplying a multidigit number by a two-digit number and for dividing a multidigit number by a one-digit number; use relationships between them to simplify computations and to check results.

Name______________________________

Multiply Greater Numbers

Estimate. Then find the product. Write the method you used.

1. $\begin{array}{r} 221 \\ \times 30 \\ \hline \end{array}$ ______

2. $\begin{array}{r} 653 \\ \times 32 \\ \hline \end{array}$ ______

3. $\begin{array}{r} 5{,}000 \\ \times\ 70 \\ \hline \end{array}$ ______

4. $\begin{array}{r} 3{,}221 \\ \times\ 23 \\ \hline \end{array}$ ______

5. $\begin{array}{r} 312 \\ \times 20 \\ \hline \end{array}$ ______

6. $\begin{array}{r} 666 \\ \times 11 \\ \hline \end{array}$ ______

7. $\begin{array}{r} 867 \\ \times 59 \\ \hline \end{array}$ ______

8. $\begin{array}{r} 9{,}000 \\ \times\ 80 \\ \hline \end{array}$ ______

9. $3,433 × 22 ______

10. 505 × $90 ______

11. 62 × 2,763 ______

12. 52 × $10.10 ______

13. 50 × $14.78 ______

14. 19 × $91.28 ______

15. $15.73 × 80 ______

16. $71.02 × 33 ______

Problem Solving and Test Prep

17. June had a party at home. June's birthday plates cost $10.97 each. If there were a total of 23 people at the party including June, how much did the plates cost?

______

18. A local store sells balloons at $29.45 a case. Frank bought 48 cases. How much did the balloons cost?

______

19. What is the best method to multiply 40 × 800?

    A mental math
    B calculator
    C paper and pencil
    D none of the above

20. Which shows the closest estimate of 61 × 829?

    A 65 × 820 = 53,300
    B 100 × 1,000 = 100,000
    C 60 × 830 = 49,800
    D 50 × 800 = 40,000

Name ____________________

Problem Solving Workshop Skill: Evaluate Reasonableness

The Bureau of Land Management sponsors wild horse adoptions in which people can bid for horses at auction. Toni wants to adopt 4 wild horses. Each horse adoption costs $185. Toni estimates that the cost of caring for each horse is about $1,000 for the year. Is it reasonable to say that it will cost Toni $4,800 to adopt and care for the 4 horses for this year?

1. What are you asked to find?

____

____

2. What operation or operations will you use to solve this problem?

____

3. Is $4,800 a reasonable estimate to adopt and care for 4 horses for the year?

____

4. How did you determine whether or not the estimate was reasonable?

____

____

____

____

Solve the Problem.

5. Each month Mario spends $5 on sports cards, $12 on food, and $9 on school supplies. Mario estimates that he spends $75 in 3 months. Is his estimate reasonable? What is the actual amount that Mario spends in 3 months?

____

6. Amy's school has 524 students. The teachers gave each student 3 pencils. Amy figured out how many pencils were handed out. Her answer was 1,048. Was her answer reasonable? Why or why not?

____

____

____

____

Name________________________________

Problem Solving Workshop Skill: Evaluate Reasonableness

Problem Solving Skill Practice

Solve the problem. Then evaluate the reasonableness of your answer. Explain.

1. Mr. Kohfeld sells farm eggs for $1.37 a carton. If he sells 1 carton to each of 4 neighbors, how much money does Mr. Kohfeld earn?

2. Vivian has the same $6.49 breakfast every day at a local grill. How much does Vivian spend in 7 days?

3. Yoshi is an athlete who has a breakfast of 1,049 calories each morning. How many calories does Yoshi consume for breakfast in 7 days?

4. Together Elise and Chris spelled 27 words correctly. Chris spelled 5 more than Elise. How many words did each student spell correctly?

Mixed Applications

5. The Miller family gives 9 sacks of feed to their farm pigs a day. How many sacks of feed do the pigs eat in a year (365 days)? How do you know your answer is reasonable?

6. Joe spent $25.87 for groceries. He bought cereal for $6.25, eggs for $5.37, pancake mix for $3.67, bacon for $7.25, and juice. How much did he spend for juice?

7. Use Data Tanya is building a wall. Given the pattern, how thick is the next stone?

8. Use Data If the finished wall is 6 stones high, what is the overall height of the wall?

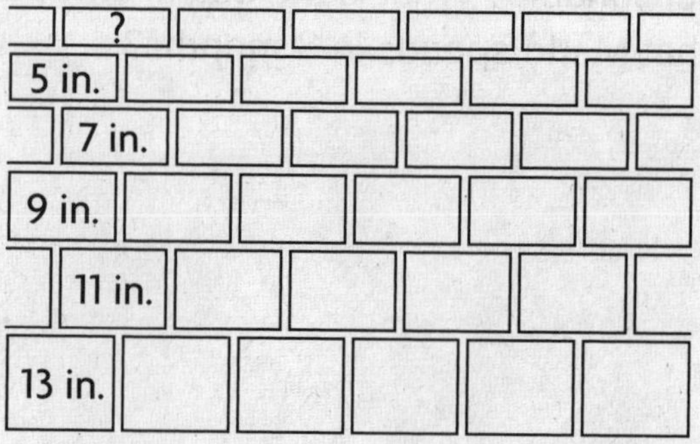

Name____ **Lesson 10.1**

Divide with Remainders

When a number cannot be divided evenly, there is an amount left over. This leftover amount is called the **remainder**.

Use counters to find the quotient and remainder.

$9\overline{)26}$

- You are dividing the divisor, 26, so use 26 counters.
- Since you are dividing 26 by 9, draw 9 circles.
  Then divide the 26 counters into 9 equal-sized groups.

- There are 2 counters in each circle, so the quotient is 2.
  There are 8 counters left over, so the remainder is 8.

So, $\begin{array}{r} 2\text{r}8 \\ 9\overline{)26} \end{array}$.

Divide. You may wish to use counters or draw a picture to help.

$7\overline{)66}$

- You are dividing the divisor, 66, so use 66 counters.
- Since you are dividing 66 by 7, draw 7 circles.
  Divide 66 counters into 7 equal-sized groups.

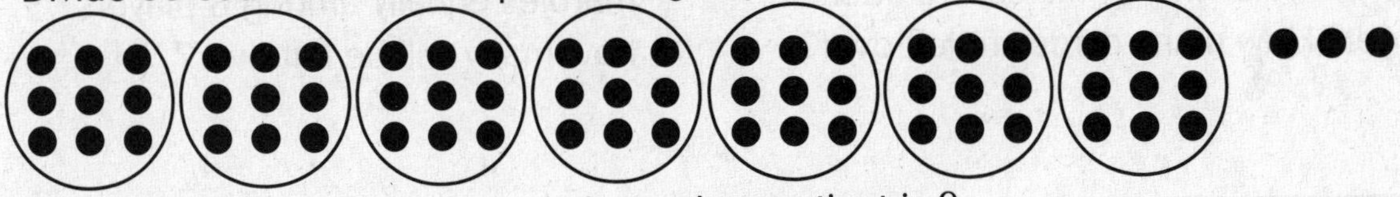

- There are 9 counters in each circle, so the quotient is 9.
  There are 3 counters left over, so the remainder is 3.

So, $\begin{array}{r} 9\text{r}3 \\ 7\overline{)66} \end{array}$.

Use counters to find the quotient and remainder.

1. $6\overline{)19}$

2. $3\overline{)14}$

Divide. You may wish to use counters or draw a picture to help you.

3. $29 \div 9$ ________

4. $31 \div 7$ ________

Name______________________________

Divide with Remainders

Use counters to find the quotient and remainder.

1. $27 \div 5 =$ ________
2. $34 \div 8 =$ ________
3. $18 \div 4 =$ ________
4. $57 \div 7 =$ ________
5. $41 \div 6 =$ ________
6. $53 \div 9 =$ ________

Divide. You may wish to use counters or draw a picture to help.

7. $26 \div 3 =$ ________
8. $34 \div 4 =$ ________
9. $50 \div 6 =$ ________
10. $9\overline{)75}$
11. $8\overline{)54}$
12. $7\overline{)60}$
13. $17 \div 3 =$ ________
14. $44 \div 5 =$ ________
15. $33 \div 3 =$ ________

Problem Solving and Test Prep

16. Five students are playing a card game using a deck of 52 cards. If the cards are divided evenly among each player, how many will each student get? How many cards are left over?

____

17. Bill made up a game using 10 each of purple, yellow, green, blue, orange, and red marbles. If Bill divides the marbles equally among 8 players, how many will be left over?

____

18. Which problem does the model describe?

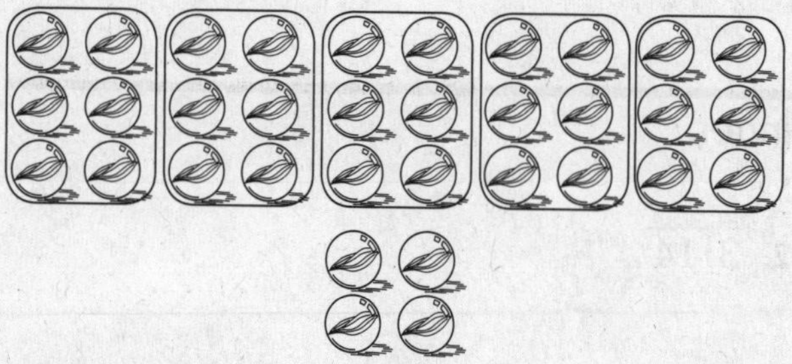

A $34 \div 5$
B $5\overline{)28}$
C $30 \div 4$
D $6\overline{)20}$

19. Which problem does the model describe?

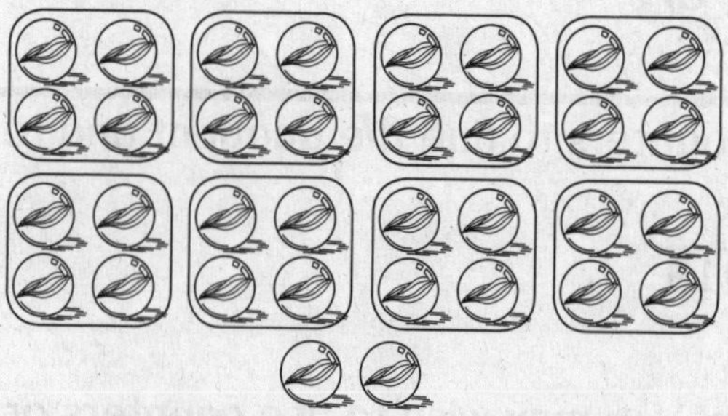

A $28 \div 6$
B $4\overline{)42}$
C $34 \div 8$
D $4\overline{)24}$

Name ______________________________ Lesson 10.2

Model 2-Digit by 1-Digit Division

You can use base ten blocks to model how to divide a 2-digit number, such as 26, by a 1-digit number, such as 3.

Use base ten blocks to find the quotient and remainder.

$7\overline{)84}$

- Show 84 as 8 tens and 4 ones.
  Then draw 7 circles, since you are dividing 84 by 7.
- Place an equal number of tens in each circle.
  If there are any tens left over, regroup them as ones.
  Now place an equal number of ones in each group.

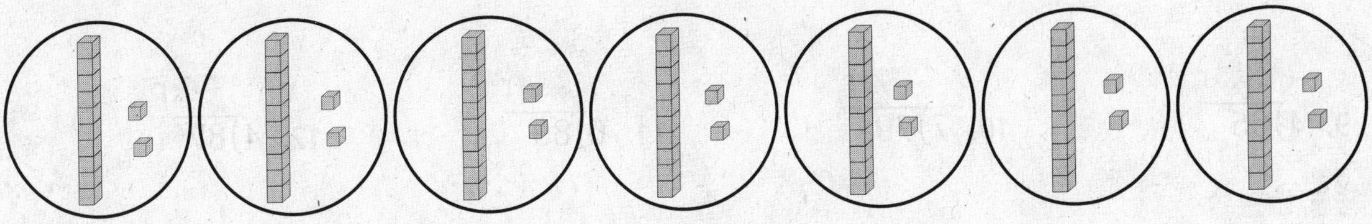

- Count the number of tens and ones in each circle to find the quotient.
  There is 1 ten and 2 ones in each circle.
  $10 + 2 = 12$, so the quotient is 12.
- There are no leftover blocks, so there is no remainder.
- So, $84 \div 7 = 12$.

Use base ten blocks to find the quotient and remainder.

1. $71 \div 8$

______

2. $35 \div 8$

______

3. $6\overline{)49}$

4. $5\overline{)28}$

Name____

Model 2-Digit by 1-Digit Division

Use base-ten blocks to find the quotient and remainder.

1. 37 ÷ 2 = ■ r ■ **2.** 53 ÷ 5 = ■ r ■ **3.** 92 ÷ 7 = ■ r ■ **4.** 54 ÷ 4 = ■ r ■

5. 56 ÷ 3 = ■ r ■ **6.** 89 ÷ 9 = ■ r ■ **7.** 78 ÷ 6 = ■ r ■ **8.** 92 ÷ 8 = ■ r ■

9. 4)65 = ■ r ■ **10.** 7)79 = ■ r ■ **11.** 6)89 = ■ r ■ **12.** 4)87 = ■ r ■

Divide. You may wish to use base-ten blocks.

13. 3)77 = ■ r ■ **14.** 2)67 = ■ r ■ **15.** 4)66 = ■ r ■ **16.** 5)67 = ■ r ■

17. 37 ÷ 2 = ■ r ■ **18.** 98 ÷ 4 = ■ r ■ **19.** 91 ÷ 6 = ■ r ■ **20.** 72 ÷ 7 = ■ r ■

21. 8)93 = ■ r ■ **22.** 6)57 = ■ r ■ **23.** 4)77 = ■ r ■ **24.** 9)59 = ■ r ■

Name ______________________

Spiral Review

For 1–4, estimate by using rounding.

1. 19,526 + 11,062

2. 8,263 – 4,829

3. 268,099 – 133,526

4. 332,185 + 398,626

For 8–10, use the data tables below.

Average High Temperature					
Month	Apr	May	Jun	Jul	Aug
Temperature	79°F	85°F	91°F	95°F	96°F

8. What is the median of the data?

Average Low Temperature					
Month	Apr	May	Jun	Jul	Aug
Temperature	55°F	55°F	61°F	64°F	70°F

9. What is the median of the data?

10. What is the mode of the data?

For 5–7, name the solid figure. Then tell how many faces.

5. ______________________

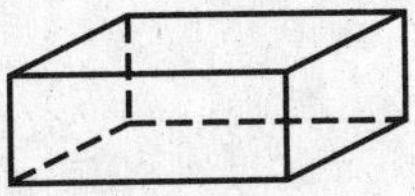

6. ______________________

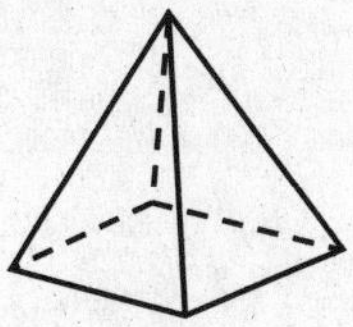

7. ______________________

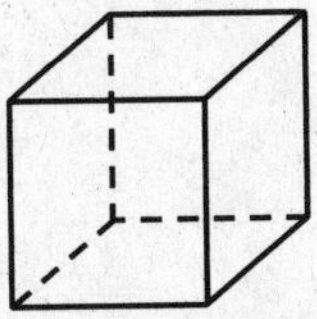

For 11–14, solve the equation.

11. $c \div 5 = 9$

12. $8 \times f = 56$

13. $32 \div x = 8$

14. $7 \times a = 49$

Name____

Record 2-Digit by 1-Digit Division

You can use long division to divide 2-digit numbers by 1-digit numbers. Sometimes the numbers will divide evenly. Other times there will be a remainder.

Choose a method. Then divide and record.

$5\overline{)85}$

Begin by dividing 8 tens by 5.
$8 \div 5 = 1$
Write 1 in the tens place.

$$\begin{array}{r} 1 \\ 5\overline{)85} \\ -5 \\ \hline 3 \end{array} \quad \text{or} \quad \begin{array}{r} 1 \\ 5\overline{)8} \\ -5 \\ \hline 3 \end{array}$$

Multiply. $1 \times 5 = 5$.
Subtract. $8 - 5 = 3$.
Compare. $3 < 5$.

Then bring down the 5 ones and divide 35 by 5.

$35 \div 5 = 7$
Write 7 in the ones place.

$$\begin{array}{r} 17 \\ 5\overline{)85} \\ -5\downarrow \\ \hline 35 \\ -35 \\ \hline 0 \end{array} \quad \text{or} \quad \begin{array}{r} 7 \\ 5\overline{)35} \\ -35 \\ \hline 0 \end{array}$$

Multiply. $7 \times 5 = 35$.
Subtract. $35 - 35 = 0$.
Compare. $0 < 5$.
So, $85 \div 5 = 17$.

Choose a method. Then divide and record.

1. $7\overline{)82}$

2. $6\overline{)74}$

3. $3\overline{)57}$

4. $7\overline{)95}$

Name____ **Lesson 10.3**

Record 2-Digit by 1-Digit Division

Divide and record.

1. $4\overline{)93}$

2. $7\overline{)75}$

3. $5\overline{)97}$

4. $49 \div 3 =$ ______

5. $61 \div 2 =$ ______

6. $95 \div 7 =$ ______

7. $9\overline{)87}$

8. $6\overline{)87}$

9. $8\overline{)99}$

ALGEBRA Complete each table.

10.

Number of Cups	16	20	24	28	32
Number of Quarts	4	5	■	■	■

11.

Number of Pints	64	72	80	88	96
Number of Gallons	8	9	■	■	■

Problem Solving and Test Prep

12. Sixty-three students signed up for golf. The coach divided them into groups with 4 students in each group. How many students were left over?

____

13. There are 6 runners on each relay team. If a total of 77 runners signed up, how many relay teams could there be?

____

14. Four students divided 85 base-ten blocks equally among them. How many base-ten blocks does each student receive?

A 20
B 21
C 22
D 24

15. Three students divided 85 base-ten rods equally among them. How many base-ten rods were left over?

A 4
B 3
C 2
D 1

Name____

Problem Solving Workshop Strategy: Compare Strategies

Ari runs a training school for pet actors. Last year he trained 3 times as many dogs as cats. If the total number of dogs and cats he trained last year is 84, how many cats did Ari train?

Read to Understand

1. What do you need to find out in this problem?

____

Plan

2. Which strategy would you choose to solve this problem: Draw a Diagram or Guess and Check?

____

____

Solve

3. Use the strategy you chose to solve the problem.

4. What is the solution to this problem?

____

Check

5. Do you think you chose the better strategy? Why or why not?

____

____

____

Solve.

6. A crate has 5 times as many apples as oranges. If there are 120 pieces of fruit in the crate, how many are oranges?

____

7. Julio has 6 times as many green marbles as red marbles. If he has 210 marbles altogether, how many are green?

____

NS 3.2 Demonstrate an understanding of, and the ability to use, standard algorithms for multiplying a multidigit number by a two-digit number and for dividing a multidigit number by a one-digit number; use relationships between them to simplify computations and to check results.

Name______________________________________ Lesson 10.4

Problem Solving Workshop Strategy: Compare Strategies

Problem Solving Strategy Practice

Choose a strategy to solve the problems.

1. Fiona's dog is 4 times as long as Rod's dog. End to end, they are 60 inches long. How long is Fiona's dog?

____

2. Davey divided 112-ounces of rabbit food equally into 7 containers. How much did each container hold?

____

3. Dina's pet cat weighs six times as much as her pet mouse. Altogether, her cat and her mouse weigh 14 pounds. How much does Dina's cat weigh?

____

4. Mel collects 91 animal fact cards. He organizes them in a book that holds 7 cards on each page. If Mel fills each page, how many pages will he use?

____

Mixed Strategy Practice

USE DATA For 5–6, use the chart.

5. Together the height of Dan's 3 dogs is 38 inches. What breeds are they?

____

6. Order the dogs in the table from shortest to tallest.

____

____

Dog Heights

Breed	Height
Bichon Frise	10 in.
Border Collie	20 in.
Chihuahua	8 in.
Irish Setter	27 in.
Labrador Retriever	24 in.
Shar-Pei	19 in.
Siberian Huskey	22 in.

7. Altogether, Haille's dog statue collection weighs 20 pounds. One statue weighs 8 pounds and the rest weigh half as much. How many dog statues does Haille have?

____

8. **Pose a Problem** Use the information from Exercise 5 to write a new problem that asks to explain the answer.

____

____

____

Name________________________ Lesson 10.5

Mental Math: Division Patterns

Use mental math to complete the pattern.

64 ÷ 8 = ☐

640 ÷ 8 = ☐

6,400 ÷ 8 = ☐

- Solve for the basic math fact. — 64 ÷ 8 = 8
- In 640 ÷ 8, there is one zero after the basic number in the dividend. So there will be one zero after the basic number in the quotient. — **640** ÷ 8 = **80**
- In 6,400 ÷ 8, there are 2 zeros after the basic number in the dividend, so there will be 2 zeros after the basic number in the quotient. — 6,**400** ÷ 8 = **800**
- So, 64 ÷ 8 = 8, 640 ÷ 8 = 80, and 6,400 ÷ 8 = 800.

Use mental math to complete the pattern.

90 ÷ 9 = ☐

900 ÷ 9 = ☐

9,000 ÷ 9 = ☐

- Solve for the basic math fact. — 90 ÷ 9 = 10
- In 900 ÷ 9, there is one zero after the basic number in the dividend. So there will be one zero after the basic number in the quotient. — **900** ÷ 9 = **100**
- In 9,000 ÷ 9, there are 2 zeros after the basic number in the dividend, so there will be 2 zeros after the basic number in the quotient. — **9,000** ÷ 9 = **1,000**
- So, 90 ÷ 9 = 10, 900 ÷ 9 = 100, and 9,000 ÷ 9 = 1,000.

Use mental math to complete the pattern.

1. 25 ÷ 5 = ☐
250 ÷ 5 = ☐
2,500 ÷ 5 = ☐

2. 63 ÷ 9 = ☐
630 ÷ 9 = ☐
6,300 ÷ 9 = ☐

3. 56 ÷ 7 = ☐
560 ÷ 7 = ☐
5,600 ÷ 7 = ☐

4. 18 ÷ 3 = ☐
180 ÷ 3 = ☐
1,800 ÷ 3 = ☐

5. 36 ÷ 4 = ☐
360 ÷ 4 = ☐
3,600 ÷ 4 = ☐

6. 30 ÷ 3 = ☐
300 ÷ 3 = ☐
3,000 ÷ 3 = ☐

Mental Math: Division Patterns

Use mental math to complete the pattern.

1. $72 \div 8 = 9$

$720 \div 8 =$ ______

$7{,}200 \div 8 =$ ______

$72{,}000 \div 8 =$ ______

2. $42 \div 7 =$ ______

______ $\div 7 = 60$

$4{,}200 \div 7 =$ ______

$42{,}000 \div 7 =$ ______

3. ______ $\div 6 = 4$

$240 \div 6 =$ ______

______ $\div 6 = 400$

$24{,}000 \div 6 =$ ______

4. $30 \div 3 =$ ______

______ $\div 3 = 100$

$3{,}000 \div 3 =$ ______

______ $\div 3 = 10{,}000$

5. ______ $\div 5 = 8$

$400 \div 5 =$ ______

______ $\div 5 = 800$

$40{,}000 \div 5 =$ ______

6. $28 \div 4 =$ ______

______ $\div 4 = 70$

$2{,}800 \div 4 =$ ______

______ $\div 4 = 7{,}000$

Use mental math and patterns to find the quotient.

7. $1{,}600 \div 4 =$ ■ ______

8. $28{,}000 \div 7 =$ ■ ______

9. $50 \div 5 =$ ■ ______

10. $900 \div 3 =$ ■ ______

11. $32{,}000 \div 4 =$ ■ ______

12. $2{,}000 \div 5 =$ ■ ______

13. $600 \div 2 =$ ■ ______

14. $3{,}500 \div 7 =$ ■ ______

Problem Solving and Test Prep

15. Maria has 4,500 stamps in her collection. She puts an equal amount of stamps into 9 books. How many stamps will be in each book?

______

16. Tex wants to put 640 stickers in his sticker book. If there are 8 stickers to a page, how many pages will Tex fill?

______

17. The theme park tickets sells for $4 each. It collects $2,000 in one day. How many tickets does the park sell in one day?

A 50

B 500

C 5,000

D 50,000

18. Dee collected $60 for selling tickets. If she sold 5 tickets, how much did each ticket cost?

A $12

B $24

C $30

D $45

Name____ **Lesson 10.6**

Mental Math: Estimate Quotients

You can use compatible numbers and rounding to estimate quotients.

Estimate the quotient.

147 ÷ 3

Use compatible numbers to estimate.

- Look at the numbers in the problem. Think of some basic math facts you know that are similar to the numbers in the problem.

  15 ÷ 3
  150 ÷ 3
  150 is close to 147

- Use mental math to divide.

  15 ÷ 3 = 5
  150 ÷ 3 = 50

So, a good estimate for 147 ÷ 3 is about 50.

Estimate the quotient.

4,522 ÷ 9

Use rounding to estimate.

- Round the dividend to the nearest thousand. Round 9 up to 10.

  4,522 rounds up to 5,000

- Rewrite the problem using the rounded numbers.

  5,000 ÷ 10

- Use mental math to divide.

  5 ÷ 1 = 5
  5,000 ÷ 10 = 500

So, a good estimate for 4,522 ÷ 9 is about 500.

Estimate the quotient.

1. 409 ÷ 4 ________

2. 212 ÷ 5 ________

3. 7,839 ÷ 2 ________

4. 311 ÷ 6 ________

5. 2,103 ÷ 4 ________

6. 731 ÷ 7 ________

7. 239 ÷ 4 ________

8. 403 ÷ 8 ________

9. 119 ÷ 3 ________

10. 5,788 ÷ 6 ________

11. 787 ÷ 4 ________

12. 273 ÷ 4 ________

Name____ **Lesson 10.6**

Mental Math: Estimate Quotients

Estimate the quotient.

1. 392 ÷ 4 ______
2. 489 ÷ 6 ______
3. 536 ÷ 9 ______
4. 802 ÷ 8 ______
5. 632 ÷ 7 ______
6. 32,488 ÷ 4 ______
7. 3,456 ÷ 5 ______
8. 7,820 ÷ 8 ______

Estimate to compare. Write <, >, or = for each ◯.

9. 276 ÷ 3 ◯ 460 ÷ 5
10. 332 ÷ 6 ◯ 412 ÷ 5
11. 527 ÷ 6 ◯ 249 ÷ 3
12. 138 ÷ 2 ◯ 544 ÷ 9
13. 478 ÷ 7 ◯ 223 ÷ 3
14. 3,112 ÷ 8 ◯ 1,661 ÷ 8

Problem Solving and Test Prep

USE DATA For 15–16, use the table.

Resting Heartbeats of Select Mammals

Mammal	Rate per 5 minutes
Human	375
Horse	240
Dog	475
Mouse	2,490

15. Which beats faster, a dog's heart in 5 minutes or a mouse's heart in 1 minute?

____

16. Which beats slower in 1 minute: a human's heart or a horse's heart?

____

17. A Common Loon's heart beats about 1,250 times in 5 minutes. What is the best estimate of the number of times its heart beats in one minute?

A 20
B 40
C 250
D 400

18. Nine equal-length Arizona Black Rattlesnakes laid in a row measure 378 inches. What is the best estimate of the length of 1 rattlesnake?

A 20
B 40
C 200
D 400

Spiral Review

For 1–5, estimate the product. Choose the method.

1. $68 \times 24 =$ ________

2. $89 \times 37 =$ ________

3. $52 \times 46 =$ ________

4. $79 \times 467 =$ ________

5. $30 \times 115 =$ ________

For 9–10, use the bar graph.

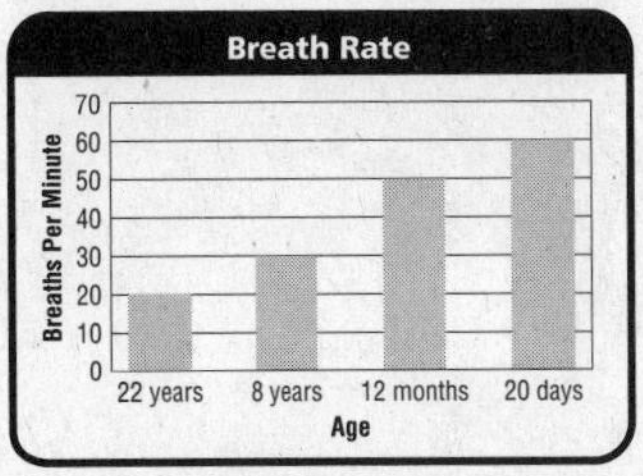

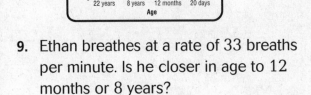

9. Ethan breathes at a rate of 33 breaths per minute. Is he closer in age to 12 months or 8 years?

____

10. Phillip breathes at a rate of 47 breaths per minute, and Safia breathes at a rate of 25 breaths per minute. Who is older?

____

For 6–8, find the length of each line segment.

6. 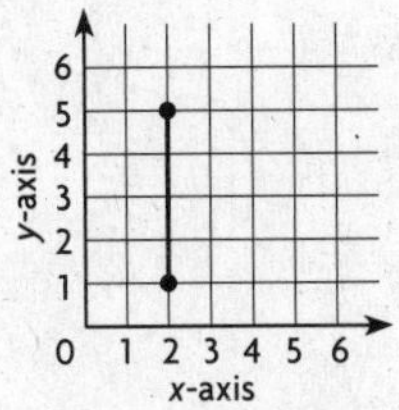

____

7 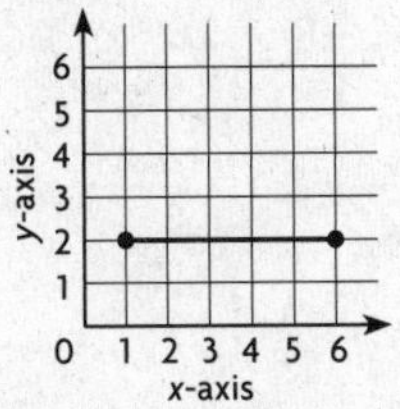

____

8. 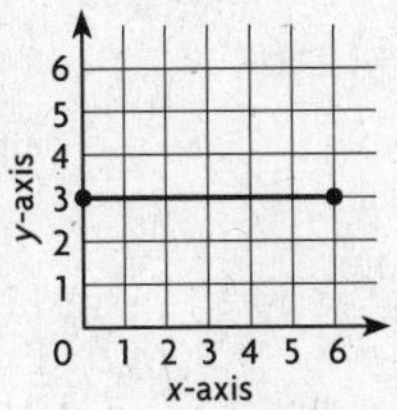

____

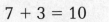

For 11–14, add to or subtract from both sides of the equation. Find the new values.

11. Add 8.

$7 + 3 = 10$ ________

12. Subtract 5.

$8 - 3 + 3 = 5 + 3$ ________

13. Add 11.

$7 - 3 - 1 = 15 - 11 - 1$

____

14. Subtract 10.

$28 + 2 + 3 = 5 + 20 + 8$

____

Name______________________________ Lesson 10.7

Place the First Digit

When you use long division, it is important to place the first digit of the quotient in the right spot. You can use estimation and place value to help you figure out where to place the first digit.

Tell where to place the first digit. Then divide.

$6\overline{)126}$

- Use compatible numbers to estimate. 126 ÷ 6 is about the same as 120 ÷ 6. — $12 \div 6 = 2$, $120 \div 6 = 20$
- 120 ÷ 6 = 20, so the first digit is in the tens place.
- Now divide and place the first digit in the quotient in the tens place, too.

$$\begin{array}{r} 21 \\ 6\overline{)126} \\ -\underline{12} \\ 06 \\ -\underline{6} \\ 0 \end{array}$$

So, the first digit is the tens place, and 126 ÷ 6 = 21.

---

809 ÷ 8

- See if the number in the hundreds place in 809 is larger than 8.
- If the number in the hundreds place is smaller than the divisor, then you begin at the next place to the right: the tens place.
- If the number in the hundreds place is larger than or equal to the divisor, then you can place the first digit in the hundreds place.
- Here, 8 is equal to 8, so the first digit will be in the hundreds place. Now divide.
- So, the first digit is in the hundreds place, and 809 ÷ 8 = 101r1.

$$\begin{array}{r} 101r1 \\ 8\overline{)809} \\ -\underline{8} \\ 00 \\ -\underline{0} \\ 09 \\ -\underline{8} \\ 1 \end{array}$$

Tell where to place the first digit. Then divide.

1. $8\overline{)286}$ **2.** $5\overline{)743}$ **3.** 536 ÷ 4 **4.** 647 ÷ 9

________ ________ ________ ________

NS 3.2 Demonstrate and understanding of, and the ability to use, standard algorithms for multiplying a multidigit number by a two-digit number and for dividing a multidigit number by a one-digit number; use relationships between them to simplify computations and to check results.

Name____

Place the First Digit

Tell where to place the first digit. Then divide.

1. $4\overline{)511}$ **2.** $7\overline{)621}$ **3.** $2\overline{)124}$ **4.** $3\overline{)423}$

______ ______ ______ ______

5. $136 \div 2$ **6.** $215 \div 5$ **7.** $468 \div 6$ **8.** $357 \div 8$

______ ______ ______ ______

Divide.

9. $3\overline{)166}$ **10.** $9\overline{)785}$ **11.** $4\overline{)334}$ **12.** $6\overline{)577}$

______ ______ ______ ______

13. $116 \div 2$ **14.** $425 \div 5$ **15.** $627 \div 7$ **16.** $436 \div 8$

______ ______ ______ ______

Problem Solving and Test Prep

17. Petra picked 135 petals from the flowers of sweet pea plants. Each flower has 5 petals. How many flowers did Petra pull petals from?

______

18. Todd wants to plant some thyme equally in 8 areas in his garden. If he has 264 plants, how many thyme plants can Todd put in each area?

______

19. In which place is the first digit in the quotient $118 \div 4$?

A ones **C** hundreds
B tens **D** thousands

20. In which place is the first digit in the quotient $1{,}022 \div 5$?

A ones **C** hundreds
B tens **D** thousands

Name____

Problem Solving Workshop
Skill: Interpret the Remainder

Guides lead groups of 9 people on biking tours in the park. There are 96 people who decided to go on the tours. How many people will be on a tour that is not full? Interpret the remainder.

1. What are you asked to find?

____

2. Divide 96 by 9. How many groups of 9 people will the guides lead on biking tours?

____

3. What is the remainder? What does it represent?

____

4. How many people are on a tour that is not full?

____

5. How can you check your answer?

____

Solve. Write a, b, or c to explain how to interpret the remainder.

a. Quotient remains the same. Drop the remainder.

b. Increase the quotient by 1.

c. Use the remainder as the answer.

6. Each van holds 9 people. There are 82 people traveling. How many vans will be completely full?

____

7. Each car in an amusement park ride must be filled with 6 people from the 28 people waiting in line. How many people will be on a car that is not full?

____

Name ______________________

Problem Solving Workshop
Skill: Interpret the Remainder

Problem Solving Skill Practice

Solve. Write *a*, *b*, or *c* to explain how to interpret the remainder.

a. Quotient stays the same. Drop the remainder.

b. Increase the quotient by 1.

c. Use the remainder as the answer.

1. The crafts teacher gave 8 campers a total of 55 beads to make necklaces. If he divided the beads equally among the campers, how many did each camper have?

____

2. In all, campers from 3 tents brought 89 logs for a bonfire. Two tents brought equal amounts but the third brought a few more. How many more?

____

3. Gene had 150 cups of water to divide equally among 9 campers. How many cups did he give each camper?

____

4. Camp leaders divided 52 cans of food equally among 9 campers. How many cans of food were left over?

____

Mixed Applications

5. Geena had 34 hot dogs. She gave 3 camp counselors 2 hot dogs each before dividing the rest equally among the 7 campers. How many hot dogs did she give each camper?

____

6. The morning of a hiking trip the temperature was 54°F. By mid-afternoon, the temperature had risen to 93°F. How much warmer was the afternoon temperature?

____

7. Pose a Problem Exchange the known for unknown information in Exercise 5 to write a new problem.

____

____

____

8. Wynn bought these camping tools: a flashlight, an axe for $15, a lantern for $12, and a camp stool for $23. If he spent $57, how much did the flashlight cost?

____

Name______________________________________ **Lesson 11.2**

Divide 3-Digit Numbers and Money

You can use multiplication to check division problems.

Divide and check.

$924 \div 8$ or $8\overline{)924}$

- Estimate to see where to place the first digit. Since there is 1 hundred, place the first digit in the hundreds place.

924 is about 900;
8 is about 9.
$900 \div 9 = 100$.

- Divide $9 \div 8$. Place the 1 in the hundreds place.
- Multiply 8×1. Place the 8 below the 9 hundreds.
- Subtract the hundreds. Bring down the 2 tens.

$$\begin{array}{r} 1 \\ 8\overline{)924} \\ -8\downarrow \\ \hline 12 \end{array} \quad \text{or} \quad \begin{array}{r} 1 \\ 8\overline{)9} \\ -8 \\ \hline 1 \end{array}$$

Repeat the steps.

- Divide $12 \div 8$. Place the 1 in the tens place.
- Multiply 8×1. Place the 8 below the 2 tens.
- Subtract the tens. Bring down the 4 ones.

$$\begin{array}{r} 11 \\ 8\overline{)924} \\ -8\downarrow \\ \hline 12 \\ -8\downarrow \\ \hline 44 \end{array} \quad \text{or} \quad \begin{array}{r} 8\overline{)12} \\ -8 \\ \hline 4 \end{array}$$

Repeat the steps.

- Divide $44 \div 8$. Place the 5 in the ones place.
- Multiply 8×5. Place the 40 below the 44 ones.
- Subtract the ones. The remainder is 4.

So, $924 \div 8 = 115$ r4.

$$\begin{array}{r} 115 \\ 8\overline{)924} \\ -8\downarrow \\ \hline 12 \\ -8\downarrow \\ \hline 44 \\ -40 \\ \hline 4 \end{array} \quad \text{or} \quad \begin{array}{r} 5 \\ 8\overline{)44} \\ -40 \\ \hline 4 \end{array}$$

Check.

Multiply the quotient by the divisor.
Add the product and the remainder.
Since the sum is equal to the dividend, your answer is correct.

$115 \times 8 = 920$
$920 + 4 = 924$

Divide and check.

1. $6\overline{)\$582}$ **2.** $7\overline{)397}$ **3.** $3\overline{)858}$

Name ____________________

Divide 3-Digit Numbers and Money

Divide and check.

1. $147 \div 5 =$ ________ **2.** $\$357 \div 7 =$ ________ **3.** $575 \div 4 =$ ________

4. $6\overline{)\$844}$ **5.** $9\overline{)874}$ **6.** $8\overline{)766}$

ALGEBRA Find the missing digit.

7. $577 \div ■ = 115\text{ r}2$ **8.** $■10 \div 2 = \$405$ **9.** $734 \div 3 = 24■\text{ r}2$ **10.** $\$572 \div 6 = ■5\text{r}2$

11. $\begin{array}{r} ■5\text{ r}8 \\ 9\overline{)593} \end{array}$ **12.** $\begin{array}{r} 145\text{ r}2 \\ 4\overline{)5■2} \end{array}$ **13.** $\begin{array}{r} 71\text{ r}4 \\ ■\overline{)572} \end{array}$ **14.** $\begin{array}{r} 69\text{ r}■ \\ 7\overline{)488} \end{array}$

Problem Solving and Test Prep

15. In all, Alfred paid $18 for 12 bundles of asparagus at a local grocery store. If the bundles were in a buy-one-get-one-free sale, how much did each bundle cost before the sale?

____

16. Eva wants to divide 122 yards of yarn into 5-yard lengths to make potholders. How many potholders can Eva make? How many yards of yarn will be left over?

____

17. Ed divided 735 football cards among 8 friends. How many cards did each friend get?

A 98

B 91

C 99

D 99r3

18. Four cans of spaghetti are on sale for $4.64. How much does one can cost?

____

Name________________________________ **Lesson 11.3**

Zeros in Division

Divide and check.

927 ÷ 9 or 9)927

Estimate to see where to place the first digit.
- Since the quotient is at least 100, place the first digit in the hundreds place.

927 is about 900;
9 is compatible with 900.
900 ÷ 9 = 100.

- Divide 9 ÷ 9. Place the 1 in the hundreds place.
- Multiply 9 × 1. Place the 9 under 9 hundreds.
- Subtract the hundreds. Bring down the 2 tens.

$$\begin{array}{r} 1 \\ 9\overline{)927} \\ -9\downarrow \\ \hline 02 \end{array} \quad \text{or} \quad \begin{array}{r} 1 \\ 9\overline{)9} \\ -9 \\ \hline 0 \end{array}$$

Repeat the steps.
- Divide 2 ÷ 9. Place the 0 in the tens place.
- Multiply 9 × 0. Place the 0 under 2 tens.
- Subtract the tens. Bring down the 7 ones.

$$\begin{array}{r} 10 \\ 9\overline{)927} \\ -9\downarrow \\ \hline 02 \\ -0\downarrow \\ \hline 27 \end{array} \quad \text{or} \quad \begin{array}{r} 0 \\ 9\overline{)2} \\ -0 \\ \hline 2 \end{array}$$

Repeat the steps.
- Divide 27 ÷ 9. Place the 3 in the ones place.
- Multiply 9 × 3. Place the 27 below the 27 tens.
- Subtract. There is no remainder.

So, 927 ÷ 9 = 103.

$$\begin{array}{r} 103 \\ 9\overline{)927} \\ -9\downarrow \\ \hline 02 \\ -0\downarrow \\ \hline 27 \\ -27 \\ \hline 0 \end{array} \quad \text{or} \quad \begin{array}{r} 3 \\ 9\overline{)27} \\ -27 \\ \hline 0 \end{array}$$

Check.

Multiply the quotient by the divisor. 103 × 9 = 927
Add the product and the remainder. 927 + 0 = 927
Since the sum is equal to the dividend, your answer is correct.

Divide and check.

1. 5)524

2. 7)910

3. 4)815

Name______________________________

Zeros in Division

Write the number of digits in each quotient.

1. $366 \div 3$ ______
2. $5\overline{)374}$ ______
3. $635 \div 7$ ______
4. $4\overline{)923}$ ______
5. $672 \div 8$ ______
6. $5\overline{)811}$ ______
7. $9 \div 921$ ______
8. $6\overline{)597}$ ______
9. $816 \div 2$ ______
10. $7\overline{)177}$ ______

Divide and check.

11. $495 \div 5 =$ ______
12. $719 \div 6 =$ ______
13. $3\overline{)735}$
14. $4\overline{)897}$
15. $210 \div 4 =$ ______
16. $103 \div$ ______ $= 14$ r5
17. ______ $\div 5 = 61$

Problem Solving and Test Prep

18. Yoshi has a collection of 702 miniature cars that he displays on 6 shelves in his bookcase. If the cars are divided equally, how many are on each shelf?

______

19. In 5 days, scouts made a total of 865 trinkets for a fundraiser. If they made the same number each day, how many did they make in 1 day?

______

20. Greta has 594 flyers in stacks of 9 flyers each. How do you find the number of stacks Greta made? Explain.

______

______

______

21. Susan has 320 slices of banana bread. She wants to fill bags with 8 slices of banana bread each. How many bags will Susan fill?

______

Name ____________________ **Week 17**

Spiral Review

For 1–3, write a fraction for the shaded part. Write a fraction for the unshaded part.

1.

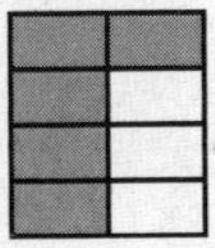

__________ __________

2.

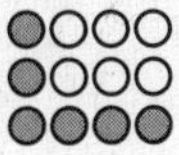

__________ __________

3.

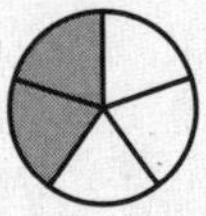

__________ __________

4. Complete the table for the equation: $y = x + 2$

Input, x	1	2	3	4	5	6	7	8
Output, y								

5. Graph the equation on the coordinate grid.

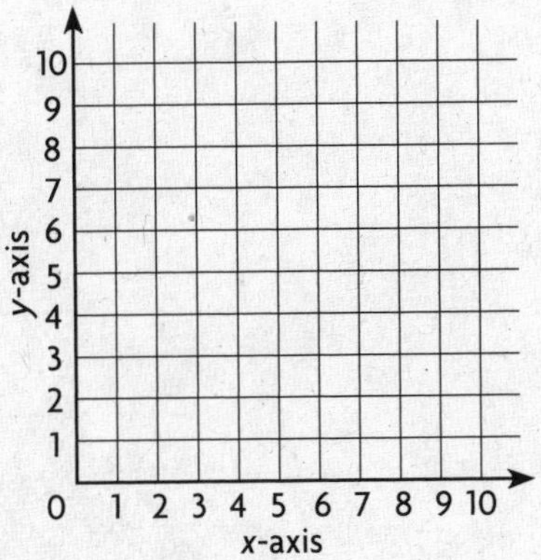

For 6–7, use the bag of numbered tiles below.

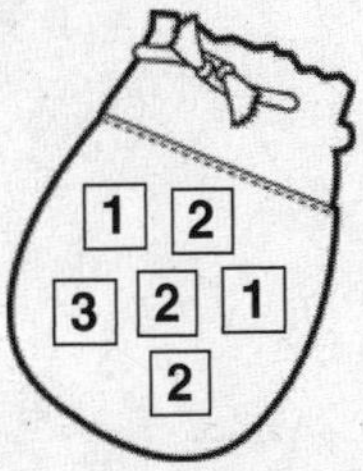

6. What are the possible outcomes for pulling one numbered tile from the bag?

____

7. What is the probability of pulling a 1 from the bag?

____

For 8–11, multiply both sides of the equation by the given number. Find the new values.

8. $(2 + 5) = (4 + 3)$; multiply by 2

____

9. $(4 \times 4) = (32 \div 2)$; multiply by 3

____

10. $(9 \div 3) = (7 - 4)$; multiply by 5

____

11. $(35 - 25) = (9 + 1)$; multiply by 6

____

Name____________________

Divide Greater Numbers

Divide. Write the method you used.

$3\overline{)2,198}$ or $2,198 \div 3$

- Use a **paper and pencil** when the answer has to be exact.
- So, $2,198 \div 3 = 732r2$.

```
   732r2
 3)2198
  -21
    09
   -9
    08
   -6
     2
```

$7\overline{)4,270}$ or $4,270 \div 7$

- Use mental math for numbers that are easy to work with. Here, the dividend contains multiples of 7, the divisor. So, use mental math to solve.

$4,270 \div 7$
$4,200 \div 7 = 600$
$70 \div 7 = 10$

- Add the quotients.

$600 + 10 = 610$

- So, $4,270 \div 7 = 610$.

Divide. Write the method used.

1. $8\overline{)3,729}$ **2.** $5\overline{)\$30.75}$ **3.** $8\overline{)4,888}$ **4.** $4\overline{)1,472}$

5. $8\overline{)\$26.08}$ **6.** $4\overline{)2,400}$ **7.** $3\overline{)\$76.56}$ **8.** $6\overline{)5,803}$

Name________________________________

Divide Greater Numbers

Divide. Write the method you used.

1. $2\overline{)643}$
2. $6\overline{)2,418}$
3. $4\overline{)6,458}$
4. $5\overline{)1,467}$
5. $3\overline{)2,483}$
6. $7\overline{)8,123}$
7. $8\overline{)7,467}$
8. $3\overline{)5,105}$
9. $7\overline{)6,111}$
10. $4\overline{)9,600}$

ALGEBRA Find the dividend.

11. ■ ÷ 3 = 178 ______
12. ■ ÷ 4 = 733 ______
13. ■ ÷ 7 = 410 ______
14. ■ ÷ 9 = 245 r5 ______
15. ■ ÷ 6 = 637 r1 ______
16. ■ ÷ 8 = 801 r4 ______

Problem Solving and Test Prep

17. Leona's team scored a total of 854 points in 7 days. Pilar's team scored a total of 750 points in 6 days. Which team scored more points each day?

______

18. Vicki has 792 seeds to put into packets. If she puts 9 seeds in each packet, how many packets will Vicki need?

______

19. Seth donated a total of $3,336 over 6 months to a charity. If he donated exactly the same amount each month, how much did Seth donate each month?

A $210
B $333
C $336
D $556

20. Joe computed that he drove 1,890 miles a year roundtrip, to and from work. If his commute is 9 miles roundtrip, how many days did Joe work?

A 210
B 333
C 336
D 556

Name ______________________________ **Lesson 12.1**

Factors and Multiples

A **factor** is a number multiplied by another number to find a product. For example, 3 and 5 are factors of 15. The number 15 is a **multiple** of 5 and 3, because $3 \times 5 = 15$.

Use arrays to find all the factors of 6.

Make an array with 6 squares.

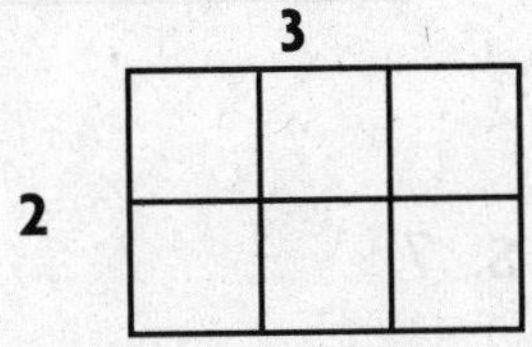

This array contains 2 rows of 3 squares each. So 2 and 3 are both factors of 6.

Make a different array with 6 squares.

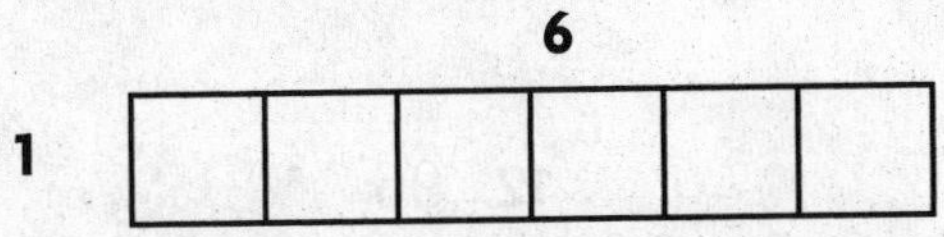

An array with 1 row of 6 contains 6 squares. So 1 and 6 are both factors of 6.

So, the factors of 6 are 1, 2, 3, and 6.

List the first ten multiples of 8.

Multiply 8 by the numbers 1 through 10.

List each product.

$1 \times 8 = 8$	$6 \times 8 = 48$
$2 \times 8 = 16$	$7 \times 8 = 56$
$3 \times 8 = 24$	$8 \times 8 = 64$
$4 \times 8 = 32$	$9 \times 8 = 72$
$5 \times 8 = 40$	$10 \times 8 = 80$

So, the first ten multiples of 8 are:
8, 16, 24, 32, 40, 48, 56, 64, 72, 80.

Use arrays to find all the factors of each product.

1. 10 **2.** 16 **3.** 21

______________ ______________ ______________

______________ ______________ ______________

List the first 10 multiples of each number.

4. 3 ______________________

____

5. 5 ______________________

____

6. 7 ______________________

____

7. 10 ______________________

____

Name____

Factors and Multiples

Use arrays to find all of the factors of each product.

1. 12 ______ ______

2. 18 ______ ______

3. 30 ______ ______

4. 21 ______ ______

List the first ten multiples of each number.

5. 11 ______ ______

6. 4 ______ ______

7. 9 ______ ______

8. 7 ______ ______

Is 8 a factor of each number? Write *yes* or *no*.

9. 16 ______

10. 35 ______

11. 56 ______

12. 96 ______

Is 32 a multiple of each number? Write *yes* or *no*.

13. 1 ______

14. 16 ______

15. 13 ______

16. 8 ______

Problem Solving and Test Prep

17. Tammy wants to make a pattern of multiples of 2 that are also factors of 16. What will be the numbers in Tammy's pattern?

____

18. Which multiples of 4 are also factors of 36?

____

19. Which multiple of 7 is a factor of 49?

A 7
B 14
C 21
D 28

20. Fred is placing 16 cups on a table in equal rows. In what ways can he arrange these cups?

____

Name______________________________

Prime and Composite Numbers

A **prime number** has only two factors: 1 and itself. The number 3 is prime because it has only 1 and 3 as its factors. A **composite number** has more than two factors. The number 15 is composite because its factors are 1, 3, 5, and 15.

Make arrays to find the factors. Write *prime* or *composite* for each number.

7

Make all the arrays you can with square tiles.

1 × 7

1 row of 7 tiles

7 × 1

7 rows of 1 tile each

List the factors.

The factors of 7 are 1 and 7.

Since 7 only has 2 factors, 1 and 7, it is *prime*.

9

Make all the array you can with 9 square tiles.

3 × 3

3 row of 3 tiles

1 × 9

1 row of 9 tiles

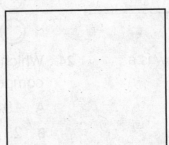

9 × 1

9 rows of 1 tile each

List the factors.

The factors of 9 are 1, 3, and 9.

Since 9 has more than 2 factors, 1, 3, and 9, it is *composite*.

Make arrays to find the factors. Write *prime* or *composite* for each number.

1. 16 ____________

2. 21 ____________

3. 19 ____________

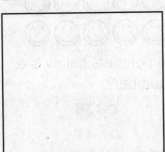

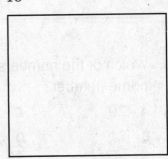

Name ____________________

Lesson 12.2

Prime and Composite Numbers

Make arrays to find the factors. Write *prime* or *composite* for each number.

1. 9 ________ 2. 17 ________ 3. 24 ________ 4. 36 ________

5. 41 ________ 6. 2 ________ 7. 27 ________ 8. 57 ________

Write *prime* or *composite* for each number.

9. 54 ________ 10. 37 ________ 11. 29 ________ 12. 40 ________

13. 45 ________ 14. 33 ________ 15. 51 ________ 16. 88 ________

17. 42 ________ 18. 11 ________ 19. 21 ________ 20. 67 ________

Problem Solving and Test Prep

USE DATA For 21–22, use the array of stickers.

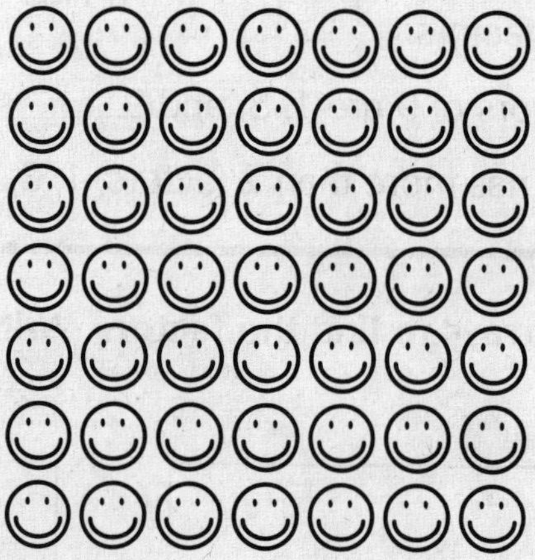

21. In what other ways could the stickers be arranged in equal rows?

____

22. Is 49 prime, or composite? Explain.

____

____

23. Which of the numbers below is a prime number?

A 20 C 24

B 21 D 29

24. Which of the numbers below is a composite number?

A 19 C 36

B 23 D 41

Name____

Lesson 12.3

Factor Whole Numbers

You can show some whole numbers as the product of two or more factors.

The model shows 24 tiles arranged in two equal arrays.

Each array has 3 rows and 4 columns.

The equation for one of these arrays is: $3 \times 4 = 12$.

Remember, there are 2 arrays representing the whole number 24.

So, the equation for the array of all 24 tiles is:

$24 = 2 \times (3 \times 4)$.

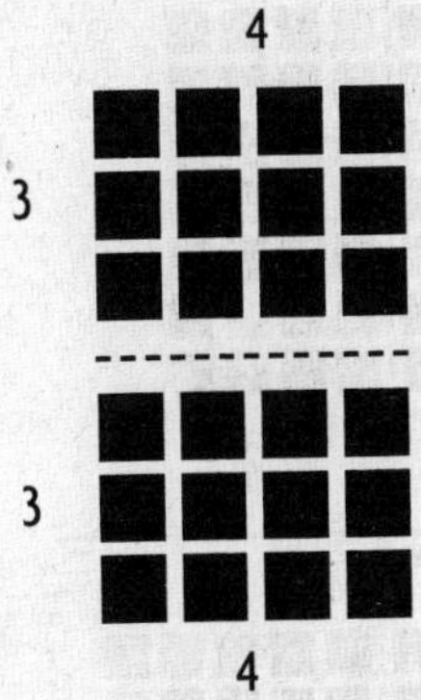

Write an equation for the arrays shown below.

1.

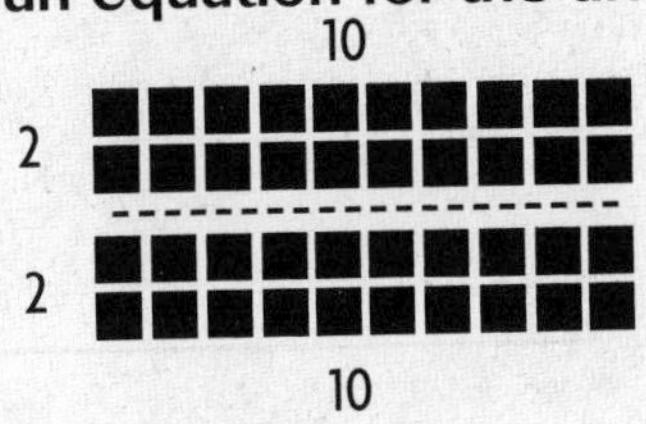

____

2.

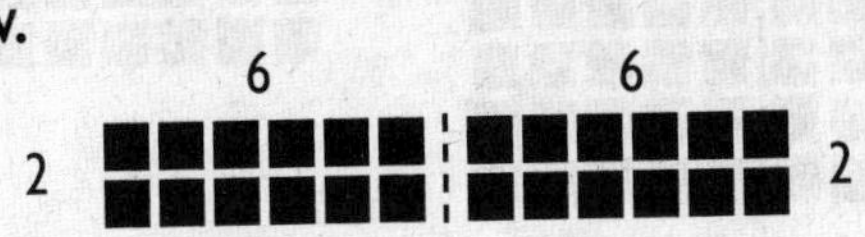

____

3.

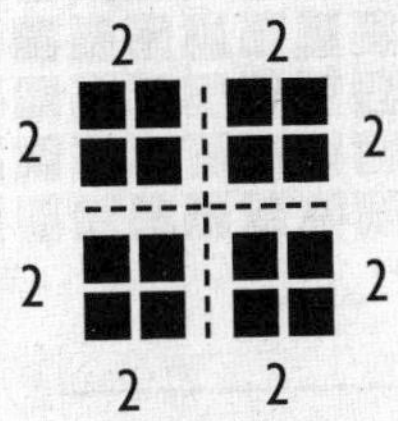

____

4.

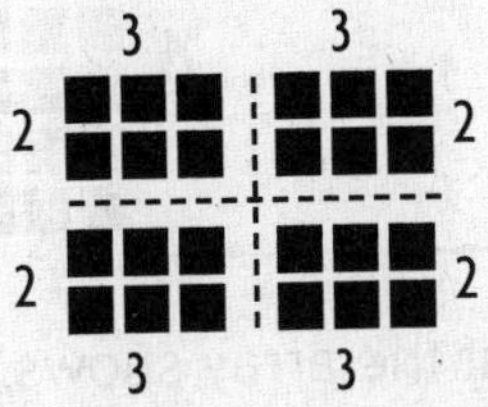

____

5.

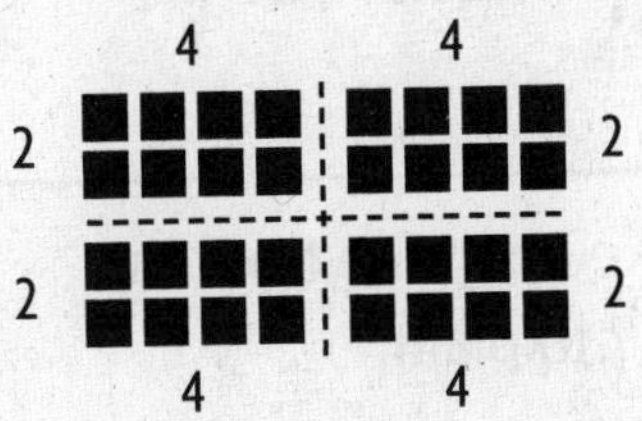

____

6.

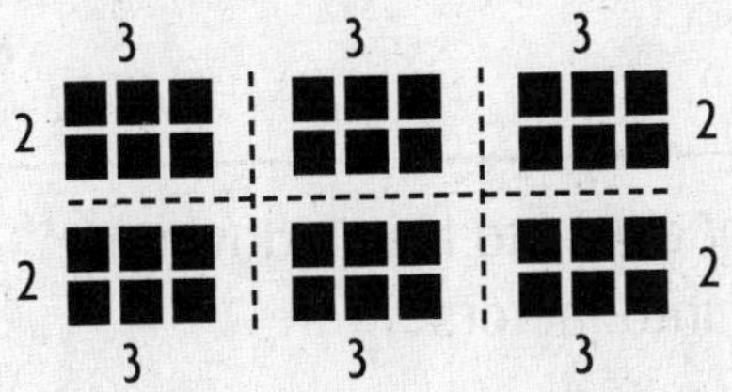

____

Name________________________________

Factor Whole Numbers

Write a multiplication equation for the arrays shown.

1. 

____

2. 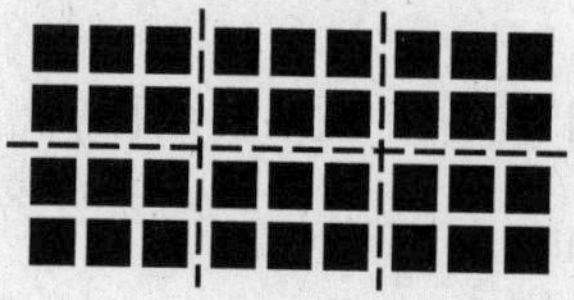

____

3. 

____

4. 

____

5. 

____

6. 

____

For 7–8, use the array on the right.

7. What are two different ways to break apart the array?

24

6

____

8. Write the equation that the array shows. ____________________

Problem Solving and Test Prep

9. What are the factors of 12?

____

10. What are the factors of 16?

____

11. Which is *not* a way to break down the number 48 into factors?

A $2 \times 4 \times 6$ **C** $3 \times 4 \times 4$

B $3 \times 6 \times 5$ **D** 3×16

12. Can you make a square from the factors of 81? Explain.

____

____

Name ______________________________ **Week 18**

Spiral Review

For 1–4, write two equivalent fractions for each.

1. $\frac{1}{5}$ ________ ________

2. $\frac{2}{6}$ ________ ________

3. $\frac{7}{14}$ ________ ________

4. $\frac{6}{7}$ ________ ________

For 8–9, use the bar graph.

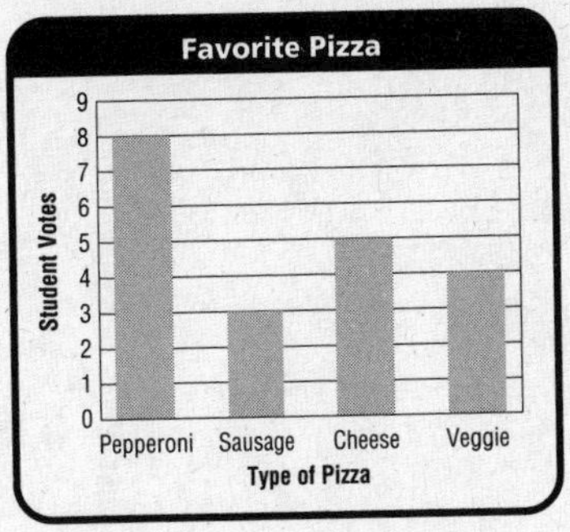

8. Which type of pizza got the most votes? ____________________

9. If three more students voted for veggie pizza, how would you show that on the graph?

____

____

For 5–7, find the length of each line segment.

5. 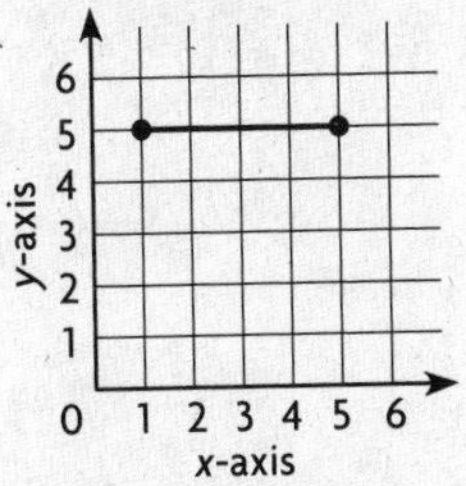

____

6. 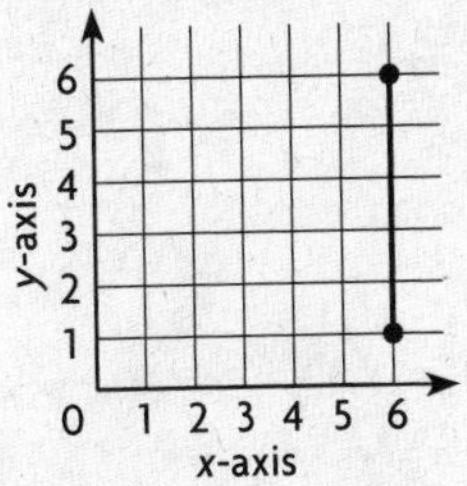

____

7. 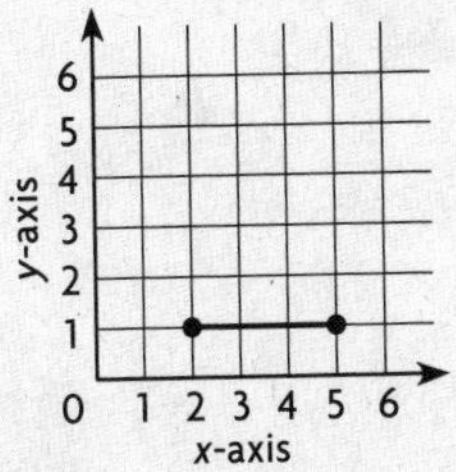

____

For 10–15, find the missing factor.

10. $8 \times \square = 40$

11. $\square \times 12 = 24$

12. $7 \times \square = 21$

13. $\square \times 3 = 18$

14. $\square \times 4 = 36$

15. $4 \times \square = 40$

Name ____________________

Find Prime Factors

Prime factors are all the factors of a number that are prime.

Make a factor tree to find the prime factors of 10.

Choose two factors of 10.
Think: 2×5

Check to see if both factors are prime.
If a factor is a composite, factor again.
Think: 2 and 5 are both prime numbers.

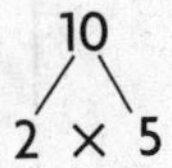

Write the factors from least to greatest.
So, $10 = 2 \times 5$.

Make a factor tree to find the prime factors of 8.

Choose two factors of 8.
Think: 2×4

Check to see if both factors are prime.
If a factor is a composite, factor again.
Think: 2 is a prime number.
4 is a composite number so I need to factor again.

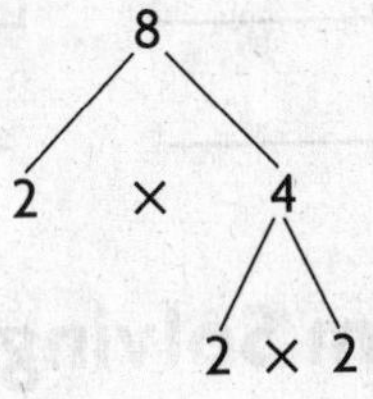

The prime factors of 8 are 2, 2, and 2.

Write the factors from least to greatest.
So, $8 = 2 \times 2 \times 2$.

Make a factor tree to find the prime factors.

1. 27 ____________

2. 15 ____________

3. 12 ____________

4. 14 ____________

5. 20 ____________

6. 42 ____________

Name________________________________

Lesson 12.4

Find Prime Factors

Make a factor tree to find the prime factors.

1. 8	**2.** 20	**3.** 35	**4.** 26	**5.** 33
________	________	________	________	________
6. 9	**7.** 54	**8.** 77	**9.** 81	**10.** 34
________	________	________	________	________

Problem Solving and Test Prep

11. I am an odd number between 11 and 21. I am the product of two prime numbers. What number am I?

____

12. I am the product of two composite numbers, and am between 11 and 17. What are the prime numbers that I am made of?

____

13. Which of these represents another way to write the product 6×8?

A $2 \times 2 \times 2 \times 2 \times 3$

B $2 \times 2 \times 3 \times 3 \times 4$

C $3 \times 3 \times 8$

D $2 \times 2 \times 3 \times 6$

14. Write all of the prime numbers greater than 20 but less than 35.

____

Name____ Lesson 12.5

Number Patterns

To find number patterns, find a rule. If a number pattern increases, try addition or multiplication. If a number pattern decreases, try subtraction or division.

Find a rule. Then find the next two numbers in your pattern.
18, 28, 38, 48, ■, ■

Look at the first two numbers, 18 and 28.
What rule changes 18 to 28?

Since 28 is greater than 18, try addition or multiplication.

Try "add 10" because 18 + 10 = 28.

Test "add 10:" 28 + 10 = 38.

The rule works.

Find the next two numbers.

Think: 48 + 10 = 58; 58 + 10 = 68.

So the rule is "add 10," and the next two numbers in the pattern are 58 and 68.

Find a rule. Then find the next two numbers in the pattern.

1. 25, 20, 15, 10, ■, ■
____

2. 13, 19, 25, 31, ■, ■
____

3. 30, 33, 32, 35, 34, 37, ■, ■
____

4. 10, 21, 32, 43, ■, ■
____

5. 2, 4, 8, 16, ■, ■
____

6. 3, 5, 10, 12, 24, 26, ■, ■
____

7. 100, 80, 60, 40, ■, ■
____

8. 7, 12, 10, 15, 13, 18, ■, ■
____

9. 800, 400, 200, 100, ■, ■
____

10. 3, 2, 12, 11, ■, ■
____

Name ____________________

Number Patterns

Find a rule. Then find the next two numbers in your pattern.

1. 108, 99, 90, 81, ☐, ☐

____

2. 2, 4, 6, 8, ☐, ☐

____

3. 2, 4, 8, 16, ☐, ☐

____

4. 85, 88, 82, 85, 79, 82, ☐, ☐

____

ALGEBRA Find a rule. Then find the missing numbers.

5. 2, 6, 10, ☐, 18, 22, 26, ☐

____

6. 545, 540, 535, ☐, 525, ☐

____

7. 600, 590, 592, 582, 584, ☐

____

8. 400, 410, 409, ☐, 418, ☐

____

Use the rule to make a number pattern. Write the first four numbers in the pattern.

9. Rule: Add 7.

Start with 14.

____

10. Rule: Subtract 6.

Start with 72.

____

11. Rule: Add 2, subtract 5.

Start with 98.

____

12. Rule: Multiply by 2, subtract 1.

Start with 2.

____

Problem Solving and Test Prep

13. Look at the following number pattern. What is the next number if the rule is multiply by 2?

3, 6, 12, ☐

14. Use the pattern 6, 9, 18, 21. What is a rule if the next number in this pattern is 42?

____

15. Which of the following describes a rule for this pattern? 3, 8, 5, 10, 7, 12

A Add 3, subtract 5

B Add 5, subtract 3

C Add 5, subtract 2

D Add 3, subtract 3

16. Which are the next two numbers in this pattern? 192, 96, 48, 24, ☐, ☐

A 10, 5

B 12, 6

C 6, 3

D 5, 2

Name____

Problem Solving Workshop Strategy: Find a Pattern

The table below shows the number of windows on each floor of a skyscraper. How many windows might be on the tenth floor?

Floor	1	2	3	4	5	6	7	8	9
Number of Windows	4	8	6	10	8	12	10	14	12

Read to Understand

1. What are you trying to find?

____

Plan

2. How would the Find a Pattern strategy help you solve the problem?

____

____

Solve

3. What's the pattern for the table? How many windows might be on the tenth floor?

____

Check

4. Does the answer make sense for the problem? Explain.

____

Find a pattern to solve.

5. Peter started an exercise program. On the first day he did 5 pushups. On the second day he did 10 pushups. On the third day he did 15 pushups. If the pattern continues, how many pushups will Peter do on the sixth day?

____

6. Jan rode her bike for 10 minutes the first day, 12 minutes the second day, and 14 minutes the third day. If the pattern continues, how many minutes will Jan ride her bike on the fifth day?

____

Name____ **Lesson 12.6**

Problem Solving Workshop Strategy: Find a Pattern

Problem Solving Strategy Practice

Find a pattern to solve.

1. A 3 by 3 array of blocks is painted so that every other row, starting with row 1, begins with a red block, and the alternate rows begin with a black block. If this pattern continues, does the 12th row begin with red or black?

____

2. What are the next three shapes in the pattern likely to be?

____

3. The first day on a March calendar is Saturday. March includes 31 days. On which day of the week will March end?

____

4. How many blocks are needed to build a stairstep pattern that has a base of 10, a height of 10, and where each step is one block high and one block deep?

____

Mixed Strategy Practice

5. USE DATA If the pattern continues, how much would each 5-inch spike cost if you buy 10,000?

____

Ralph's Hardware Builder's Sale

Spike	10	100	1,000
5-inch	10 cents ea.	8 cents ea.	6 cents ea.
10-inch	15 cents ea.	13 cents ea.	10 cents ea.
15-inch	20 cents ea.	16 cents ea.	12 cents ea.

6. Jules bought 5 pet turtles for $2 each. How much money did Jules spend on turtles in all?

____

7. Dorothy bought gloves with a $20 dollar bill. The gloves cost $6. How much change did Dorothy receive?

____

Name____________________

Collect and Organize Data

You can use a frequency table to organize data.

Use the population frequency table to answer the question.

Washington Elementary School Population	
Grade	**Number of Students**
K	45
1	42
2	54
3	58
4	41
5	55
6	50

If 2 sixth graders move away and 5 fifth graders enroll, how many fifth and sixth grade students will there be in all?

- Adjust the number of sixth graders. Two move away, so you subtract 2. $50 - 2 = 48$.
- Adjust the number of fifth graders. Five enroll, so you add five. $55 + 5 = 60$.
- Add the number of fifth and sixth graders.

  $48 + 60 = 108$

So, there will be 108 fifth and sixth graders in all.

For 1–2, use the frequency table above.

1. How many students are in second and third grades combined?

   ____

2. How many more students are there in sixth grade than in Kindergarten?

   ____

For 3–5, use the Favorite Drawing Tool frequency table below. Tell whether each statement is true or false. Explain.

Favorite Drawing Tool	
Drawing Tool	**Frequency**
Colored pencils	3
Crayons	8
Markers	7

3. More students chose markers than colored pencils.

   ____

   ____

   ____

4. More students chose colored pencils than crayons.

   ____

   ____

   ____

5. A total of 20 students responded to the survey about favorite drawing tools.

   ____

   ____

Name________________________________

Lesson 13.1

Collect and Organize Data

For 1–2, use the Favorite Snacks table. Tell whether each statement is true or false.

Students' Favorite Snacks	
Snack	**Votes**
Apple	12
Banana	7
Carrots	8
Celery	4

1. More students chose carrots than bananas.

____

2. More students chose carrots and celery than apples and bananas.

____

For 3–5, use the Sports Participation table.

Sports Participation		
Sport	**Boys**	**Girls**
Golf	12	19
Softball	18	17
Tennis	9	11
Volleyball	13	12

3. How many more boys participate in volleyball than tennis?

____

4. How many more girls participate in golf than in tennis?

____

5. How many more boys and girls together play softball than volleyball?

____

Problem Solving and Test Prep

USE DATA For 6–7, use the Sports Participation table above.

6. Which is the most popular sport for girls? for boys?

____

7. Who has the largest overall participation in sports: girls or boys?

____

8. How many people were surveyed?

A 186

B 194

C 196

D 200

Favorite Sport	Votes
Golf	37
Softball	63
Tennis	52
Volleyball	44

9.

Favorite Pet					
Pet	Dog	Cat	Bird	Turtle	Fish
Votes	卌 II	卌	III	I	II

How many people were surveyed?

A 18 B 3 C 5 D 7

Name ______________________

Spiral Review

For 1–4, name the number represented by each letter.

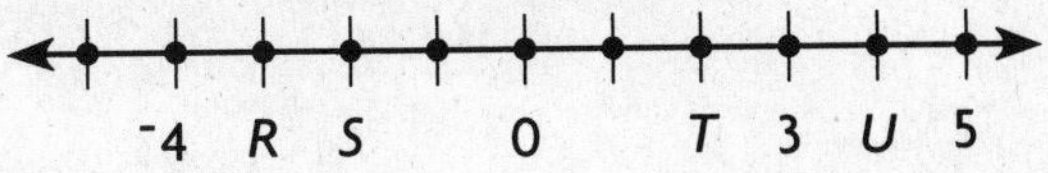

1. *R* ______

2. *S* ______

3. *T* ______

4. *U* ______

For 5–6, use the relationship between plane and solid figures to solve.

5. For her drawing, Melody traces around the bottom of a cone. What plane figure is Melody creating?

____

6. Sam wants to make the following design by stamping shapes onto paper using only 1 solid figure.

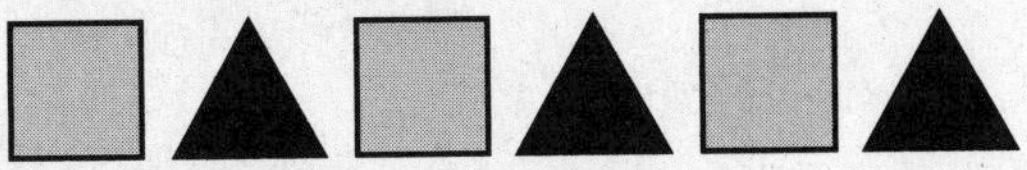

To make both of the plane figures in his design, what solid figure should Sam use?

____

For 7–9, use the spinner below. Tell whether each event is *likely, unlikely,* or *impossible*.

7. The pointer will land on 2.

____

8. The pointer will land on 3.

____

9. The pointer will land on 5.

____

For 10–15, find the product.

10. $(3 \times 3) \times 2$

____

11. $(6 \times 2) \times 2$

____

12. $3 \times (2 \times 2)$

____

13. $9 \times (1 \times 4)$

____

14. $8 \times (2 \times 4)$

____

15. $1 \times (9 \times 9)$

____

Name____ **Lesson 13.2**

Make and Interpret Venn Diagrams

You can use a **Venn diagram** to sort and describe data. Remember the section where circles overlap shows what data the circles have in common.

What label should you use for Section C?

- Section A has multiples of 2.
- Section B has multiples of 2 and 3.
- So, the label for section C should be multiples of 3.

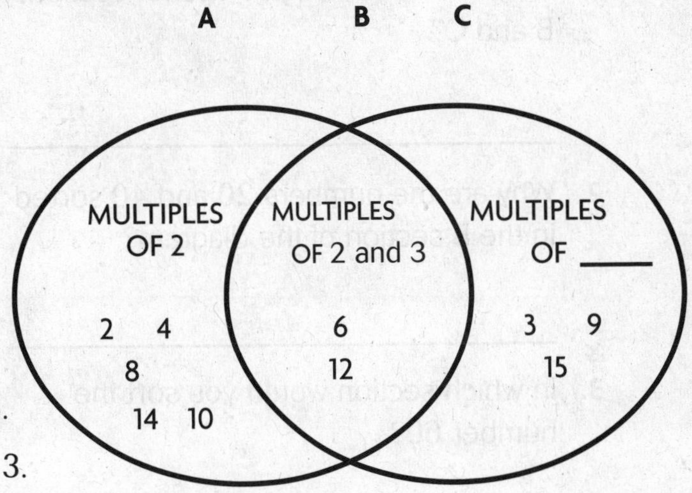

Why are the numbers 6 and 12 sorted in the B section of the diagram?

- The number 6 is a multiple of both 2 and 3.
- The number 12 is a multiple of both 2 and 3.

So, the numbers 6 and 12 are in the B section because they are multiples of both 2 and 3.

For 1–2, use the Venn diagram at the right.

1. What label should you use for section B?

____

____

2. What are two numbers that could go in section B?

____

____

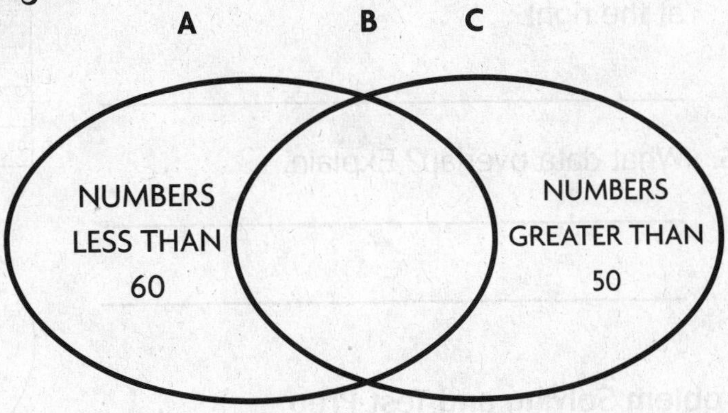

For 3–4, use the Venn diagram below.

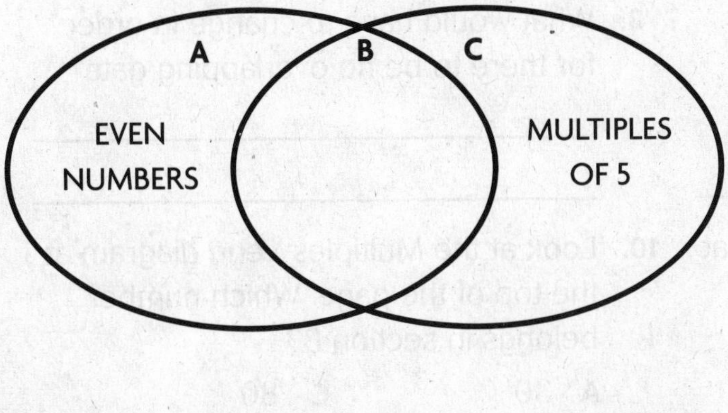

3. What are two numbers that could go in section B?

____

____

4. Can the number 15 go in section B? Explain.

____

____

Name____ **Lesson 13.2**

Make and Interpret Venn Diagrams

For 1–4, use the Multiples Venn diagram.

Multiples

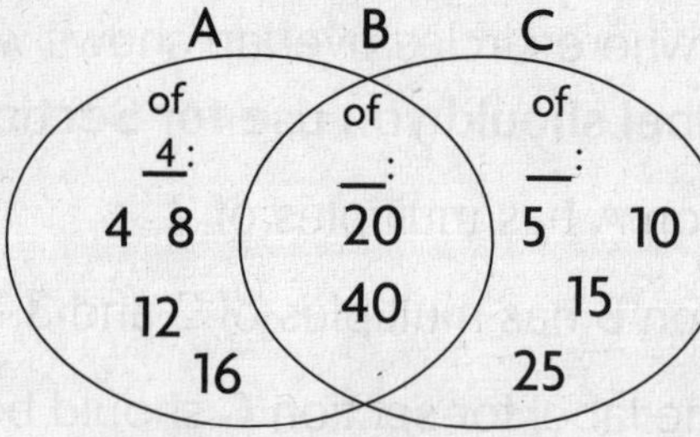

1. What labels should you use for sections B and C?

____

2. Why are the numbers 20 and 40 sorted in the B section of the diagram?

____

3. In which section would you sort the number 60?

____

4. **Reasoning** If section A were multiples of 45 and section C were multiples of 71, would section B contain a number less than 100? Explain.

____

____

____

For 5–6, use the Breakfast Choices table.

5. Show the results in the Venn diagram at the right.

____

6. What data overlap? Explain.

____

____

Breakfast Choices

Food	Student Names
Cereal	Jane, Mani, Liddy, Steve, Ana
Fruit	Ben, Cecee, Beth
Both	Dave, Raiza

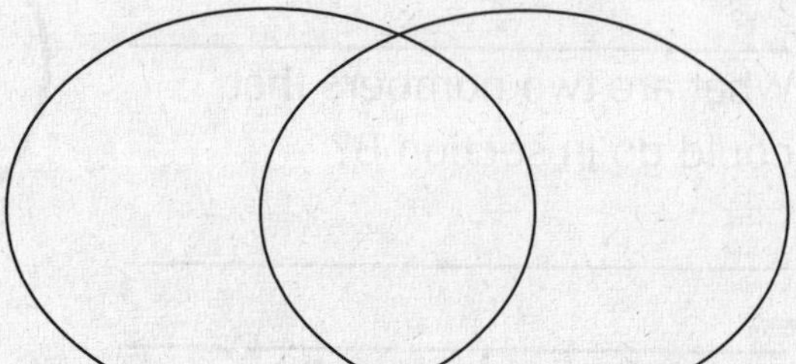

Problem Solving and Test Prep

USE DATA For 7–8, use the Breakfast Choices table.

7. How many students chose either cereal or fruit?

____

8. What would have to change in order for there to be no overlapping data?

____

____

9. Look at the Multiples Venn diagram at the top of the page. Which number belongs in section C?

A 22 **C** 204

B 28 **D** 250

10. Look at the Multiples Venn diagram at the top of the page. Which number belongs in section B?

A 30 **C** 80

B 50 **D** 65

Name____ **Lesson 13.3**

Find Mode and Median

The **median** is the middle number in a set of data. The **mode** is the number that occurs the most often in a set of data.

Rainfall						
Month	Apr	May	Jun	Jul	Aug	Sep
Number of Inches	9	6	8	6	4	9

Find the mode.

- Find the number or numbers that appear most often.

  The numbers 6 and 9 both appear twice.

So, there are two modes: 6 and 9.

Find the median.

- List the numbers in order from least to greatest.

  4, 6, 6, 8, 9, 9

- Find the number in the middle of the list.

  Both 6 and 8 are the middle numbers. So, add the two middle numbers and divide by 2.

  $6 + 8 = 14$
  $14 \div 2 = 7$

So, the median is 7.

Find the median and mode.

1.

Games Played				
Month	Jan	Feb	Mar	Apr
Games	6	3	6	9

median:_____ mode:_____

2.

Books Read					
Month	Aug	Sep	Oct	Nov	Dec
Books	8	9	2	6	2

median:_____ mode:_____

3.

Science Club					
Age	7	8	9	10	11
Frequency	5	2	4	6	4

median: _____mode: _____

4.

Thunderstorms				
Month	May	Jun	Jul	Aug
Storms	5	5	7	8

median:_____ mode:_____

5.

Days Practiced				
Month	Sep	Oct	Nov	Dec
Days	10	3	6	10

median:_____ mode:_____

6.

Apples Eaten				
Day	Mon	Tues	Wed	Thur
Apples	2	2	3	1

median:_____ mode:_____

SDAP 1.2 Identify the mode(s) for sets of categorical data and the mode(s), median, and any apparent outliers for numerical data sets.

Name ______________________________

Find Mode and Median

Find the median and mode.

1.

Rainfall					
Month	**Nov**	**Dec**	**Jan**	**Feb**	**Mar**
Inches	1	2	7	9	2

Median __________

Mode __________

2.

Animals Fed						
Days	**Mon**	**Tues**	**Wed**	**Thurs**	**Fri**	**Sat**
Number	5	12	5	9	11	7

Median __________

Mode __________

USE DATA For 3–5, use the graphs.

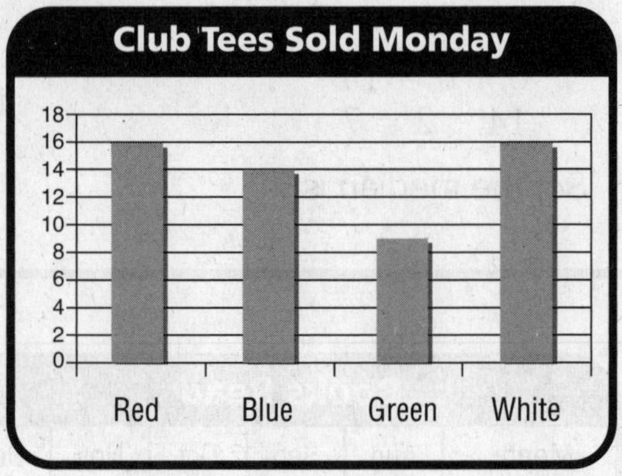

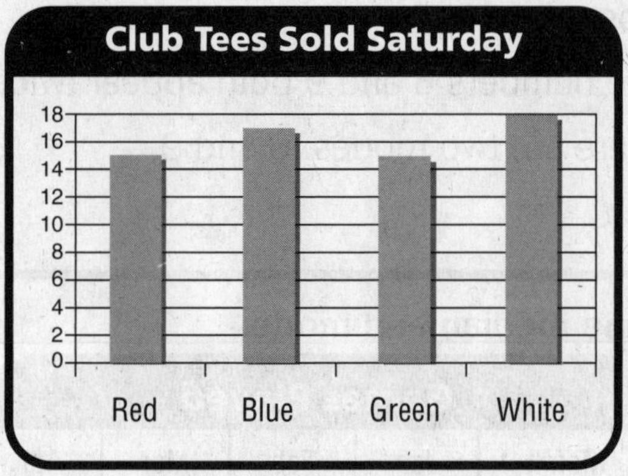

3. What is the difference between the median of tees sold on Monday and the median of tees sold on Saturday? __________

4. What is the range for both days combined? __________

5. How many more tees were sold on Saturday than on Monday?

____

Problem Solving and Test Prep

6. What is the median of the following set of test scores?

4, 7, 10, 9, 9, 5, 7, 8, 8, 9

A 5

B 8

C 9

D 10

7. Look at the Club Tees Sold on Saturday from the bar graph above. What would be the mode if one more white tee was sold?

A 9

B 10

C 15

D 18

Name______________________________ **Lesson 13.4**

Read Line Plots

You can use a **line plot** to graph data. The **range** is the difference between the lowest and highest values in the data. A **clump** is where data is concentrated in one place. A **hole** is where there is no data. An **outlier** is a number separated from the rest of the numbers.

The Number of Hours line plot shows the results of a tourist survey at NASA. Use the line plot to answer the questions below.

What is the range of the hours spent visiting?

- Find the highest number of hours spent visiting on the line plot.

  8 is the highest number.

- Find the lowest number of hours on the line plot.

  1 is the lowest number.

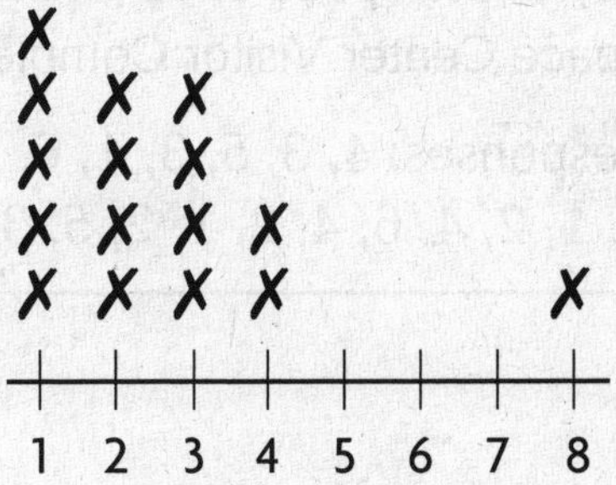

Number of Hours

- Find the difference between the highest and lowest numbers.

  8 − 1 = 7.

  So, the range is 7.

For how long did most tourists visit NASA? How can you tell?

- Look at the line plot to see which hour has the most responses.

  So, most people visited for 1 hour because the line plot has the most Xs at 1.

For 1–2, use the Hours Playing Sports line plot.

1. What is the range of the data?

   ____

2. How long did most respondants play sports? How can you tell?

   ____

   ____

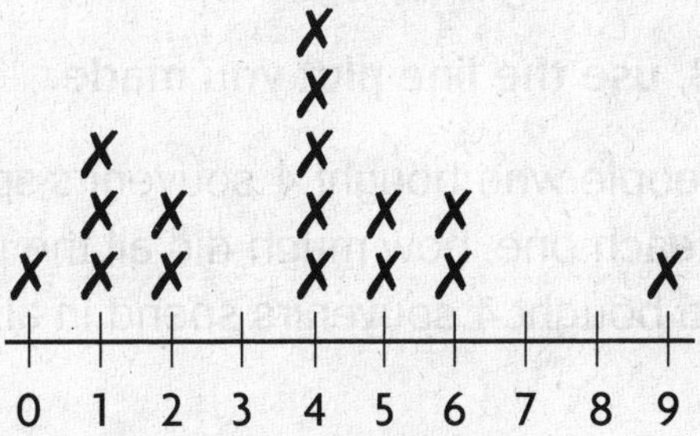

Hours Playing Sports

For 3–4, use the Books Read in October line plot.

3. What is the outlier in the data?

   ____

4. What is the range of the data?

   ____

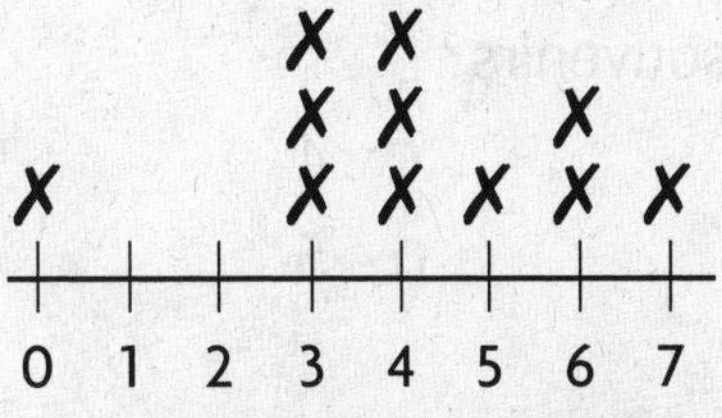

Books Read in October

Name____

Read Line Plots

For 1–4, use the Tourist Souvenir Survey data.

1. Use the data below to fill in the tally table and line plot.

Tourist Souvenir Survey

Question: How many souvenirs did you buy at the Space Shop while at Kennedy Space Center Visitor Complex?

Responses: 4, 3, 5, 3, 1, 6, 5, 2, 5, 1, 9, 6, 1, 2, 4, 6, 4, 2, 1, 2, 5, 3, 4, 1, 6

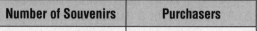

Tourist Souvenir-Buying Results

Number of Souvenirs	Purchasers
1	
2	
3	
4	
5	
6	
7	
8	
9	

Tourist Souvenir-Buying Results

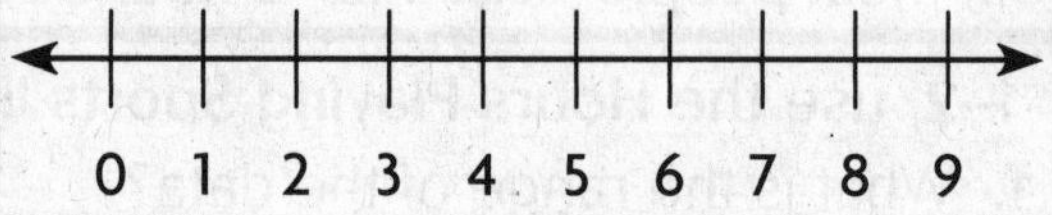

2. Is there an outlier in the data? Explain.

____

____

3. What is the range of souvenirs bought?

____

4. How many souvenirs were purchased in all?

____

Problem Solving and Test Prep

For 5–8, use the line plot you made.

5. If people who bought 4 souvenirs spent $2 for each one, how much did all the people who bought 4 souvenirs spend in all?

____

6. What does the outlier in the data suggest about souvenir buying?

____

____

7. What is the median for the data collected on the souvenirs?

A 2 **C** 4
B 3 **D** 5

8. What is the mode for the data collected on the souvenirs?

A 1 **C** 3
B 2 **D** 4

Name ____________________

Choose a Reasonable Scale

The **interval** of a graph is the difference between one number and the next on the scale. Graphs use different intervals based on the data being graphed.

Choose 5, 10, or 100 as the most reasonable interval for the set of data.

Set A: 200, 350, 100, 250, 500

The numbers are all in the hundreds. An interval of 5 or 10 would be too small to fit all the data on the graph. An interval of 100 would show the data clearly on the graph.

So, an interval of 100 would be most reasonable for Set A.

Set B: 25, 79, 50, 45, 90

The numbers are all less than 100, so an interval of 100 would be too big. An interval of 5 would be too small to fit all the data. An interval of 10 would show the data clearly on the graph.

So, an interval of 10 would be most reasonable for Set B.

Choose 5, 10, or 100 as the most reasonable interval for the set of data.

1. 180, 195, 183, 210, 200

____

2. 8, 13, 12, 6, 9

____

3. 19, 13, 11, 7, 5

____

4. 35, 65, 72, 21, 83

____

5. 90, 130, 200, 150, 310

____

6. 560, 620, 720, 890, 360

____

7. 8, 1, 3, 7, 4, 6

____

8. 93, 66, 25, 75, 71

____

9. 52, 60, 34, 88, 69

____

10. 757, 283, 891, 362, 549

____

SDAP 1.0 Students organize, represent, and interpret numerical and categorical data and clearly communicate their findings.

Name________________________________ Lesson 13.5

Choose a Reasonable Scale

For 1–2, choose 5, 10, or 100 as the most reasonable interval for each set of data.

1. 35, 55, 77, 85, 20, 17 ______________

2. 125, 200, 150, 75, 277, 290 ______________

For 3–6, use the Favorite Summer Sport graph.

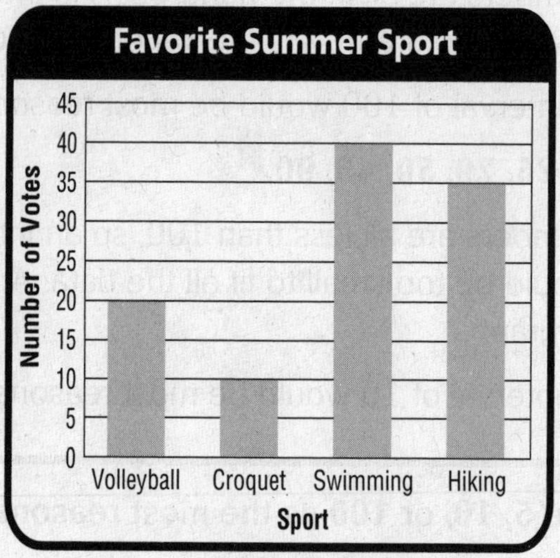

3. What are the scale and the interval used in the graph?

____

4. How would the length of the bars change if the interval were 10?

____

5. How many votes were cast?

____

6. How many more votes did swimming get than croquet and volleyball combined?

____

Problem Solving and Test Prep

USE DATA For 7–10, use the Favorite Winter Sport graph.

7. What is the least favorite winter sport?

____

8. How many fewer people voted for sledding than skiing and ice skating combined?

____

9. What is the interval on the Winter Sport graph?

A 5 **C** 15

B 10 **D** 20

10. What is the scale of the Winter Sport graph?

A 0–80 **C** 0–100

B 0–50 **D** 0–20

Name ____________________ **Week 20**

Spiral Review

For 1–3, write each fraction as a decimal. You may draw a picture.

1. $\frac{1}{2}$ ______

2. $\frac{1}{10}$ ______

3. $\frac{3}{4}$ ______

For 4–6, describe the lines. Write *intersecting* or *parallel*.

4. 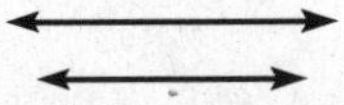

____

5. 

____

6. 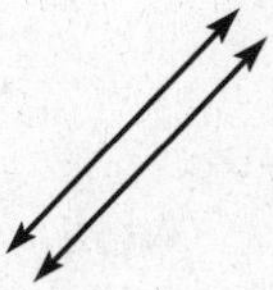

____

For 7–10, list the possible outcomes for each.

7. Elena rolls a cube numbered 1–6.

____

8. Jimena flips a coin.

____

9. Trang spins a spinner.

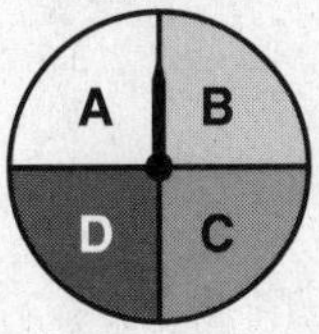

____

10. Chris pulls a tile.

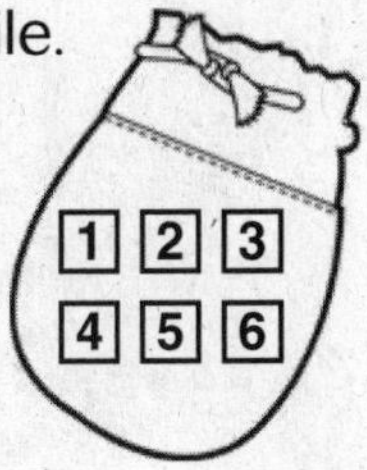

____

For 11–15, follow the order of operations to find the value of each expression.

11. $(4 + 7) + 3 \times 2$

____

12. $(3 + 2 + 7) \div 4$

____

13. $(24 - 18) \times (2 + 5)$

____

14. $(19 - 12) + 4 \times 5$

____

15. $2 \times 3 \div (4 - 2)$

____

Name ____________________

Problem Solving Workshop Skill: Make Generalizations

The chart to the right shows the maximum target heart rates for different ages while exercising.

Target Heart Rates

Age	Beats per Minute
25	195
45	175
65	155
85	135

Miyu is 25 years old. She does aerobics for a half hour and then measures her heart rate. If it is 180 beats per minute, should Miyu slow her exercising, or can she continue at the same pace? Explain your answer.

1. What are you asked to find?

2. Look at the information in the chart. Make a generalization.

3. Find a relationship between Miyu and the information in the chart. What conclusion can you draw about Miyu?

4. What is the answer to this problem?

5. What can you do to check your answer to this problem?

For 6–7, use the table above. Make a generalization. Then solve the problem.

6. What if the graph showed a maximum target heart rate of 145? For what age would that rate be?

7. Sid is 45 years old. While exercising, his heart beats 190 times per minute. Should Sid slow down or continue exercising?

Name____ **Lesson 13.6**

Problem Solving Workshop
Skill: Make Generalizations

Problem Solving Skill Practice

USE DATA For 1–3, use the Adult Weight Range chart. Make a generalization. Then solve the problem.

1. Complete the sentences.

Known Information

The chart shows the healthy weight range for adults. Adults who are ___ tall should weigh between ___ and 184 pounds. Healthy adults who weigh between 139 and 174 pounds may be about ___ tall. An adult who is 5′7″ should weigh between ___ and ___.

- Minimum weights increase in ___-pound increments.
- Maximum weights increase in ___-pound increments.

Height (ft, in)	Adult Weight Ranges (in pounds)	
	Minimum	Maximum
5′7″	127	159
5′8″	131	164
5′9″	135	169
5′10″	139	174
5′11″	143	179
6′0″	147	184

2. Kosi is 5′9″ tall. What is a healthy weight range for Kosi?

____

3. Gwen is a healthy adult who weighs 135 pounds. According to the chart, what might be Gwen's range in height?

____

Mixed Applications

For 4–7, use the Adult Weight Range chart.

4. How much greater is the weight range of a healthy adult who is 6′0″ tall than one who is 5′7″?

____

5. Gino weighs 180 pounds. About how much more does Gino weigh than Tu who is at maximum weight for 5′9″?

____

6. If the pattern continues, what will be the weight range for an adult who is 6′1″ tall?

____

7. **Pose a Problem** Look at exercise 4. Change the numbers to make a new problem.

____

____

____

Name____

Interpret Bar Graphs

You can use **bar graphs** to compare data. In vertical bar graphs, the bars run from the bottom of the graph to the top of the graph. In horizontal bar graphs, the bars run from left to right.

Use the Camp Choices bar graph.

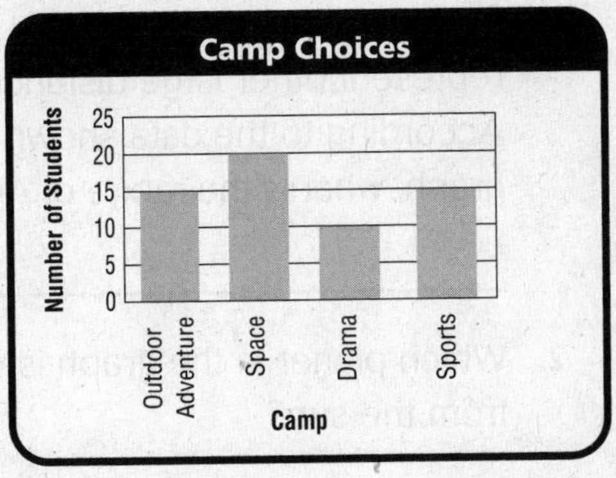

Which camp was chosen by the fewest students?

The shortest bar is the one for drama. So, drama camp was chosen by the fewest number of students.

How many students chose Space Camp?

The bar for Space Camp goes up to the line for 20. So, 20 students chose Space Camp.

For 1–2, use the Favorite Fruit bar graph.

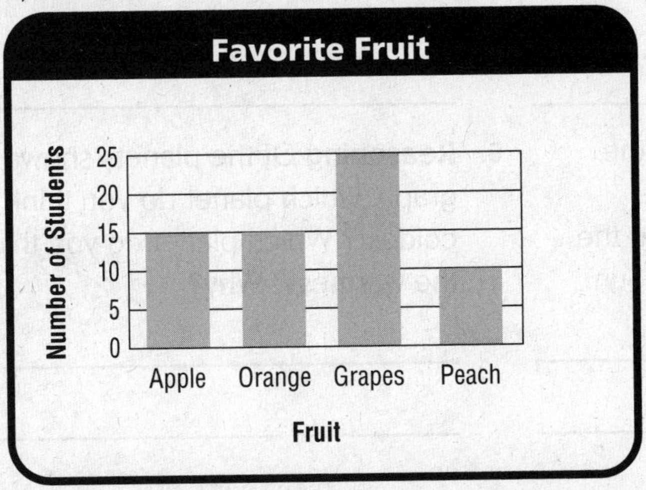

1. Which fruit do most students prefer?

____

2. How many students prefer oranges?

____

For 3–4, use the Favorite Sports bar graph.

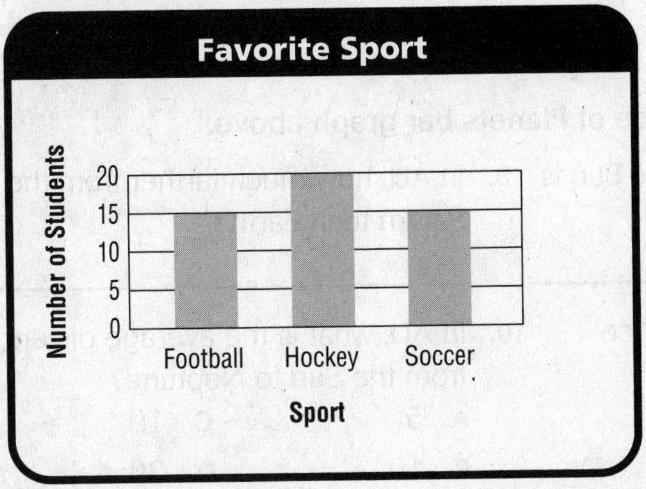

3. Which sports received the same number of votes?

____

4. How many votes did hockey receive?

____

Name______________________________

Interpret Bar Graphs

For 1–6, use the Distance of Planets bar graph.

1. An Astronomical Unit (AU) is the average distance between the Earth and the sun. Scientists use Astronomical Units to help represent other large distances. According to the data shown in the graph, what is the range of AU shown?

____

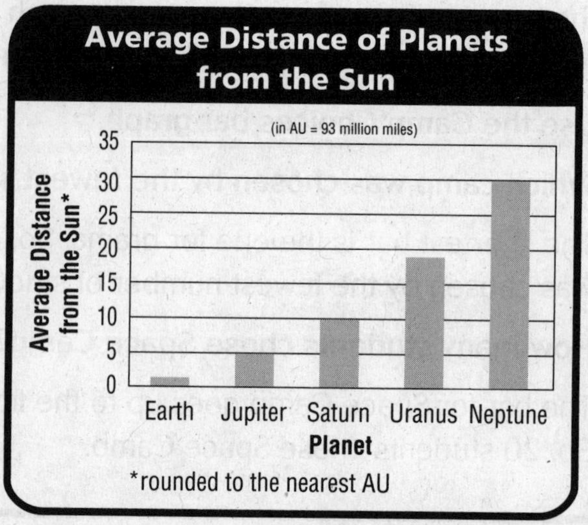

2. Which planet in the graph is farthest from the sun?

____

3. Which planet is 6 times farther away from the Sun than Jupiter?

____

4. Which planet's distance from the Sun is the median of the data?

____

5. List the names of the planets in the graph in order from the greatest average distance from the Sun to the least average distance from the Sun.

____

____

____

6. **Reasoning** Of the planets shown in the graph, which planet do you think is the coldest? Which planet do you think is the warmest? Why?

____

____

____

Problem Solving and Test Prep

USE DATA For 7–10, use the Distance of Planets bar graph above.

7. In AU, how much farther from the Sun is Uranus than Jupiter?

____

8. In AU, how much farther from the Sun is Saturn than Earth?

____

9. In AU, what is the average distance from the Sun to Uranus?

A 5 **C** 19

B 10 **D** 30

10. In AU, what is the average distance from the Sun to Neptune?

A 5 **C** 19

B 10 **D** 30

Name____

Make Bar and Double-Bar Graphs

You can use **double-bar graphs** to compare similar kinds of data. Double-bar graphs have two bars for each category.

Use the data in the table to make a double-bar graph.

Favorite Time of Day

Time	Boys	Girls
Morning	6	8
Afternoon	20	17
Evening	33	31

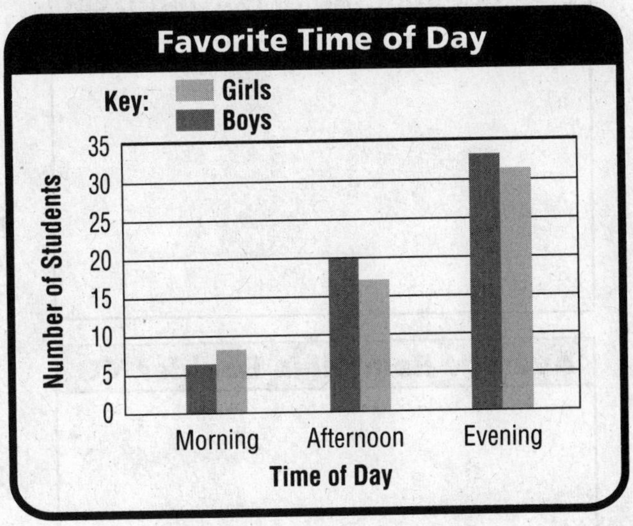

- Make a bar graph and decide on a title, labels, and a scale.
- The title and labels will be the same as in your table.
- Draw a bar for each time of day for the boys.
- Draw a bar for each time of day for the girls. Be sure to use a different color.
- Insert a key to show what each color represents.

For 1–2, use the Rainfall graph below.

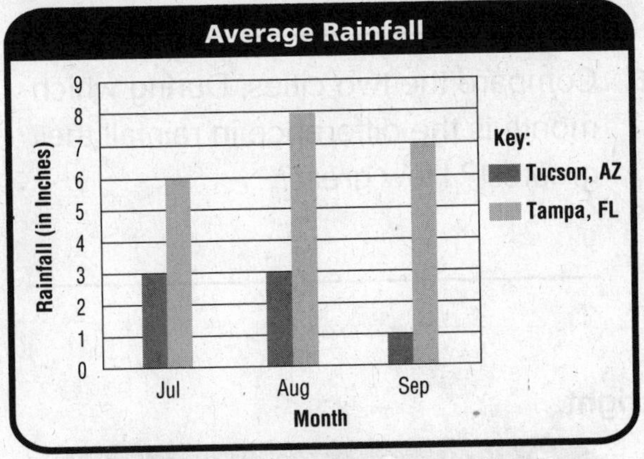

1. How much more rain does Tampa get than Tuscon in July?

____

2. During which month does Tampa get 7 inches of rainfall?

____

For 3, use the Victories table below.

3. Use the table below to make a double-bar graph in the space below.

Victories

Team	Apr	May	Jun
Jays	8	3	9
Robins	7	12	8

Name______________________________________

Make Bar and Double-Bar Graphs

Use the data in the table to make two bar graphs. Then make a double-bar graph. Use the space provided below.

1.

Average Rainfall in Portland, OR

Average Rainfall in Boulder, CO

2.

Average Rainfall

Average Rainfall (in inches)			
City	**Jan**	**Feb**	**Mar**
Portland, OR	6	5	5
Boulder, CO	1	1	2

For 3–6, use the graphs you made.

3. Which city gets the most rainfall from January through March?

____

4. During which month does Boulder get the most rainfall?

____

5. Which city has a greater range of inches of rainfall in the three months?

____

6. Compare the two cities. During which month is the difference in rainfall the greatest? How great?

____

For 7–8, use the Favorite Sports graph at the right.

7. What is the range of the data?

____

8. How many more girls than boys like soccer the most?

____

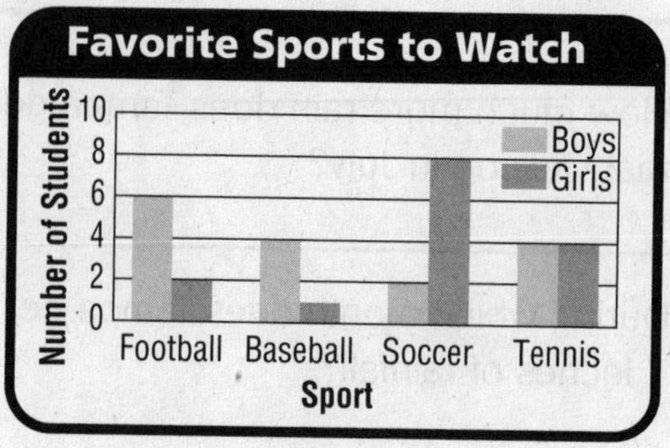

Name______________________________

Interpret Circle Graphs

You can use **circle graphs** to show data in different categories. Each color in the graph stands for a category. Each category is divided into sections to show the data.

Use the Favorite Breakfast graph.

Which type of breakfast received the fewest votes?

- The category with the fewest votes is represented by the smallest section.
- In the graph shown, the smallest section is eggs.
- So, the type of breakfast with the fewest number of votes is eggs.

Which type of breakfast received the most number of votes?

- The category with the most votes is represented by the largest section.
- In the graph shown, the largest section is cereal.
- So, the type of breakfast with the most of votes is cereal.

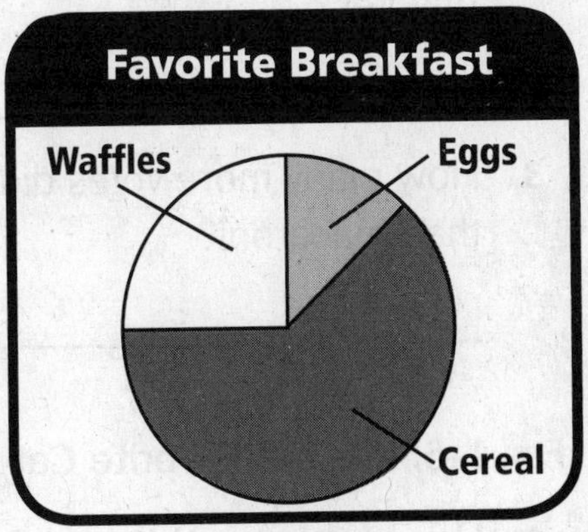

For 1–2, use the Favorite Subject graph.

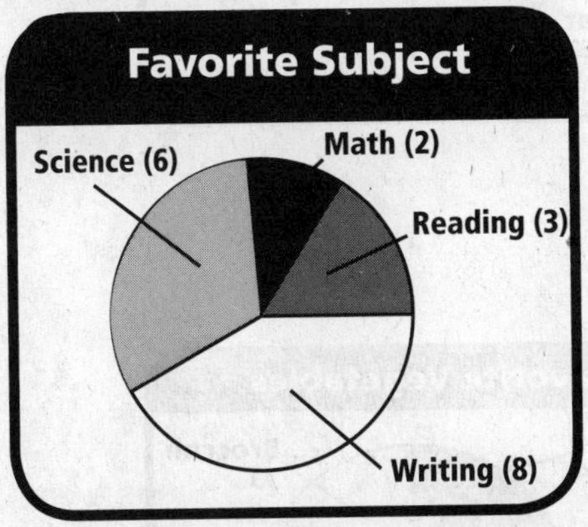

1. Which subject received the most votes?

____

2. How many students prefer science?

____

For 3–4, use the Favorite Animal graph.

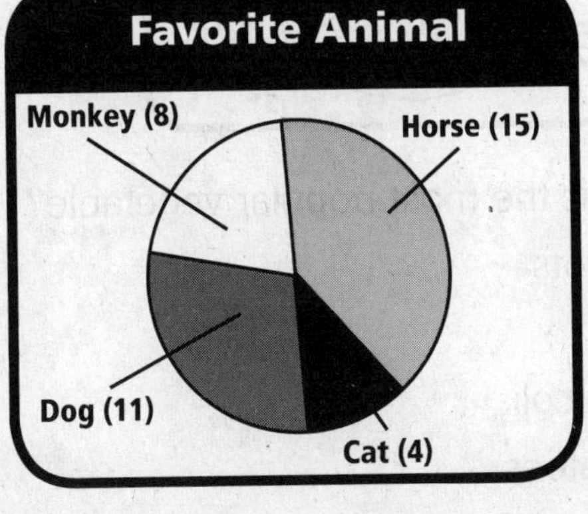

3. How many people said that dogs were their favorite animal?

____

4. How many votes did horses receive?

____

Name_______________________________

Interpret Circle Graphs

For 1–3, use the Favorite Lunch Entree graph.

1. How many votes were counted?

____

2. Which lunch has the greatest number of votes?

____

3. How many more votes did pizza get than macaroni?

____

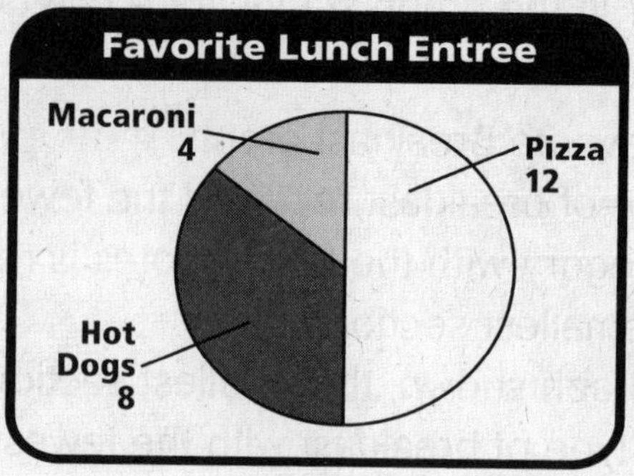

For 4–5, use the Favorite Cat graph.

4. How many people voted for a Favorite Cat?

____

5. Which cat is the favorite among voters?

____

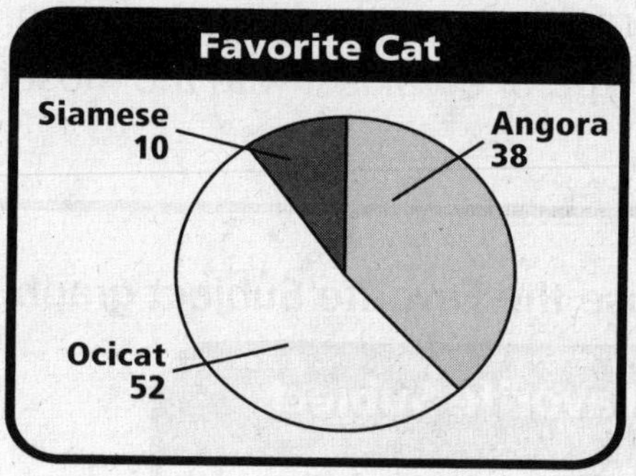

Problem Solving and Test Prep

For 6–9, use the Favorite Vegetables graph.

6. Which vegetable received the fewest votes?

____

7. How many more people voted for carrots than potatoes?

____

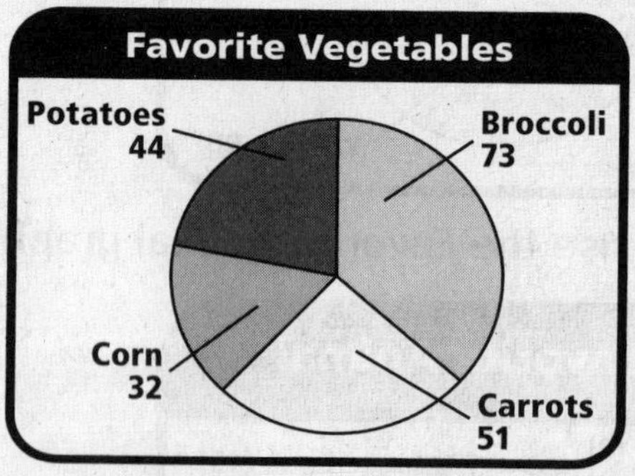

8. How many people voted in all?

   A 200
   B 150
   C 125
   D 100

9. Which is the most popular vegetable?

   A carrots
   B corn
   C broccoli
   D potatoes

Spiral Review

For 1–2, draw conclusions to solve the problem.

1. Jill lives 5.6 miles from school. Gretchen lives 5.1 miles from school. Henry lives 5.9 miles from school. Who lives closest to school?

____

2. Binder's Office Supply sells notebook binders at 2 for $10.00. The Office sells the same binders for $6.00 each. Which store has the better price?

____

For 3–5, classify each triangle. Write *isosceles, scalene,* or *equilateral*. Then write *right, acute,* or *obtuse*.

3. _______________

____

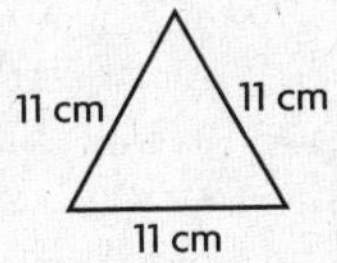

4. _______________

____

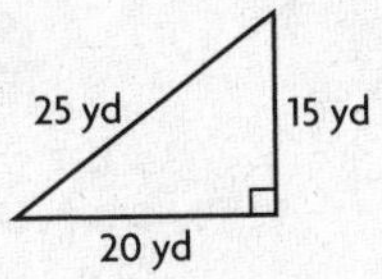

5. _______________

____

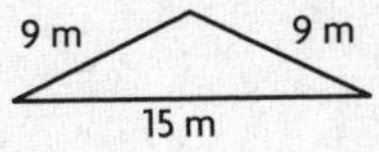

For 6–7, complete the Venn diagram below.

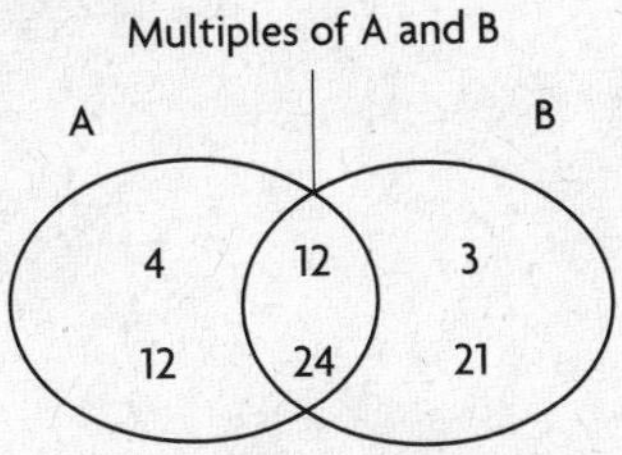

6. What labels should you use for section A and section B?

____

____

7. In which section would you sort the number 48?

____

For 8–10, find a rule. Write the rule as an equation. Use the equation to extend the pattern.

8.

Input	a	14	28	42	56
Output	b	21	35		

____

9.

Input	x	100	80	60	40
Output	y	90	70		

____

10.

Input	k	104	78		
Output	m	91	65	39	13

____

Name____

Algebra: Graph Ordered Pairs

An **ordered pair** names a point on a grid. An ordered pair has two numbers in parentheses, such as (6, 8). The first number tells how far to move horizontally across the grid. The second number tells how far to move vertically up the grid.

Use the grid.

Write the ordered pair for point *C*.

- Start at (0,0).
- Find point *C* on the grid.
  Point *C* is located 9 units to the right. So the first number in the ordered pair is 9.
- Point *C* is located 2 units up. So the second number in the ordered pair is 2.
- So, the ordered pair for *C* is (9, 2).

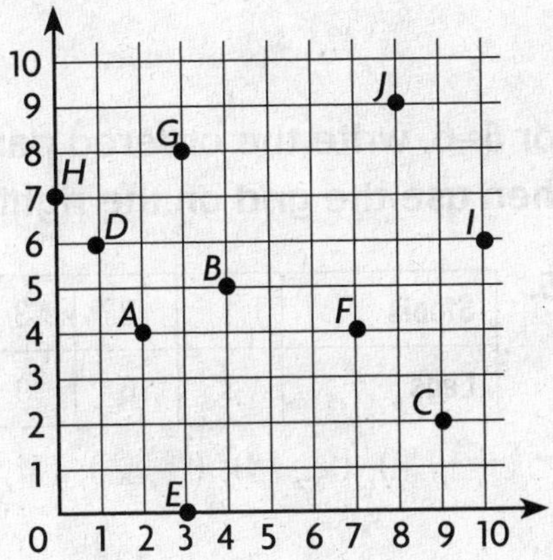

Write the ordered pair for point *D*.

- Find point *D* on the grid.
  Point *D* is located 1 unit to the right and 6 units up.
- So, the ordered pair for point *D* is (1, 6).

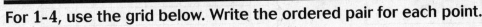

For 1-4, use the grid below. Write the ordered pair for each point.

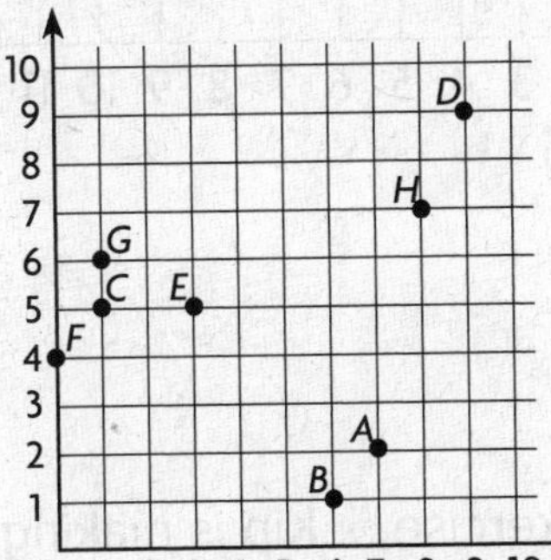

1. *E* ____________

2. *G* ____________

3. *H* ____________

4. *B* ____________

For 5-8, use the grid below. Write the ordered pair for each point.

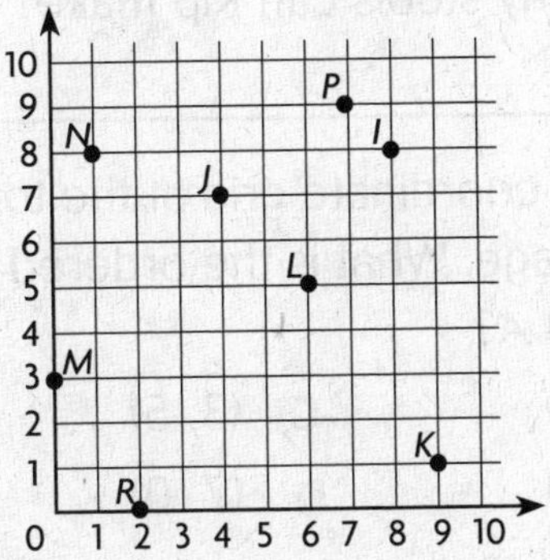

5. *J* ____________

6. *P* ____________

7. *L* ____________

8. *I* ____________

MG 2.0 Students use two-dimensional coordinate grids to represent points and graph lines and simple figures.

Name________________________________ **Lesson 14.4**

Algebra: Graph Ordered Pairs

For 1–4, use the grid at the right. Write the ordered pair for each point.

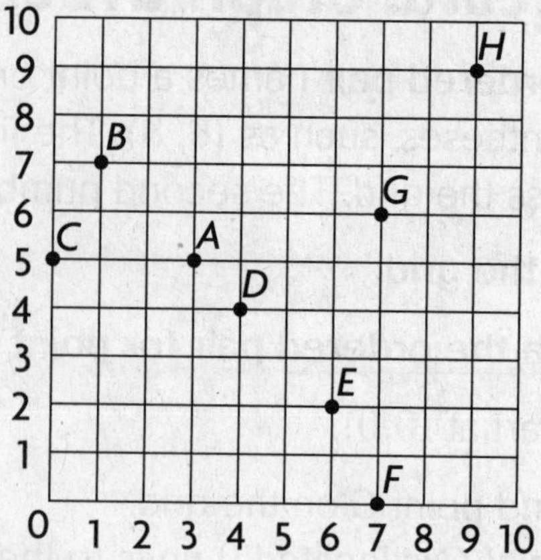

1. *C* (■, ■)
2. *G* (■, ■)
3. *D* (■, ■)
4. *B* (■, ■)

For 5–6, write the ordered pairs for each table. Then use the grid on the right to graph the ordered pairs.

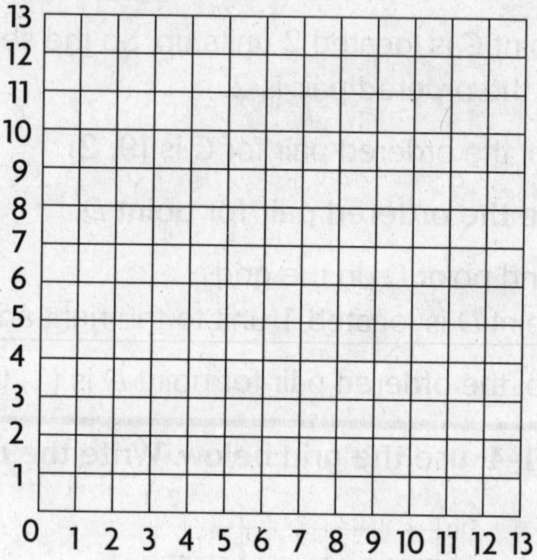

5.

Stools	1	2	3	4
Legs	3	6	9	12

(■, ■), (■, ■), (■, ■), (■, ■)

6.

Section	3	4	5	6
Pages	7	8	9	10

(■, ■), (■, ■), (■, ■), (■, ■)

Problem Solving and Test Prep

7. Look at Exercise 6. Fabio is making a book in which the sections have increasing numbers of pages. How many pages will section 10 have?

____

8. Look at Exercise 5. Kip is making three-legged stools. If he has enough seats to make stools using 24 legs, how many stools can Kip make?

____

9. Use the coordinate grid at the top of the page. What is the ordered pair for point *F*?

A (6, 2) **C** (3, 5)
B (9, 9) **D** (7, 0)

10. Use the coordinate grid at the top of the page. What is the ordered pair for point *A*?

A (6, 2) **C** (3, 5)
B (9, 9) **D** (7, 0)

Name____

Interpret Line Graphs

Line graphs use line segments to show how data changes over time. **Trends** refer to parts of the line graph where data stays the same, increases, or decreases over a period of time.

At what year was the tree 30 inches?

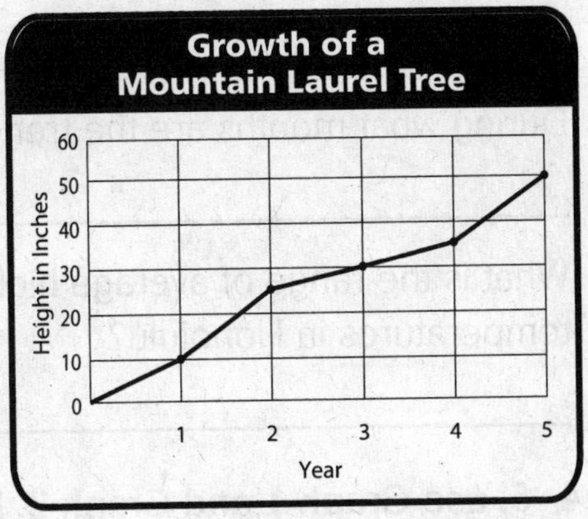

- First, find the number 30 on the scale. Follow the 30 to the right until you reach a point on the graph. Follow the point down to see which year the point represents.
- The point for 30 inches is above the year 3. So, in year 3 the plant was 30 inches.

Between which years did the tree grow more slowly? How can you tell?

- Look at the trends on the graph.
- Between years 2 and 4, the graph line gets flatter.
- This means there was the least amount of change during this time.

So, between years 2 and 4, the tree grew more slowly.

For 1–4, use the graphs below.

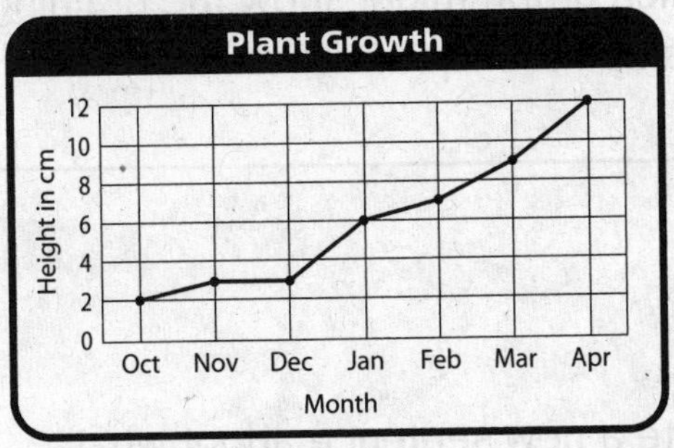

1. How tall was the plant in December?

____

2. During which 2 months did the plant's height stay the same?

____

3. During which 3 months did the team win the same number of games?

____

4. How many games did the team win in April?

____

Name ____________________

Lesson 14.5

Interpret Line Graphs

For 1–3, use the Average Highs in Honolulu graph.

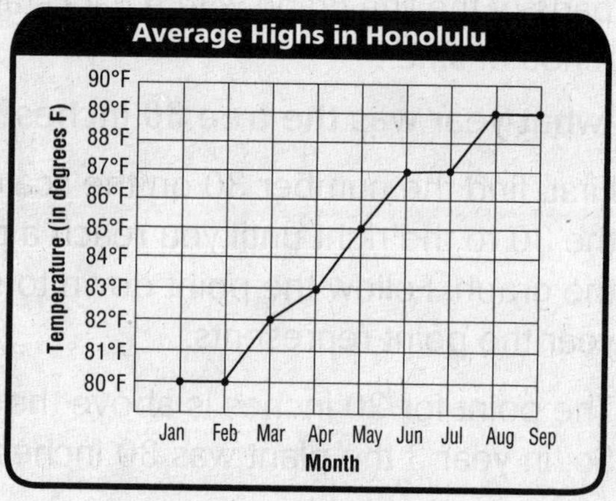

1. What is the highest average temperature in Honolulu?

____

2. During what months are the trends level?

____

3. What is the range of average high temperatures in Honolulu?

____

For 4–5, use Graph 1 and Graph 2. Explain your choice and write a label for the left side of each graph.

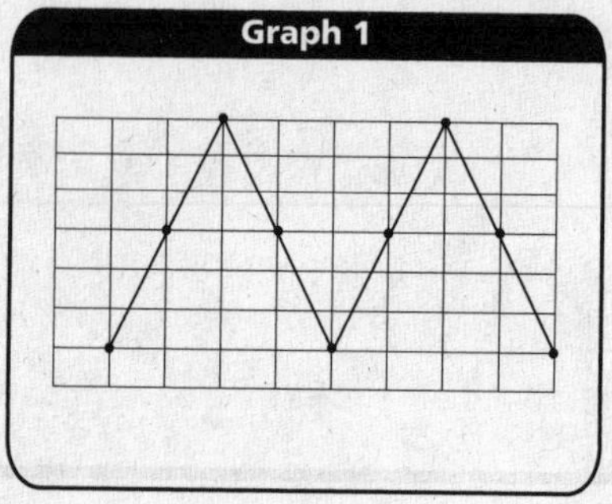

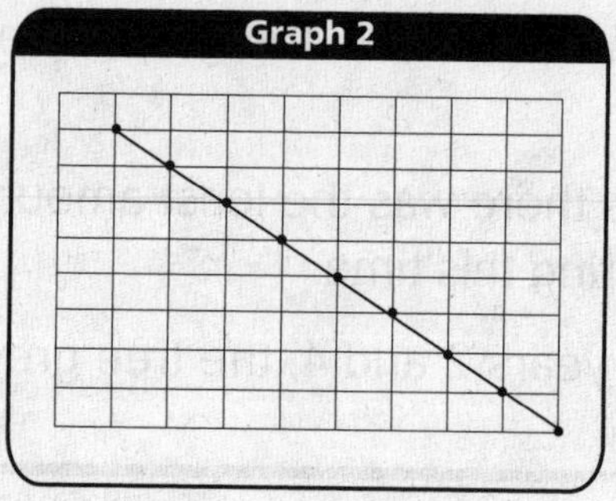

4. Which graph might show someone repeatedly going up and down stairs?

____

5. Which graph might show the draining of a bathtub filled with water?

____

Problem Solving and Test Prep

For 6–9, use Graph 1 and Graph 2.

6. Write a new sentence about what Graph 1 might show.

____

7. Write a new sentence about what Graph 2 might show.

____

8. Graph 1 shows that the data are following what trend?

   A increasing **C** staying the same
   B decreasing **D** none of these

9. Graph 2 shows that the data are following what trend?

   A increasing **C** staying the same
   B decreasing **D** none of these

Name____

Make Line Graphs

Use the data in the chart to make a line graph.

My Plant's Growth				
Week	1	2	3	4
Height (in inches)	6	9	13	15

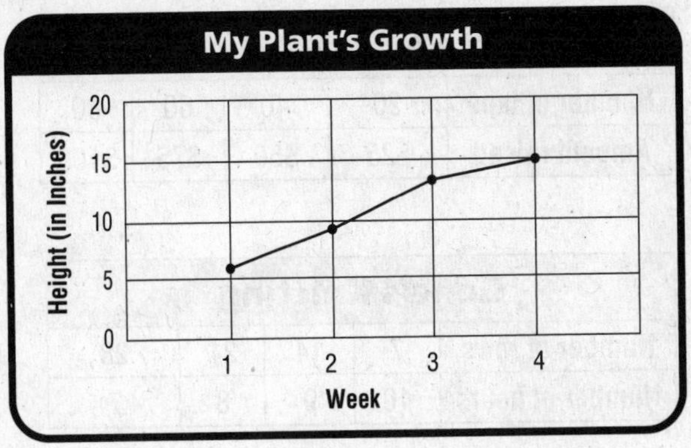

- Write a title for the graph: My Plant's Growth.
- Choose a scale and interval. Since the smallest number is 6 and the greatest number is 15, you can use a scale of 5. Write the label and scale numbers along the left side of the graph.
- Write the labels for the weeks along the bottom of the graph.
- Plot the points. Then draw line segments to connect the points from left to right.

For 1–2, use the data to make line graphs in the boxes below.

1.

Haley's Afternoon Bike Ride				
Time in Min.	30	60	90	120
Distance in Miles	9	16	21	27

2.

Bianca's Writing Progress				
Time in Minutes	15	30	45	60
Total Pages	1	3	9	11

Name____

Make Line Graphs

For 1–2, use the data below to make line graphs at the right.

1.

Bette's Bike-a-Thon				
Number of laps	20	40	60	80
Amount raised	$25	$50	$75	$100

2.

Gene's Knitting				
Number of rows	7	14	21	28
Number of hours	10	9	8	7

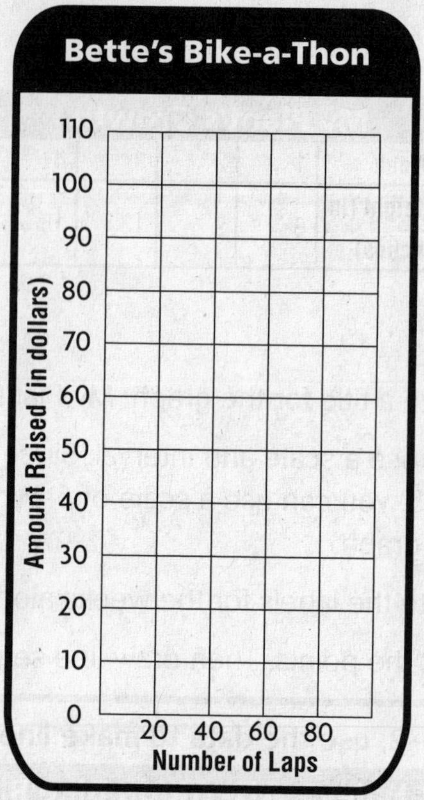

3. Look at your graph from Exercise 1. Suppose the trend continues. What amount will Bette raise if she bikes 100 laps?

____

4. What would be a better interval to show the different amounts raised in Bette's Bike-a-Thon?

____

5. Look at your graph from Exercise 2. How many rows has Gene knitted so far in all?

____

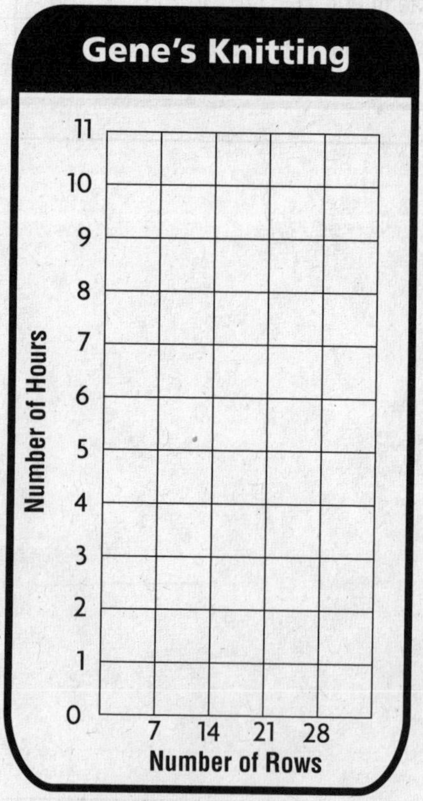

6. How many hours has it taken Gene to knit all the rows so far?

____

7. **Reasoning** Explain what is happening to Gene's speed as he continues to knit.

____

____

____

8. By what interval does the time Gene spends knitting decrease?

____

Name____

Problem Solving Workshop Strategy: Make a Graph

Manuel's class voted for their favorite places to visit. Make a double-bar graph to show the data. Then find which place had the greatest difference in votes between girls and boys.

Places We'd Like to Visit		
Place	**Girls**	**Boys**
Lake	3	1
Ocean	5	4
National Park	2	7
Amusement Park	6	8

Read to Understand

1. What are you asked to find?

____

Plan

2. How can making a graph help you solve this problem?

____

Solve

3. Complete the graph below. Write the answer to the problem on the line.

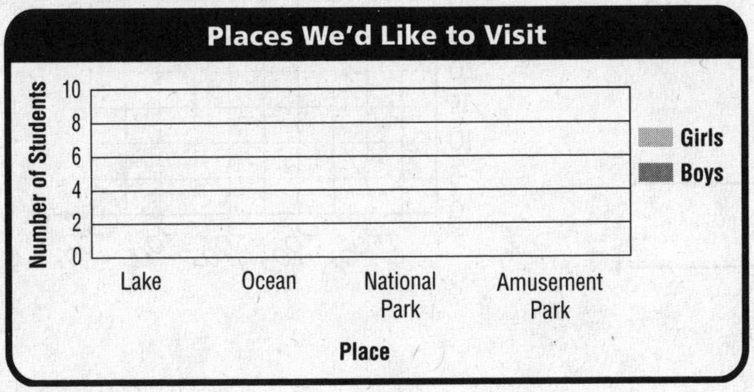

____

____

Check

4. How can you check to see that your answer is correct?

____

Use the data table to make a graph.

5.

Winter Snowfall	
Month	**Inches of Snow**
Dec	25
Jan	30
Feb	15

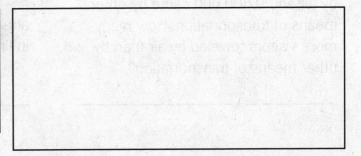

Name____

Problem Solving Workshop Strategy: Make a Graph

Problem Solving Strategy Practice

For 1–3, use the Tourist Spending table.

Tourist Spending in Florida	
Year	**$ in billions**
2001	49
2002	50
2003	52
2004	57
2005	62

1. How can you make a visual display of the data? Use the space provided at the right and make either a line graph or bar graph to display the data. Label your data.

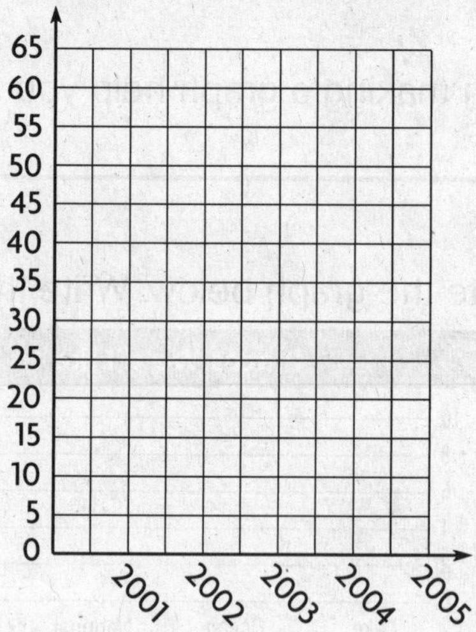

2. What trend in the data does your graph show?

____

3. What if you wanted to add that in 2000 visitors spent $45 billion. Would that change the trend in the data ?

____

____

Mixed Strategy Practice

4. Look at the graph you made above. What is the range of the data?

____

5. **Write Math** Explain why you chose one graph type over the other.

____

____

6. In the first quarter of 2004, 11,800,000 visitors came to Florida by air and 9,800,000 came by other means of transportation. How many more visitors traveled by air than by other means of transportation?

____

7. Jeff, Sean, Ida, and Paul are in line to buy souvenirs. Neither Sean nor Paul are first. Jeff is second and Paul is ahead of Sean. In what order are they in line?

____

Name ____________________

Spiral Review

For 1–4, round each number to the nearest tenth and each money amount to the nearest dollar.

1. 8.92 ________
2. \$92.25 ________
3. 21.21 ________
4. 56.79 ________

For 5–6, draw each of the following in circle *R* below.

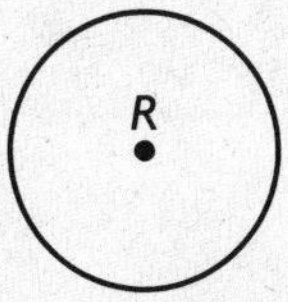

5. radius *RS*
6. diameter *QS*

For 7–8, use the data chart below.

Average Rainfall	
Jan	2 inches
Feb	4 inches
Mar	7 inches
Apr	7 inches
May	5 inches

7. What is the median of the data?

____

8. What is the mode of the data?

____

For 9–14, complete to make the equation true.

9. $11 - 4 = 5 + 6 - \square$
10. $17 + \square = 9 + 8 + 1$
11. $12 - \square + 5 = 5 + 5$
12. $5 + \square = 5 + 10 + 1$
13. $12 - 6 + 9 = 12 + \square$
14. $\square + 1 = 11 + 5 + 1$

Name____ Lesson 14.8

Choose an Appropriate Graph

You can use a variety of graphs to present data.
The graph you use depends on the type of data you are showing.

<table>
<tr><td colspan="3">bar graph or double-bar graphs
Use to show and compare data about different categories or groups. Example: cars sold in June, July, and August</td><td colspan="3">circle graphs
Use to compare parts of a group to a whole group. Example: allowance money spent</td></tr>
<tr><td colspan="2">line graphs
Use to show how data changes over time. Example: growth of a tree over one year</td><td colspan="2">line plots
Use to show the frequency of data along a number line. Example: number of students who have 1, 2, or 3 siblings</td><td colspan="2">pictographs
Use to show and compare data about different categories or groups. Example: number of shirts, shoes, and pants in a closet</td></tr>
</table>

Choose the best type of graph or plot for the data. Explain your choice.

the total number of inches of rain over 5 days

- This data focuses on how the amount of rain changed or stayed the same over the course of 5 days.
- A line graph would be the best type of graph for this data because a line graph shows how data changes over time.

For 1–6, choose the best type of graph or plot for the data.

1. the types of shoes that were most often purchased in April

____

2. how a family spent $100

____

3. how much a vine grew each year

____

4. how many home runs a team hit at each of 5 games

____

5. the number of students playing different sports at recess

____

6. the temperature in a small town over 3 months

____

Name ____________________

Lesson 14.8

Choose an Appropriate Graph

For 1–6, choose the best type of graph or plot for the data.

1. how Drew spends one afternoon

____

2. the amount of rainfall per month in a given town

____

3. favorite toys chosen by boys and girls in a day care

____

4. number of students who make A's in three different grading periods

____

5.

Mileage Traveled					
Miles	75	30	30	90	120
Day	Mon	Tues	Wed	Thurs	Fri

____

6.

Bird Wingspan			
Bird	Hummingbird	Crane	Goose
Inches	4	84	54

____

Problem Solving and Test Prep

For 7–8, use the line and bar graphs below.

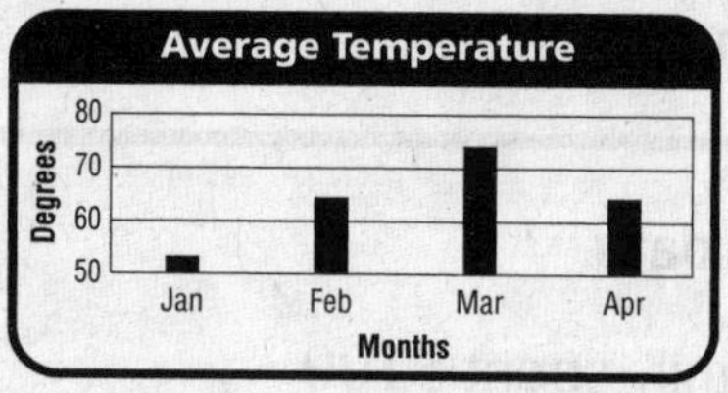

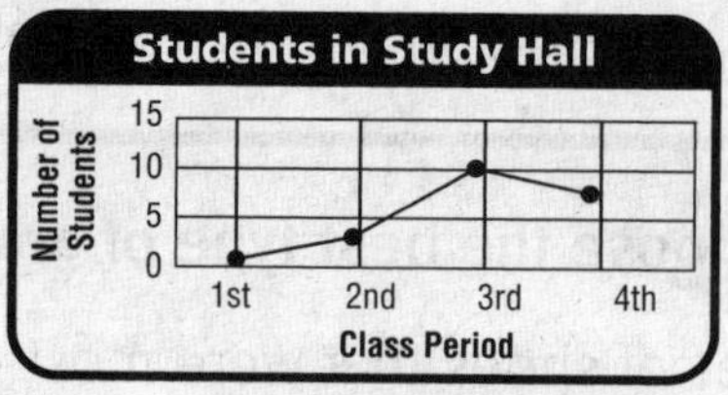

7. What graph would be a better choice to show the average temperature over several months?

____

8. What graph would be a better choice to show the number of students in study hall?

____

9. Which type of graph or plot would best display the numbers of four car models sold in a weekend?

A bar graph **C** line graph

B circle graph **D** line plot

10. Which type of graph or plot would best display how many hours students in class spent on homework over 5 days?

A bar graph **C** line graph

B circle graph **D** line plot

Name____

Temperature

The customary unit for measuring temperature is **degrees Fahrenheit (°F)**.
The metric unit for measuring temperature is **degrees Celsius (°C)**.

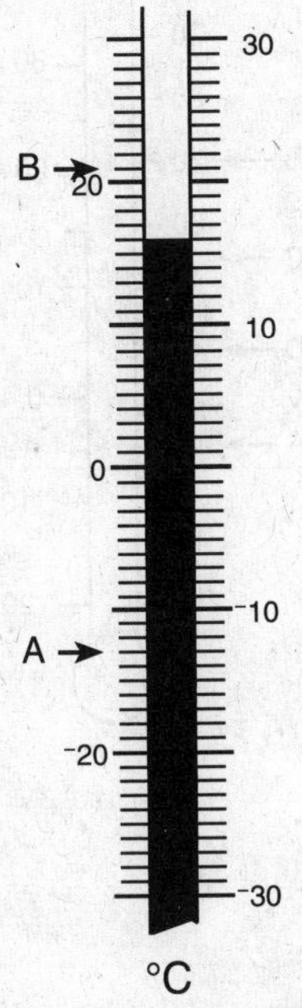

Use the thermometer to find the temperature shown by each letter.

A

- The thermometer uses degrees Celsuis.
- The letter A is below the number 0, so it is a negative temperature.
- A is at the line for ⁻13°Celsius.

So, the temperature at the letter A is ⁻13°C.

B

- The thermometer uses degrees Celsius.
- The letter B is at the line for 21°Celsius.

So, the temperature at the letter B is 21°C.

Use the thermometer below to find the temperature shown by each letter.

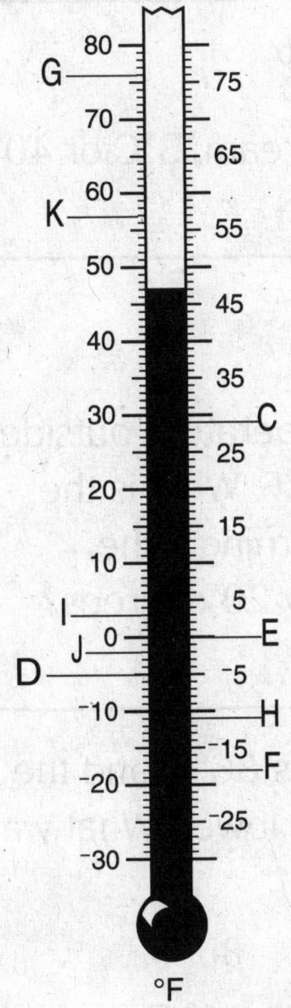

1. C ________ **2.** D ________ **3.** E ________

4. F ________ **5.** G ________ **6.** H ________

7. I ________ **8.** J ________ **9.** K ________

Name ________________________________

Temperature

Use the thermometer to find the temperature shown by each letter.

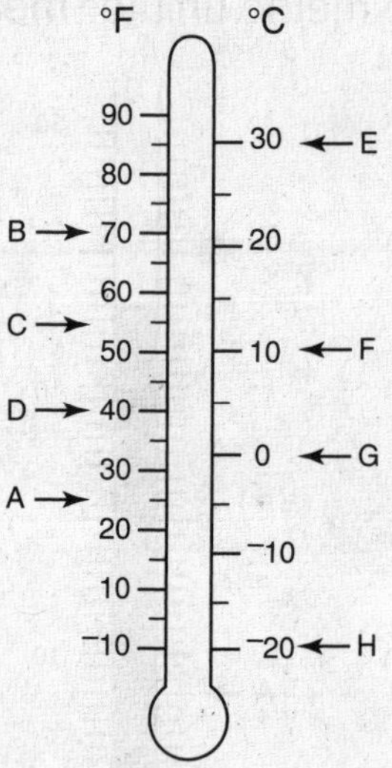

1. A ________ **2.** B ________ **3.** C ________ **4.** D ________

5. E ________ **6.** F ________ **7.** G ________ **8.** H ________

Write the temperature. Then estimate to the nearest 5 degrees.

9.

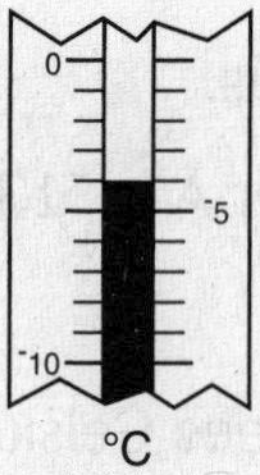

10.

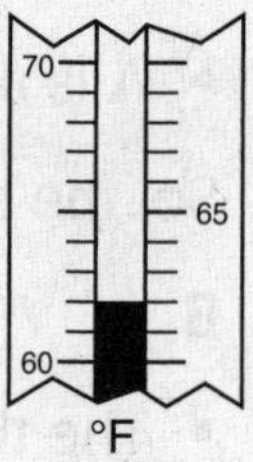

________ ________

Use a thermometer to find the change in temperature.

11. 20°F to 5°F ________ **12.** 13°F to 72°F ________ **13.** $^-8$°C to 35°C ________ **14.** 63°C to 42°C ________

Choose the better estimate.

15. hot tea: 30°C or 95°C ________ **16.** a lake: 50°F or 100°F ________ **17.** ice cream: 3°C or 40°C ________

Problem Solving and Test Prep

18. Order from greatest to least:
100°C; $^-45$°C; $^-12$°C, 32°C

____

19. Bruce measures the temperature outside one morning and it is 45°F. What is the temperature the next morning if the temperature increases by 20 degrees?

____

20. The temperature dropped from 15°C to $^-8$°C. What was the change in the temperature?

____

21. The high temperature was 86° F and the low temperature was 22° lower. What was the low temperature?

A 22° F **C** 86° F

B 64° F **D** 66° F

Name____ **Lesson 15.2**

Explore Negative Numbers

Negative numbers are less than zero, such as $^{-}3$ or $^{-}16$.
Negative numbers have a small minus ($^{-}$) sign next to them.
Positive numbers are numbers greater than zero, such as 5 and 12. Zero is neither a positive nor a negative number.

Name the numbers represented by points *D* and *E* on the number line.

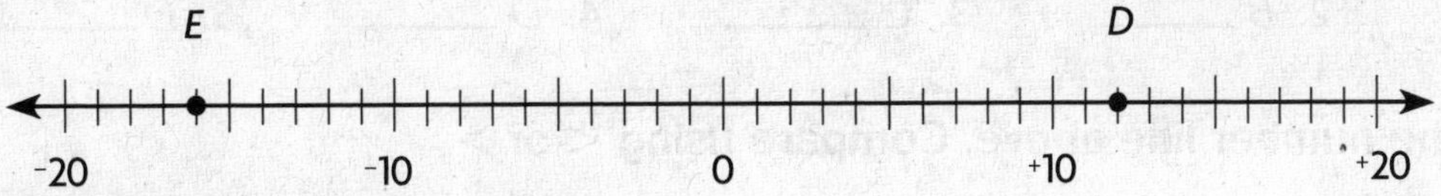

- First find the letter on the number line.
  Then, count from 0 to see which number the letter is above.
- Point *D* is located at 12. It is on the side with plus signs before the numbers. When there is a plus sign, the number is positive. Point *D* is at $^{+}12$ or positive 12.
- Point *E* is located at $^{-}16$. The minus sign ($^{-}$) in front of the numbers on that side show that the number is negative. Point *E* is at $^{-}16$ or negative 16.

So, point *D* is at $^{+}12$ and point *E* is at $^{-}16$.

Name the number represented by each letter on the number line below.

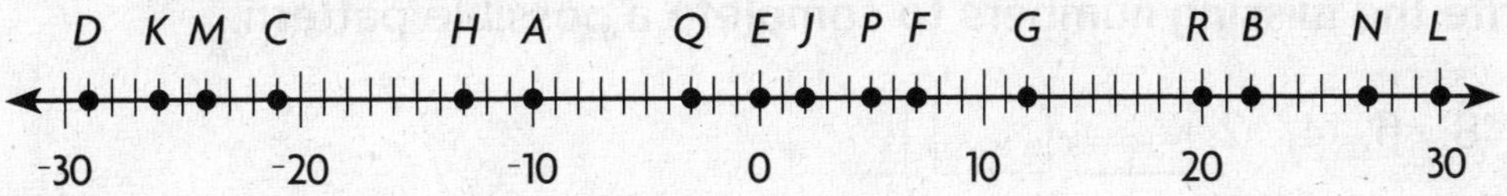

1. *A* ________	**2.** *B* ________	**3.** *C* ________	**4.** *D* ________
5. *E* ________	**6.** *F* ________	**7.** *G* ________	**8.** *H* ________
9. *J* ________	**10.** *K* ________	**11.** *L* ________	**12.** *M* ________
13. *N* ________	**14.** *P* ________	**15.** *Q* ________	**16.** *R* ________

Name______________________________

Explore Negative Numbers

Name the number represented by each letter.

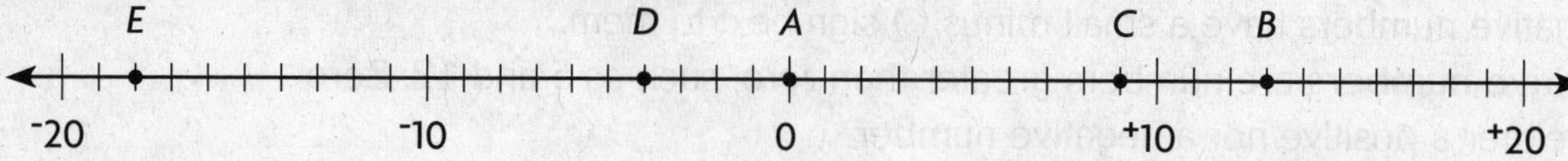

1. *A* ______ **2.** *B* ______ **3.** *C* ______ **4.** *D* ______ **5.** *E* ______

For 6–9, use the number line above. Compare using < or >.

6. $^{-}6 \bigcirc ^{-}8$ **7.** $^{+}5 \bigcirc ^{+}11$ **8.** $^{+}3 \bigcirc ^{-}4$ **9.** $^{-}15 \bigcirc ^{+}15$

Write a positive or negative number to represent each situation.

10. Fred spends all of his $8 allowance.

____

11. Mary buys 12 roses.

____

12. Ruth wins the game by 15 points.

____

13. Sam and Saya eat the last 7 apples.

____

ALGEBRA Write the missing numbers to complete a possible pattern.

14. $^{+}12, ^{+}10, ^{+}8, ^{+}6, ^{+}4, ^{+}2,$ ☐, ☐, ☐

15. $^{+}9, ^{+}7, ^{+}5, ^{+}3, ^{+}1, ^{-}1$ ☐, ☐, ☐

Problem Solving and Test Prep

16. Hans answers 10 questions right on the test. Is this a positive or negative number?

____

17. Martin loses five points to Marci. Is this a positive or negative situation for Marci? for Martin?

____

____

18. Order from least to greatest: $^{+}5, ^{-}2, ^{+}3, ^{-}6$

A $^{-}2, ^{+}3, ^{+}5, ^{-}6$

B $^{+}3, ^{+}5, ^{-}2, ^{-}6$

C $^{-}2, ^{-}6, ^{+}3, ^{+}5$

D $^{-}6, ^{-}2, ^{+}3, ^{+}5$

19. Use the number line above to find which number sentence is *false*.

A $^{+}6 < ^{+}1$

B $^{-}1 > ^{-}4$

C $^{-}1 < ^{+}4$

D $^{+}6 > ^{-}4$

Name____

Problem Solving Workshop Strategy: Act it Out

A special tool called an auger is used to drill through ice so that people can ice fish. The ice is 28 inches thick and the tool can reach 8 inches below the ice. How many inches long is the auger?

Read to Understand

1. What are asked to find?

____

Plan

2. How can acting it out help you solve this problem?

____

____

____

Solve

3. Solve the problem. Describe how you used the "Act it Out" strategy.

____

____

____

4. Write your answer in a complete sentence.

____

Check

5. How can you check that your answer is correct?

____

____

Act it out to solve.

6. In the afternoon, the temperature is 13ºF. Later that day, the temperature drops 7 degrees. By 10:00 PM, it drops 5 more degrees. What is the temperature at 10:00 PM?

____

7. Kari buys a ball for $11.99, a glove for $3.45, and a pair of shoelaces for $1.36. He gives the cashier $20.00. How much change should Karl get back?

____

Name____

Problem Solving Workshop Strategy: Act It Out

Practice Solving Strategy Practice

Act it out to solve.

1. Sally wants to go swimming this afternoon if the temperature is above 85°F. This morning the temperature was 92°F. By noon, it was 76°F. Did Sally go swimming? If not, how many degrees too cool was it?

____

2. The record high in Fresno, California, in August was 112°F on August 13, 1996. The record low was 49°F on August 30, 1966. What is the difference in temperature between the two extremes?

____

3. Rena lines up four model race cars. The red one is ahead of the green one. The blue one is first. The yellow one is directly behind the red one. What color race car is last?

____

4. Jim leaves home with $40. At the mall, he bought goggles for $9.98, swimming trunks for $19.95, and lunch for $8.45. How much money did Jim have left after shopping?

____

Mixed Strategy Practice

5. Make a bar graph that shows long-track speed records.

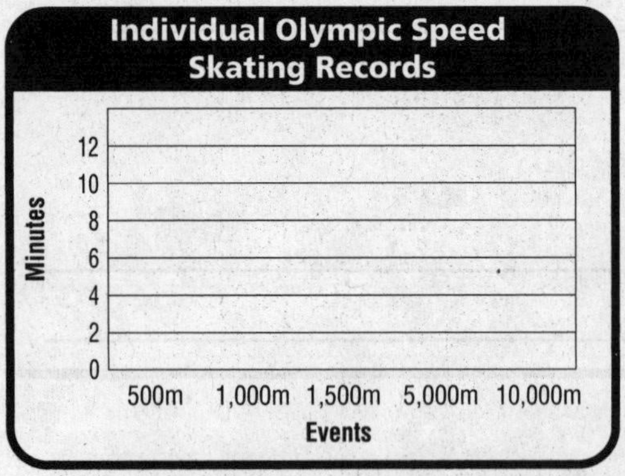

Individual Olympic Speed Skating Records (Rounded Times)

Event	Long-track
500 m	34 sec
1,000 m	1 min 7 sec
1,500 m	1 min 44 sec
5,000 m	6 min 15 sec
10,000 m	12 min 59 sec

6. An athlete can run 1,000 meters in 1 minute 4 seconds. How long would it take the athlete to run 5,000 meters?

____

7. Willa measures the temperature on Monday. By Wednesday, it had increased to 13°C. The temperature changed 21°C. What was the temperature on Monday?

____

Spiral Review

For 1–4, tell where to place the first digit. Then divide.

1. $5\overline{)343}$ ______________

2. $366 \div 2$ ______________

3. $4\overline{)599}$ ______________

4. $168 \div 2$ ______________

For 5–6, tell whether the two figures are *congruent* or *not congruent*.

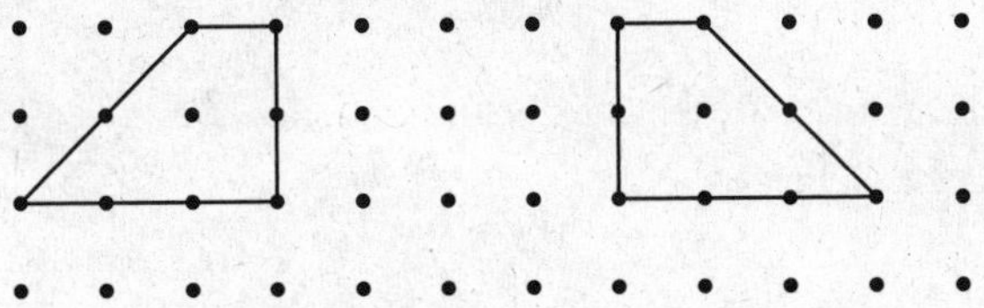

5. ______________________

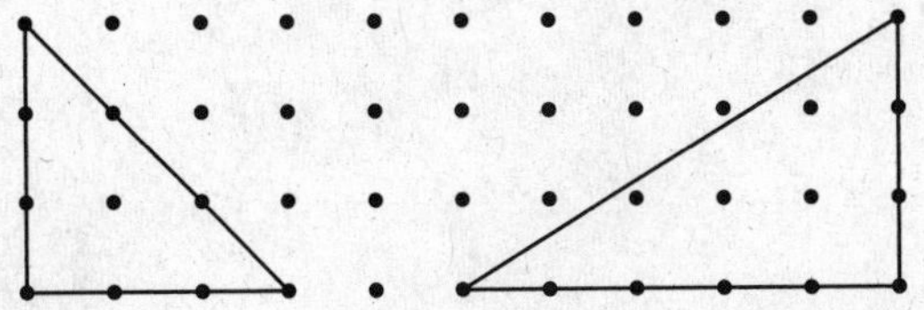

6. ______________________

7. Below is a chart that tells how long it takes Freddy to run up to 5 miles. Make a line graph using the data below.

Mile	1	2	3	4	5
minutes	8	16	24	32	40

8. How long would it take Freddy to run 6 miles?

____

For 9–11, use the rule and equation to make an input/output table.

9. add 3; $x + 3 = y$

Input, x	1	2	3	4	5
Output, y					

10. subtract 4, $x - 4 = y$

Input, x	10	9	8	7	6
Output, y					

11. add 25, $x + 25 = y$

Input, x	5	10	15	20	25
Output, y					

Name____

Use a Coordinate Plane

A **coordinate plane** is a grid formed by a horizontal line called the ***x*-axis** and a vertical line called the ***y*-axis**. You can plot points on a coordinate plane using **ordered pairs**. The first number in an ordered pair, the **x-coordinate**, tells how far to move right or left along the *x*-axis. The second number, the ***y*-coordinate**, tells how far to move up or down the *y*-axis.

Write the point for each ordered pair on the coordinate plane.

(${}^{+}2$, ${}^{-}3$)

- Start at (0, 0). The first number in the ordered pair, ${}^{+}2$, is positive. So move two units to the right.
- The second number is ${}^{-}3$. The number is negative, so move down three units. You end at point *E*.

So, the point for (${}^{+}2$, ${}^{-}3$) is *E*.

(0, ${}^{+}5$)

- Start at (0, 0). Since the first number is 0, do not move to the right or left.
- The second number is ${}^{+}5$, so move 5 units up. You end at point *C*.

So, the point for (0, ${}^{+}5$) is *C*.

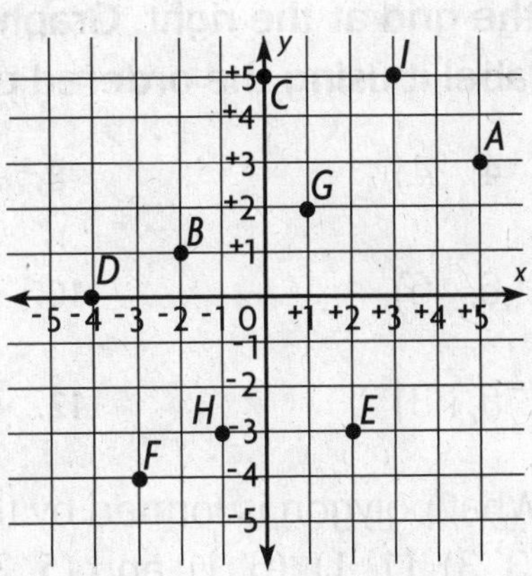

Use the coordinate plane below. Write the point for each ordered pair.

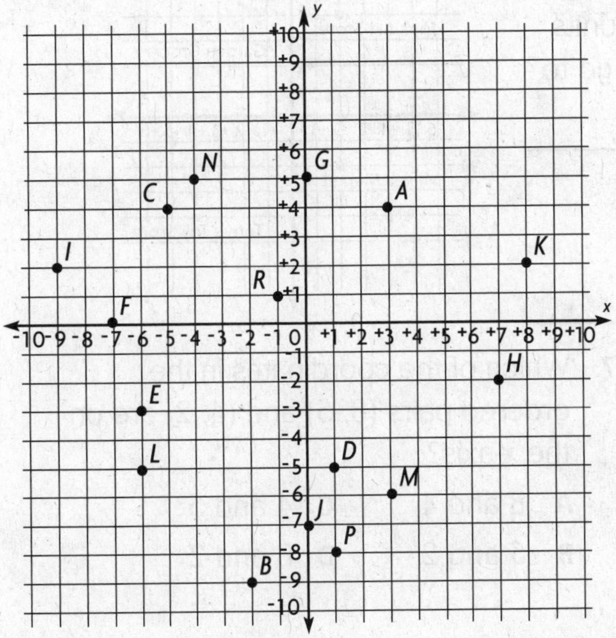

1. (${}^{-}2$, ${}^{-}9$) ____________

2. (${}^{+}3$, ${}^{+}4$) ____________

3. (0, ${}^{+}5$) ____________

4. (${}^{+}1$, ${}^{-}5$) ____________

5. (${}^{-}6$, ${}^{-}5$) ____________

6. (${}^{+}3$, ${}^{-}6$) ____________

7. (${}^{-}9$, ${}^{+}2$) ____________

8. (0, ${}^{-}7$) ____________

Name____

Lesson 15.4

Use a Coordinate Plane

Write the point for each ordered pair on the coordinate plane at the right.

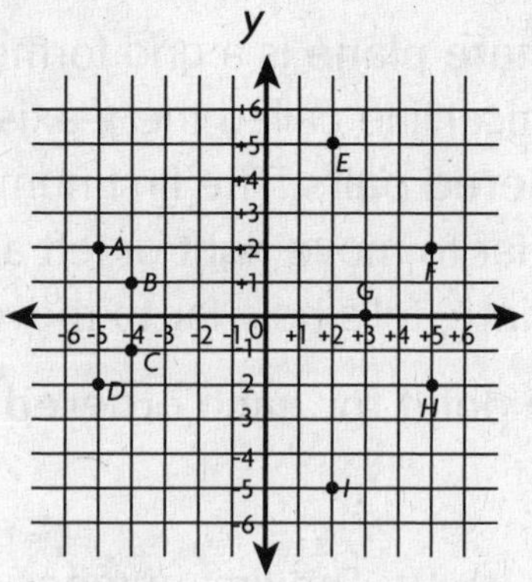

1. $(^{+}2, ^{+}5)$ = ______ 2. $(^{-}4, ^{-}1)$ = ______

3. $(^{-}5, ^{+}2)$ = ______ 4. $(^{+}3, 0)$ = ______

5. $(^{+}5, ^{+}2)$ = ______ 6. $(^{+}5, ^{-}2)$ = ______

Use the grid at the right. Graph each point and label it using the ordered pair.

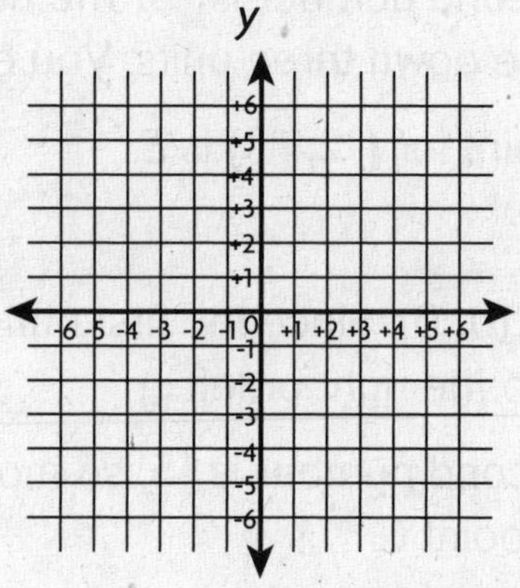

7. $(^{+}4, ^{+}4)$ 8. $(^{+}2, ^{-}1)$

9. $(^{-}6, ^{+}6)$ 10. $(^{+}1, ^{-}5)$

11. $(^{-}5, ^{+}3)$ 12. $(^{-}1, ^{-}6)$

13. What polygon is formed by the points: (3, 3), (1, 1), (6, 1), and (5, 3)?

____

Problem Solving and Test Prep

For 14–15, use the map at the right.

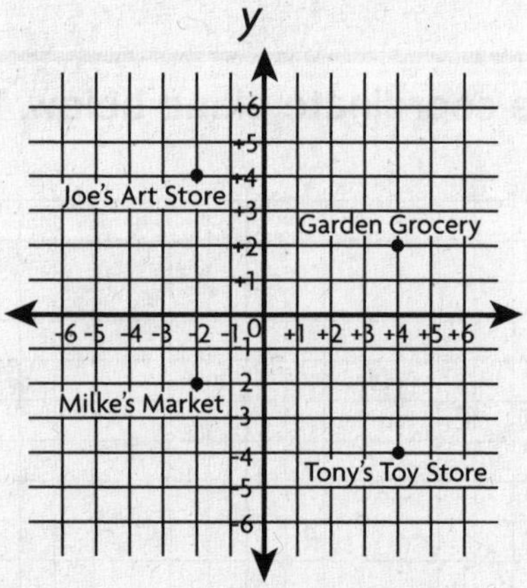

14. Kevin leaves Garden Grocery. He walks 6 units left and 2 units up. What store does Kevin go to?

____

15. Jill leaves Tony's Toy Store to go to Milke's Market. Describe Jill's path.

____

16. How many units above the origin is the point (4, 7)?

A 11 **C** 4

B 7 **D** 3

17. Which of the coordinates in the ordered pairs (8, 5) and (4, 2) are on the *x*-axis?

A 8 and 4 **C** 4 and 5

B 5 and 2 **D** 8 and 2

Name____ **Lesson 15.5**

Length on a Coordinate Plane

You can find the length of a line segment on a coordinate plane by finding the distance between the two points.

Graph the ordered pairs and connect the points. Find the length of the line segment.

(2, 2) and (6, 2)

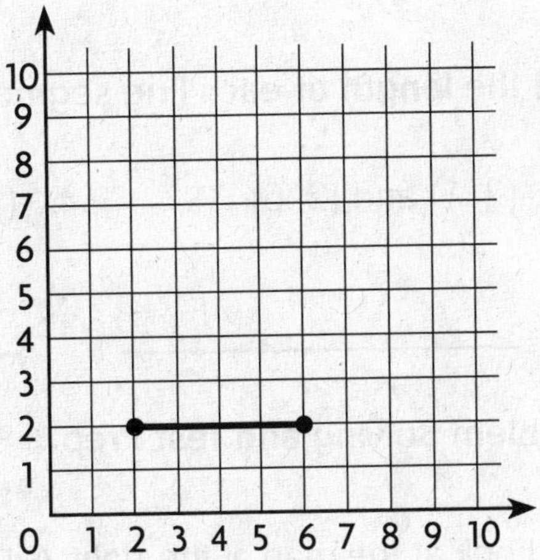

- Graph the ordered pairs on the grid.
- Draw a straight line to connect the two points.
- Find the length of the line segment.

Count: Count the units from one point to the other. To move from (2, 2) to (6, 2), you must move 4 units to the right. This means that the distance between the two points is 4 units.

Subtract: Subtract to find the difference between the two *x*-coordinates. The *x*-coordinates are 6 and 2. $6 - 2 = 4$.

- So, the line segment is 4 units long.

Find the length of each line segment.

1.

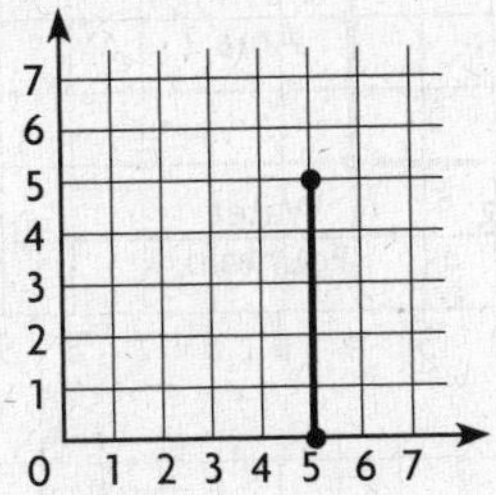

____

2.

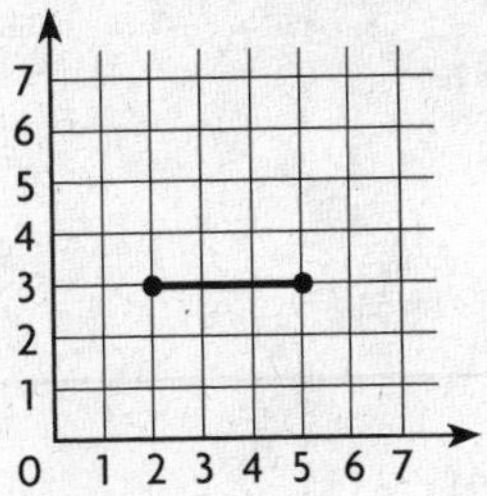

____

Graph the ordered pairs and connect the points. Find the length of each line segment.

3. (3, 2) and (3, 4)

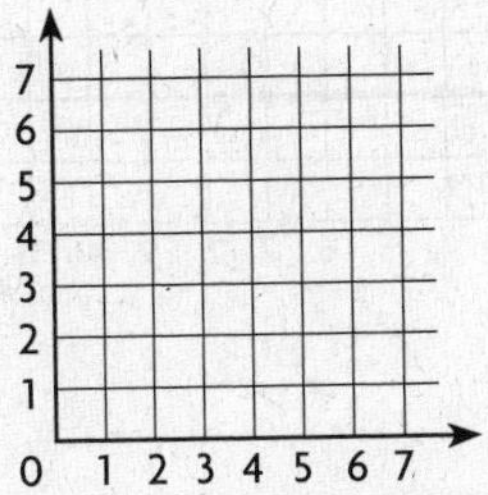

____

4. (1, 4) and (4, 4)

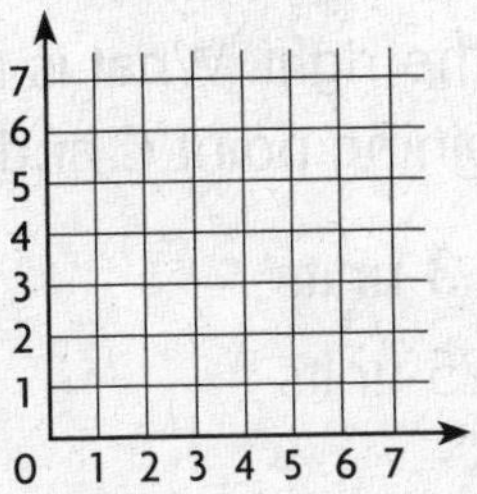

____

MG 2.2 Understand that the length of a horizontal line segment equals the difference of the *x*-coordinates. **MG 2.3** Understand that the length of a vertical line segment equals the difference of the *y*-coordinates.

Name____________________

Length on a Coordinate Plane

Graph the ordered pairs on the grid at the right and connect the points. Find the length of each line segment.

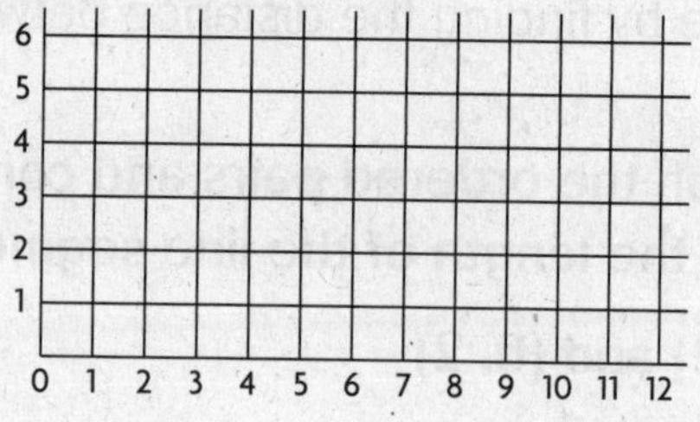

1. (7, 2) and (7, 6) ____________________

2. (10, 1) and (10, 2) ____________________

Find the length of each line segment.

3. (2, 1) and (2, 6) ____________________

4. (1, 3) and (5, 3) ____________________

5. (4, 1) and (4, 3) ____________________

Problem Solving and Test Prep

6. Look at the map at the right. Gil and Zelda bike from Start through Resting Area 1 to the water fountain. Gil continues to Resting Area 3. How much farther does Gil bike than Zelda?

Pearson Park Bike Path
Each unit = 1 kilometer

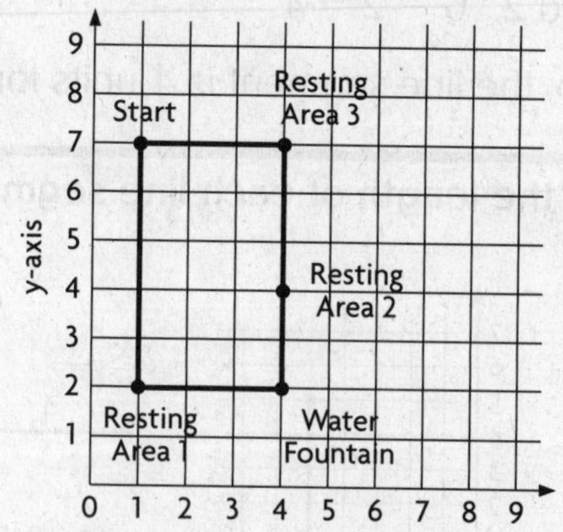

____

7. Look at the graph at the right. How can you find the number of units from point *C* to point *D*?

A add: 2 + 7 **C** subtract: 7 − 3

B add: 3 + 7 **D** subtract: 7 − 2

8. Look at the graph at the right. What is the length of the line segment joining point *C* and point *D*?

A 4 units **C** 3 units

B 2 units **D** 5 units

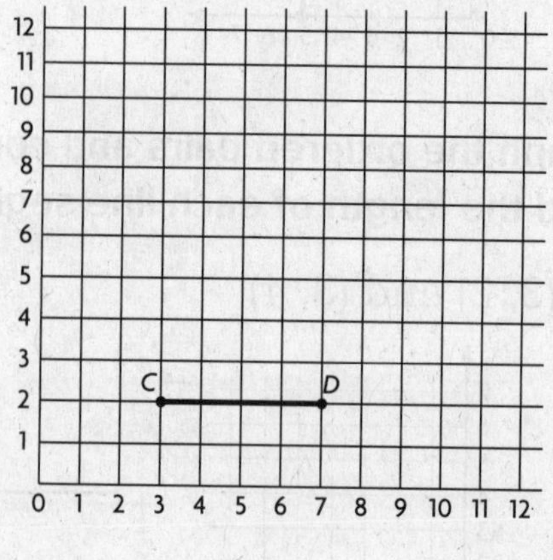

Name____ **Lesson 15.6**

Use an Equation

A **function table** shows a relationship between an input value, *x*, and an output value, *y*. The two numbers , *x* and *y*, are related by a **function**, or an equation that tells the rule of how *y* is connected to *x*. You can use a function table to find a second number when the first number in an equation is given.

Use the equation to complete the function table.

$y = (x \div 2) - 1$

Input, *x*	4	6	8	10
Output, *y*	■	■	■	■

- **Substitute each *x* value in the top row of the chart for *x* in the equation.**

$x = 4$	$x = 6$	$x = 8$	$x = 10$
$y = (x \div 2) - 1$	$y = (x \div 2) - 1$	$y = (x \div 2) - 1$	$y = (x \div 2) - 1$
$y = (4 \div 2) - 1$	$y = (6 \div 2) - 1$	$y = (8 \div 2) - 1$	$y = (10 \div 2) - 1$
$y = 2 - 1$	$y = 3 - 1$	$y = 4 - 1$	$y = 5 - 1$
$y = 1$	$y = 2$	$y = 3$	$y = 4$

So, if *x* is 4, 6, 8, and 10, then *y* is 1, 2, 3, and 4.

Use the equation to complete the function table.

1. $y = 2x + 2$

Input, *x*	1	2	3	4
Output, *y*	■	■	■	■

2. $y = 4x - 1$

Input, *x*	2	4	6	8
Output, *y*	■	■	■	■

3. $y = (x \div 3) + 2$

Input, *x*	120	90	30	3
Output, *y*	■	■	■	■

4. $y = 5x + 3$

Input, *x*	1	3	5	7
Output, *y*	■	■	■	■

5. $y = 3x - 2$

Input, *x*	2	4	6	8
Output, *y*	■	■	■	■

6. $y = (x \div 2) + 8$

Input, *x*	40	30	20	10
Output, *y*	■	■	■	■

AF 1.5 Understand that an equation such as $y = 3x + 5$ is a prescription for determining a second number when a first number is given.

Name________________________________ **Lesson 15.6**

Use an Equation

Use the equation to complete each function table.

1. $y = 2x + 4$

Input, x	1	2	3	4
Output, y	■	■	■	■

2. $y = (2x - 3) + 6$

Input, x	2	4	6	8
Output, y	■	■	■	■

3. $y = 3x - 4$

Input, x	3	6	9	12
Output, y	■	■	■	■

4. $y = (x - 5) + 6$

Input, x	5	10	15	20
Output, y	■	■	■	■

Does the ordered pair make the equation $y = 5x + 3$ true? Write *yes* or *no*.

5. (1, 8) ____________ **6.** (10, 43) ____________ **7.** (3, 9) ____________ **8.** (4, 23) ____________

9. (10, 53) ____________ **10.** (6, 33) ____________ **11.** (2, 7) ____________ **12.** (5, 28) ____________

Problem Solving and Test Prep

For 13–14, use the table.

13. How much does it cost to rent a sandcastle kit and a towel for 1, 2, or 3 hours?

____

14. Does it cost more to rent a towel and umbrella for 3 hours or a sandcastle kit and fins for 2 hours?

____

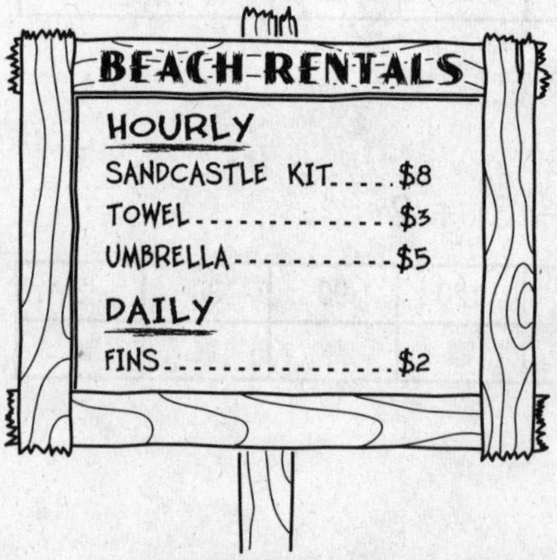

15. What is the value of y if $x = 3$ in the equation $y = 2x - 2$?

A 8 **C** 4

B 6 **D** 2

16. The difference of $x - y$ equals 4. If $x = 18$, what equation can be used to find the value of y?

A $18 - y = 4$ **C** $4 - y = 18$

B $18 + y = 4$ **D** $y = 4 \times 18$

Name____ **Lesson 15.7**

Graph Relationships

You can graph an equation on a coordinate grid. First use the equation to create ordered pairs. Then graph the ordered pairs.

Make a table using the values of 1 through 10 for *x*. Then graph the equation on a coordinate grid.
$y = 2x$.

- Multiply each x value by 2 to get the y value.

Input, *x*	1	2	3	4	5	6	7	8	9	10
Output, *y*	2	4	6	8	10	12	14	16	18	20

- After completing the table, write the ordered pairs: (1, 2), (2, 4), (3, 6), (4, 8), (5, 10), (6, 12), (7, 14), (8, 16), (9, 18), (10, 20).
- Plot the ordered pairs on the graph. The first number tells how far to move across the graph. The second number tells how far to move up the graph.
- Use a ruler to connect the points on the graph.

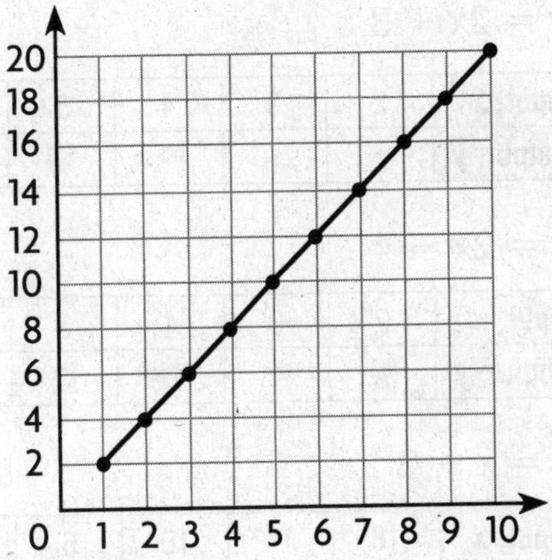

Make a table using the values of 1 through 10 for *x*. Then graph the equation on a coordinate grid.

1. $y = 2x - 1$

Input, *x*	1	2	3	4	5	6	7	8	9	10
Output, *y*										

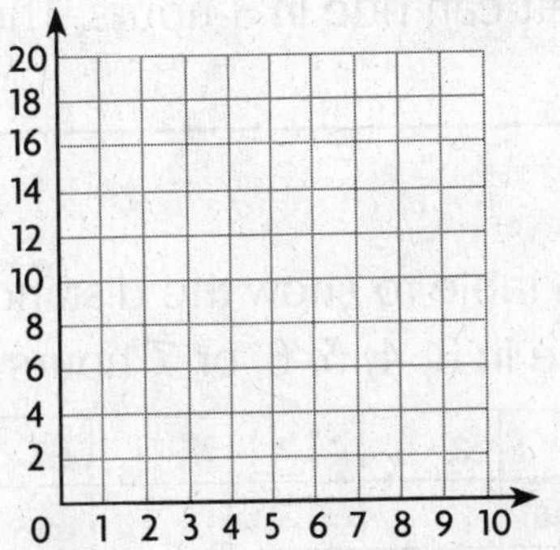

2. $y = x + 4$

Input, *x*	1	2	3	4	5	6	7	8	9	10
Output, *y*										

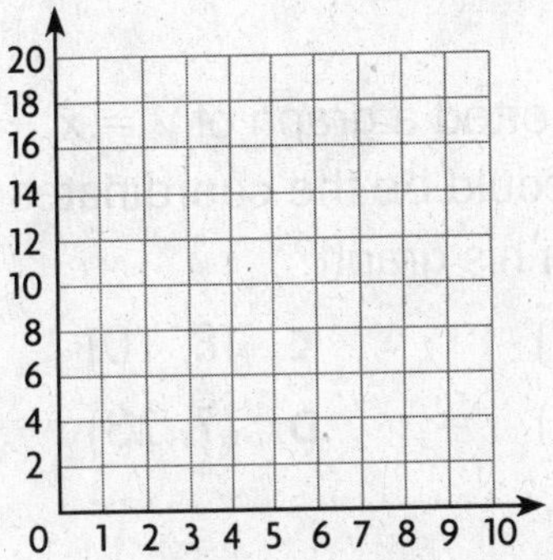

Name ____________________

Lesson 15.7

Graph Relationships

Complete each table.
Then graph the equation on the coordinate grid.

1. $y = 2x$

Input, x	5	6	7	8	9
Output, y	■	■	■	■	■

2. $y = 2x + 3$

Input, x	2	3	4	5	6
Output, y	■	■	■	■	■

3. $y = 2x - 4$

Input, x	2	3	4	5	6
Output, y	■	■	■	■	■

4. $y = x \div 2$

Input, x	18	14	10	6	2
Output, y	■	■	■	■	■

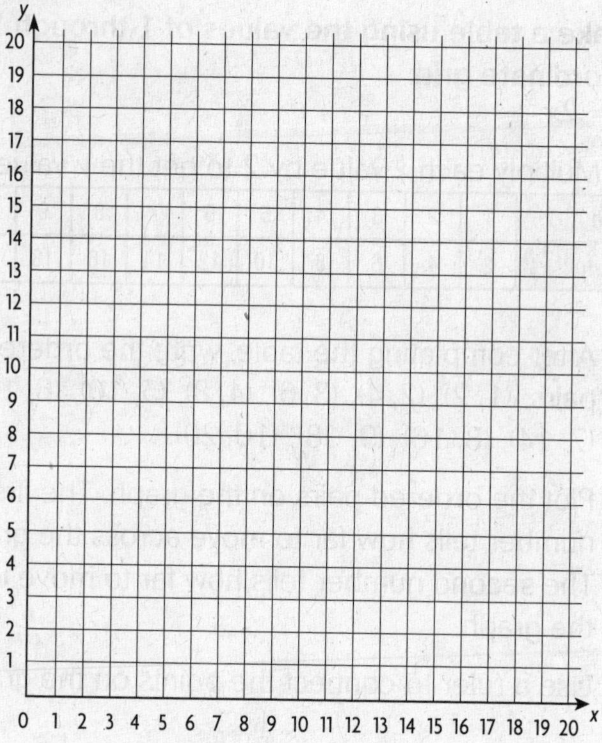

Problem Solving and Test Prep

For 5–6, use the table.

5. Write an equation to show the distance that Matt can ride in 3 hours. Then solve.

distance (d) = rate (r) × time (t)

____

Biking Speeds	
Student	**Miles per hour (r)**
Jo	6
Matt	8
Carl	12

6. Make a table to show the distance Carl can bike in 3, 4, 5, 6, or 7 hours.

Time	■	■	■	■	■
Distance	■	■	■	■	■

7. Darla plotted a graph of $y = x - 6$. Which could be the coordinates of a point on his graph?

A (2, 5)
B (7, 1)
C (8, 10)
D (7, 13)

8. Jim plotted a graph of $y = 2x - 15$. Which could be the coordinates of a point on his graph?

A (8, 31)
B (24, 30)
C (14, 18)
D (15, 15)

Name ______________________________

Spiral Review

For 1–4, find all the factors of each product. You may use arrays.

1. 18 ______________________________

2. 25 ______________________________

3. 36 ______________________________

4. 12 ______________________________

For 8–10, use the data below to find the probability of each event. Write your answers as fractions.

Tiles In a Bag	
Color Tile	Number of Tiles
Red	12
Blue	9
Yellow	4
Green	6

8. Pulling a red tile ________

9. Pulling a green tile ________

10. Pulling a blue tile ________

For 5–7, name the top view, front view, and side view of each solid figure.

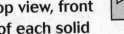

5. a cone ______________________________

6. a sphere ______________________________

7. a triangular pyramid ______________________________

For 11–15, use the multiplication properties and mental math strategies to find the product.

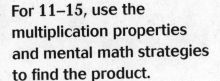

11. $3 \times (0 \times 4) =$ ________

12. $(5 \times 2) \times 7 =$ ________

13. $(4 \times 3) \times 5 =$ ________

14. $(9 \times 1) \times 4 =$ ________

15. $0 \times (4 \times 8) =$ ________

Name____

Identify Linear Relationships

You can graph linear relationships on a coordinate plane.

Write a rule. Graph the ordered pairs.

Input, x	2	4	6	8
Output, y	1	2	3	4

- Study the data in the table. Find the pattern. The output, y, is always half as much as the input, x. So the rule is:

  $y = x \div 2$.

- Write the ordered pairs for the table. The x is the first number in the pair and the y is the second number in the pair.

  (2,1), (4,2), (6,3), and (8,4).

- Graph the ordered pairs on a coordinate grid.
- Draw a line to connect the points.

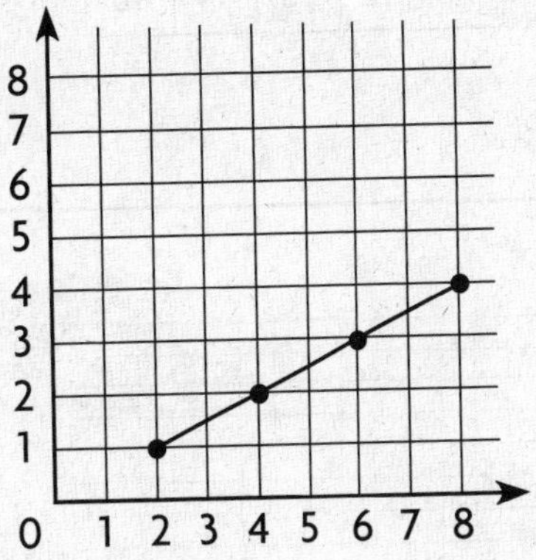

Write the rule. Graph the ordered pairs.

1.

Input, x	1	2	3	4	5
Output, y	3	4	5	6	7

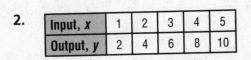

____

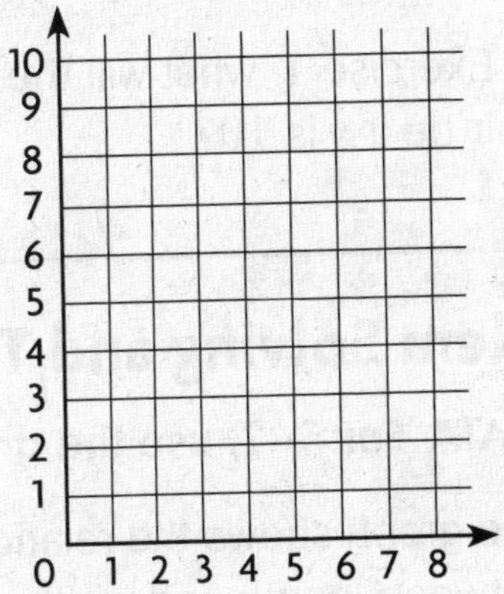

2.

Input, x	1	2	3	4	5
Output, y	2	4	6	8	10

____

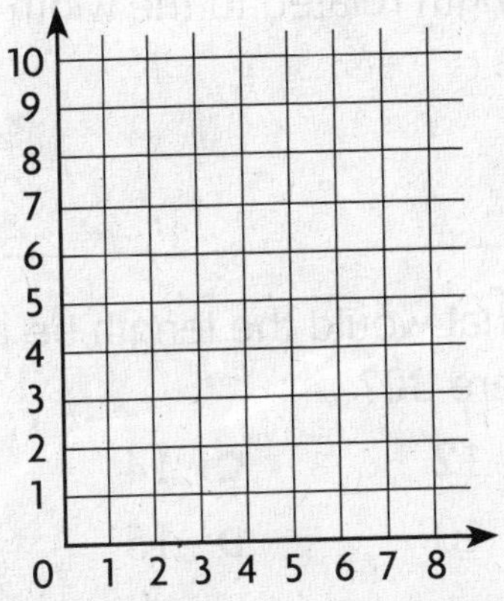

Name____

Identify Linear Relationships

Write a rule. Graph the ordered pairs.

1.

Input, *x*	1	2	3	4	5
Output, *y*	11	12	13	14	15

____

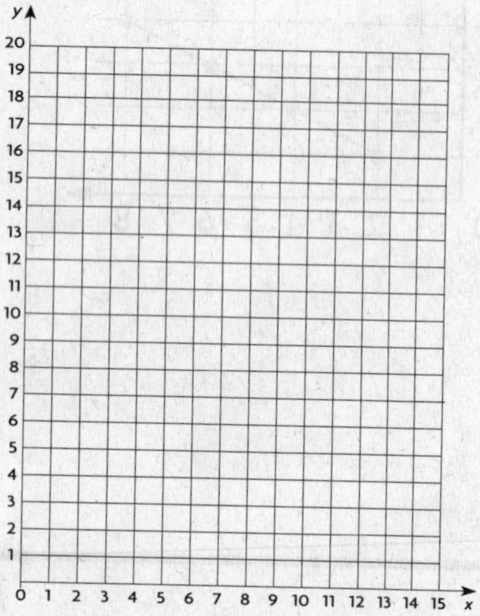

2.

Input, *x*	20	15	10	5
Output, *y*	15	10	5	0

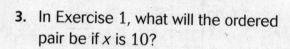

____

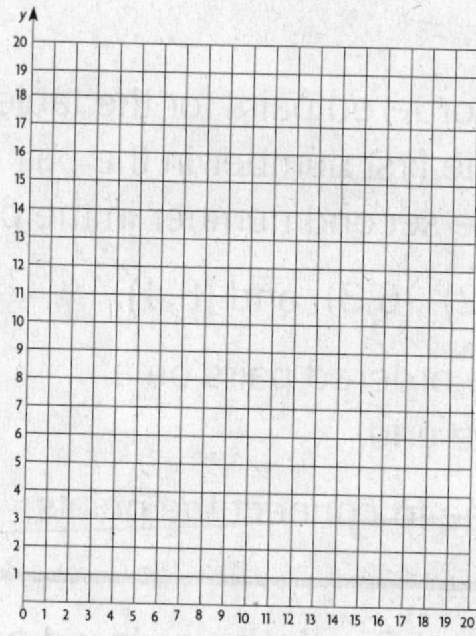

3. In Exercise 1, what will the ordered pair be if x is 10?

____

4. In Exercise 2, what will the ordered pair be if x is 30?

____

Problem Solving and Test Prep

USE DATA For 5–7, use the graph.

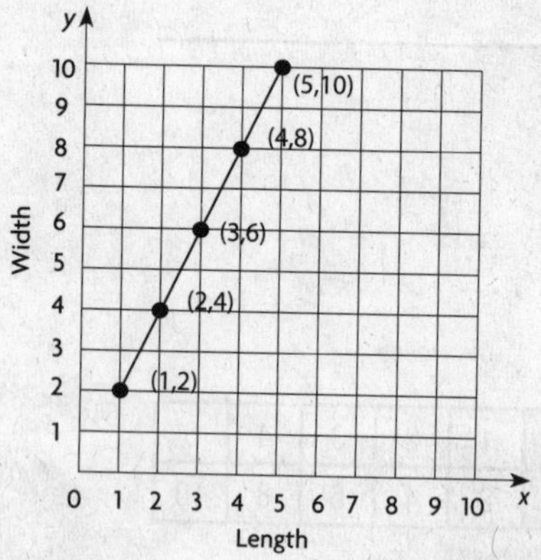

5. The graph shows the relationship between length and width. How is the length related to the width?

____

6. What would the length be if the width were 30?

A 12 **C** 14

B 13 **D** 15

7. What would the width be if the length were 10?

A 20 **C** 10

B 15 **D** 5

Name ______________________________

Read and Write Fractions

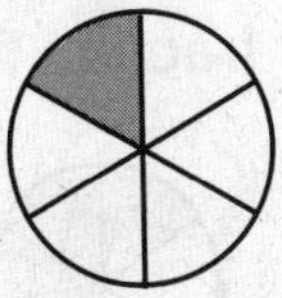

A **fraction** is a number that names part of a whole or part of a set. You use fractions every day. You can eat a fraction of a banana. You can read a fraction of a book.

Write a fraction for the shaded part.

Find the **denominator**, or bottom number.

The denominator is the total number of equal parts in the figure. The circle is divided into six equal parts, so 6 is the denominator.

Find the **numerator**, or top number.

The numerator is the number of shaded parts. There is one part shaded, so 1 is the numerator.

numerator → $\frac{1}{6}$, or one sixth of the circle is shaded.
denominator →

Write a fraction for the unshaded part.

Find the **denominator,** or bottom number.

The **denominator** is the total number of equal parts in the figure. The circle is divided into six equal parts, so 6 is the denominator.

Find the **numerator**, or top number.

The **numerator** is the number of unshaded parts. There are five unshaded parts, so 5 is the numerator.

numerator → $\frac{5}{6}$, or five sixths of the circle is unshaded.
denominator →

Write a fraction for the shaded part. Write a fraction for the unshaded part.

1.

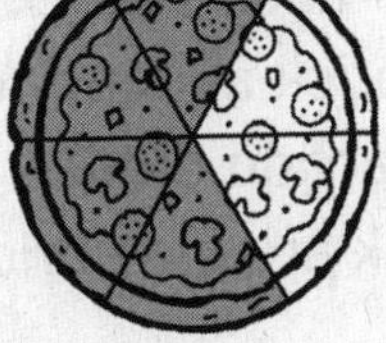

shaded ________

unshaded ________

2.

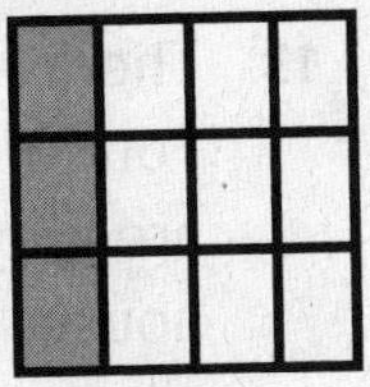

shaded ________

unshaded ________

3.

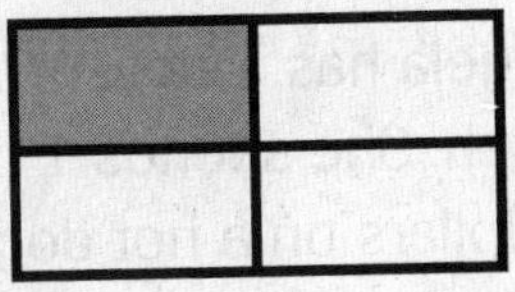

shaded ________

unshaded ________

4.

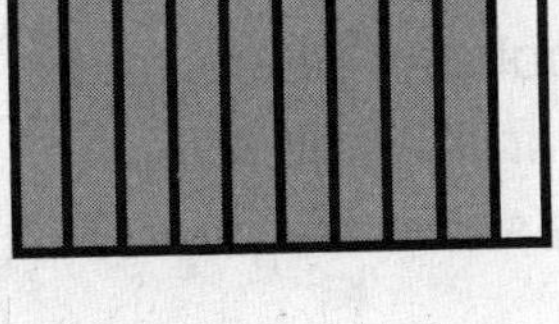

shaded ________

unshaded ________

5.

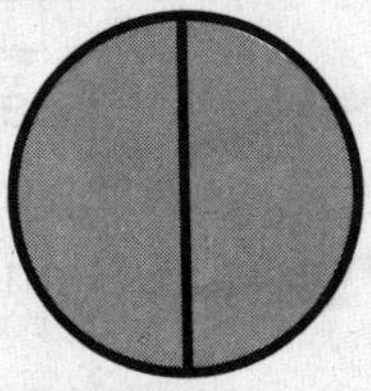

shaded ________

unshaded ________

6.

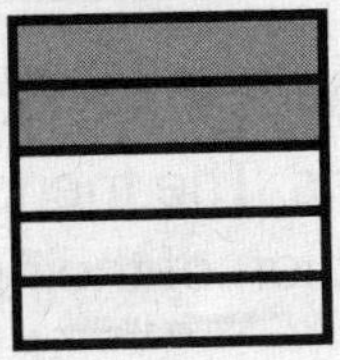

shaded ________

unshaded ________

Name____

Read and Write Fractions

Write a fraction for the shaded part. Write a fraction for the unshaded part.

1. 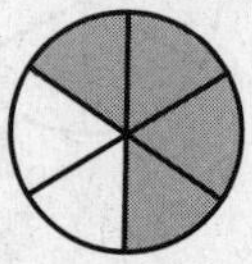

2. 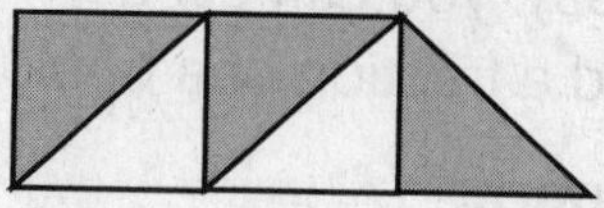

3. 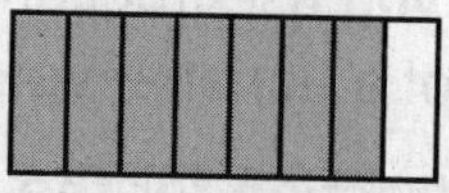

______________ ______________ ______________

Draw a picture and shade part of it to show the fraction. Write a fraction for the unshaded part.

4. $\frac{5}{6}$ **5.** $\frac{4}{10}$ **6.** $\frac{3}{7}$ **7.** $\frac{3}{5}$

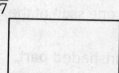

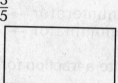

Write the fraction for each.

8. one eighth **9.** seven tenths **10.** four out of five **11.** two divided by three

__________ __________ __________ __________

Problem Solving and Test Prep

12. Angela has 5 dollars to spend on lunch. She spends 1 dollar on a soda, 3 dollars on a hot dog, and 1 dollar on a bag of pretzels. What fraction of Angela's money does she spend on a hot dog?

____

13. There are 9 houses on Zach's block. Four of them are red brick and the rest are gray brick. What fraction of the houses on Zach's block are gray brick?

____

14. Three friends cut a pizza into eight equal parts. The friends eat 3 pieces. What fraction of their pizza is left?

A $\frac{1}{8}$ **C** $\frac{3}{5}$

B $\frac{3}{8}$ **D** $\frac{5}{8}$

15. Melissa buys 3 apples, 4 pears, and 2 bananas from a fruit stand. What fraction of Melissa's fruit are pears?

A $\frac{3}{9}$ **C** $\frac{2}{9}$

B $\frac{4}{9}$ **D** $\frac{9}{9}$

Name______________________________ **Lesson 16.2**

Model Equivalent Fractions

Equivalent fractions are two or more fractions that name the same amount. There are several ways to model equivalent fractions.

Write two equivalent fractions for the model.

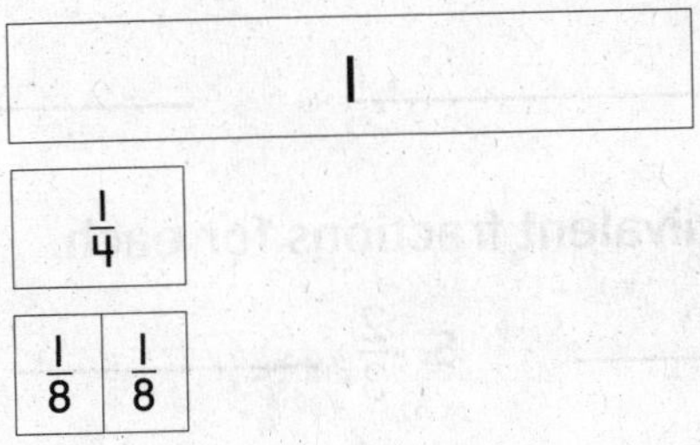

- Line up one $\frac{1}{4}$ bar with the bar for 1 to show $\frac{1}{4}$.
- Line up $\frac{1}{8}$ bars to show the same amount as $\frac{1}{4}$.
- $\frac{2}{8}$ shows the same amount as $\frac{1}{4}$.

So, $\frac{1}{4}$ and $\frac{2}{8}$ are equivalent fractions.

Write two equivalent fractions for the number lines.

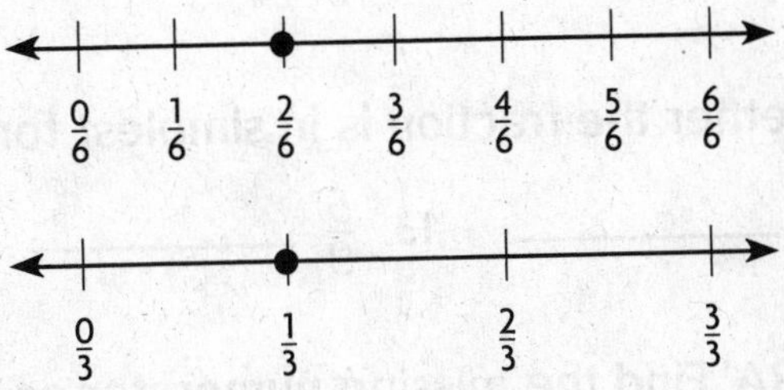

- Both number lines start at 0 and end at 1.
- When the number lines are lined up, the fraction $\frac{2}{6}$ lines up with the fraction $\frac{1}{3}$.

So, $\frac{2}{6}$ and $\frac{1}{3}$ are equivalent fractions.

Write two equivalent fractions for each model.

1.

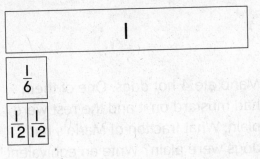

____

2.

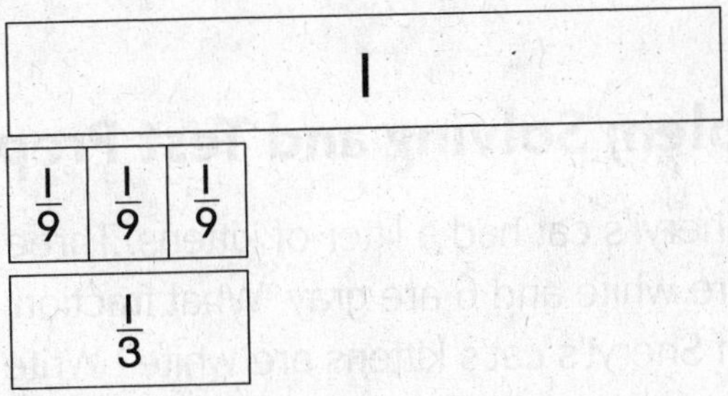

____

3.

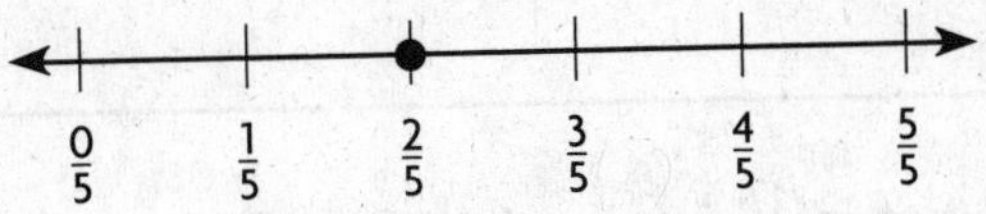

____

4.

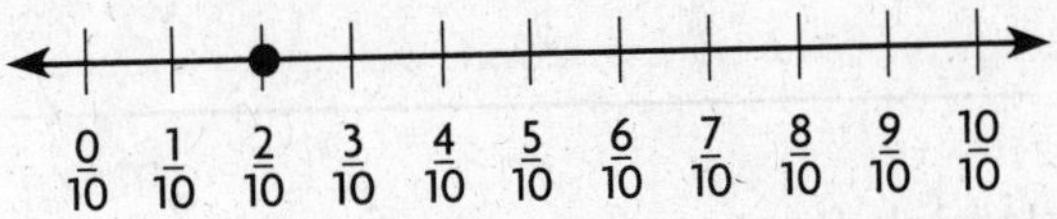

____

Name______________________________

Lesson 16.2

Model Equivalent Fractions

Write two equivalent fractions for each model.

1. 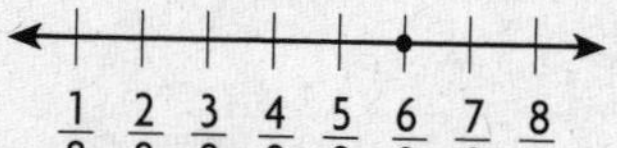

____

2. 

____

3. 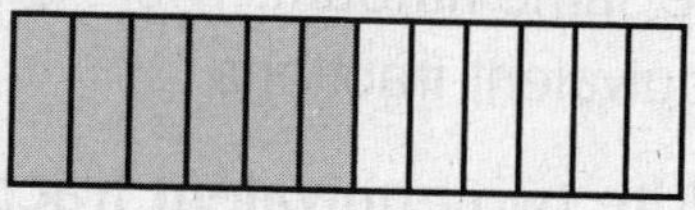

____

Write two equivalent fractions for each.

4. $\frac{1}{5}$ ______________ **5.** $\frac{2}{3}$ ______________ **6.** $\frac{3}{12}$ ______________ **7.** $\frac{6}{8}$ ______________

Tell whether the fractions are equivalent. Write *yes* or *no*.

8. $\frac{2}{9}, \frac{4}{16}$ ________ **9.** $\frac{2}{6}, \frac{8}{24}$ ________ **10.** $\frac{1}{7}, \frac{2}{14}$ ________ **11.** $\frac{6}{12}, \frac{2}{3}$ ________

Tell whether the fraction is in simplest form. If not, write it in simplest form.

12. $\frac{12}{16}$ ______________ **13.** $\frac{5}{9}$ ______________ **14.** $\frac{18}{20}$ ______________ **15.** $\frac{3}{14}$ ______________

ALGEBRA Find the missing numerator or denominator.

16. $\frac{2}{8} = \frac{\blacksquare}{24}$ **17.** $\frac{6}{16} = \frac{\blacksquare}{8}$ **18.** $\frac{7}{9} = \frac{28}{\blacksquare}$ **19.** $\frac{2}{5} = \frac{20}{\blacksquare}$

Problem Solving and Test Prep

20. Sheryl's cat had a litter of kittens. Three are white and 6 are gray. What fraction of Sheryl's cat's kittens are white? Write this amount in simplest form.

____

21. Mario ate 4 hot dogs. One of them had mustard on it and the rest were plain. What fraction of Mario's hot dogs were plain? Write an equivalent fraction for this amount.

____

22. Which fraction is equivalent to $\frac{2}{5}$?

A $\frac{3}{6}$

B $\frac{2}{8}$

C $\frac{4}{10}$

D $\frac{5}{15}$

23. What is $\frac{15}{40}$ in simplest form?

A $\frac{1}{4}$

B $\frac{5}{5}$

C $\frac{3}{8}$

D $\frac{1}{3}$

Name____ **Lesson 16.3**

Compare Fractions

You can use a number line to compare fractions with different, or unlike, denominators.

Use number lines to compare.

$\frac{1}{6}$ ● $\frac{1}{5}$

- Draw two equal-sized number lines and divide each number line according to the fractions' denominators.

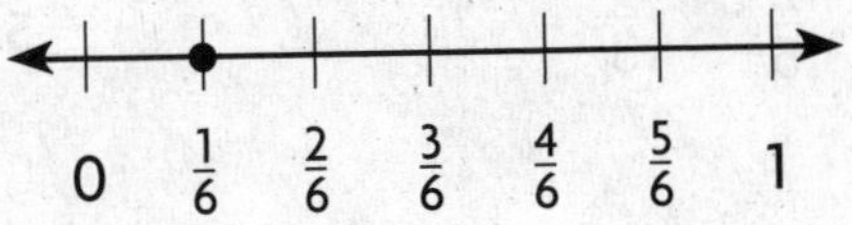

- In the fraction $\frac{1}{6}$, the denominator is 6, so you would divide your number line into six parts. Locate $\frac{1}{6}$ on the number line.
- In the fraction $\frac{1}{5}$, the denominator is 5, so you would divide your number line into 5 parts. Locate $\frac{1}{5}$ on the number line.

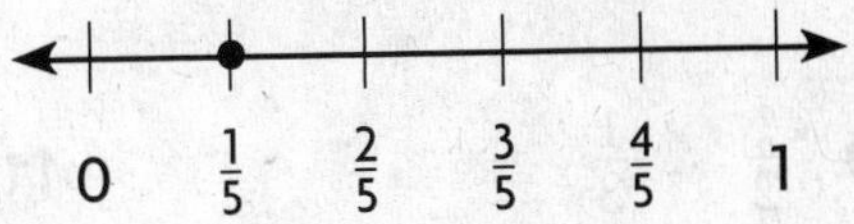

Since $\frac{1}{6}$ is farther to the left than $\frac{1}{5}$, that means $\frac{1}{6} < \frac{1}{5}$.

Use number lines to compare. Write $<$, $>$, or $=$ for each ●.

1. $\frac{5}{8}$ ● $\frac{1}{3}$

2. $\frac{2}{4}$ ● $\frac{3}{6}$

3. $\frac{3}{4}$ ● $\frac{9}{10}$

4. $\frac{5}{12}$ ● $\frac{3}{8}$

NS1.9 Identify on a number line the relative position of positive fractions, positive mixed numbers, and positive decimals to two decimal places.

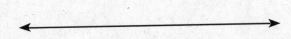

Name________________________________

Compare Fractions

Model each fraction to compare. Write <, >, or = for each ●.

1. $\frac{6}{9}$ ● $\frac{8}{9}$

2. $\frac{4}{5}$ ● $\frac{2}{3}$

3. $\frac{1}{5}$ ● $\frac{1}{8}$

4. $\frac{2}{6}$ ● $\frac{1}{3}$

5. $\frac{2}{4}$ ● $\frac{3}{5}$

6. $\frac{3}{8}$ ● $\frac{5}{8}$

7. $\frac{3}{5}$ ● $\frac{3}{4}$

8. $\frac{1}{3}$ ● $\frac{5}{8}$

9. $\frac{3}{8}$ ● $\frac{3}{4}$

10. $\frac{1}{2}$ ● $\frac{1}{3}$

11. $\frac{5}{6}$ ● $\frac{5}{8}$

12. $\frac{3}{8}$ ● $\frac{4}{8}$

Use number lines to compare.

13. $\frac{2}{3}$ ● $\frac{3}{4}$

14. $\frac{5}{9}$ ● $\frac{4}{8}$

15. $\frac{4}{10}$ ● $\frac{2}{5}$

16. $\frac{3}{10}$ ● $\frac{3}{8}$

17. $\frac{4}{12}$ ● $\frac{1}{5}$

18. $\frac{4}{16}$ ● $\frac{6}{12}$

19. $\frac{1}{5}$ ● $\frac{3}{10}$

20. $\frac{2}{3}$ ● $\frac{6}{9}$

21. $\frac{3}{4}$ ● $\frac{6}{8}$

22. $\frac{2}{6}$ ● $\frac{2}{9}$

23. $\frac{5}{8}$ ● $\frac{1}{3}$

24. $\frac{2}{4}$ ● $\frac{4}{10}$

25. $\frac{3}{7}$ ● $\frac{4}{7}$

26. $\frac{2}{6}$ ● $\frac{2}{8}$

27. $\frac{5}{9}$ ● $\frac{9}{12}$

Name ______________________________

Spiral Review

For 1–4, estimate. Then find the sum or difference.

1. $5{,}998 + 216$ ______________

2. $654 - 328$ ______________

3. $8{,}212 - 2{,}093$ ______________

4. $1{,}527 + 1{,}633$ ______________

For 5–7, find the area and perimeter of each figure.

5.
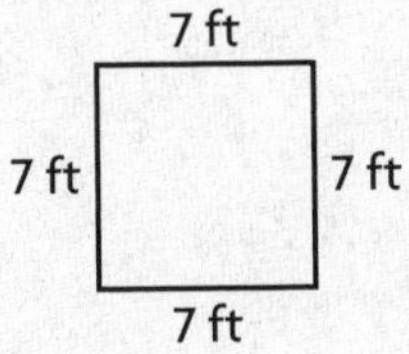

A = ______________

P = ______________

6.
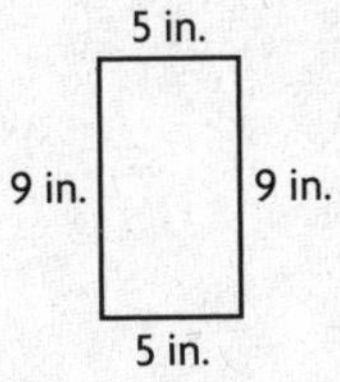

A = ______________

P = ______________

7.
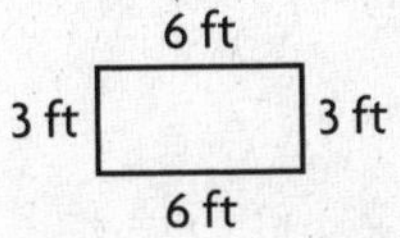

A = ______________

P = ______________

8. Audrey is a ski instructor. For her uniform, she was given a red ski jacket and a green jacket. She was also given red ski cap and a green cap. Make an organized list of the possible combinations of clothes she was given.

Color of Ski Jackets	Color of Ski Caps

9. How many possible combinations are there?

____

For 10–12, use the equation to complete each function table.

10. $y = 2x + 4$

Input, x	2	4	6	8	10
Output, y					

11. $(x \div 4) + 3 = y$

Input, x	8	12	16	20	24
Output, y					

12. $5x - 5 = y$

Input, x	2	3	4	5	6
Output, y					

Name________________________________ **Lesson 16.4**

Order Fractions

You can use fraction bars or a number line to order a group of fractions.

Order the fractions $\frac{12}{12}$, $\frac{3}{8}$, and $\frac{1}{6}$ from greatest to least.

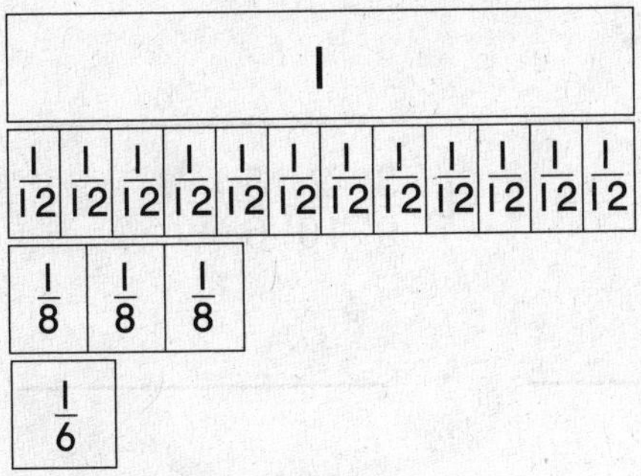

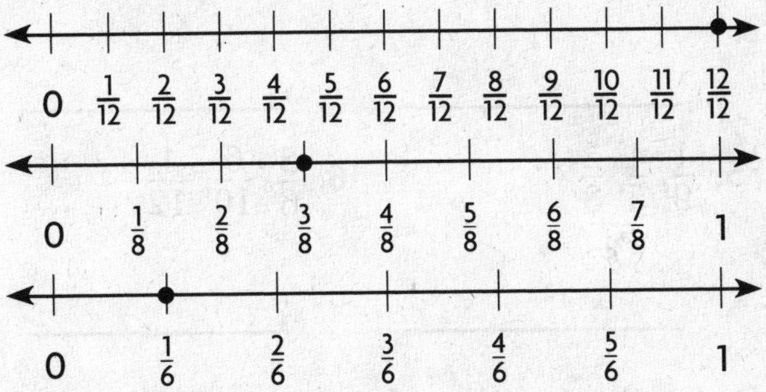

Use fraction bars.

Start with a fraction bar for 1.

Line up fraction bars for $\frac{12}{12}$, $\frac{3}{8}$, and $\frac{1}{6}$.

Arrange the fraction bars in order from longest to shortest.

So, the order from greatest to least is $\frac{12}{12}$, $\frac{3}{8}$, $\frac{1}{6}$.

Use number lines.

Draw three equal sized number lines.

Locate $\frac{12}{12}$, $\frac{3}{8}$, and $\frac{1}{6}$ each on a number line.

The fraction farthest to the right is the greatest fraction.

So, the order from greatest to least is $\frac{12}{12}$, $\frac{3}{8}$, $\frac{1}{6}$.

Order the fractions from greatest to least.

1. $\frac{3}{4}$, $\frac{1}{4}$, $\frac{9}{10}$ ____________

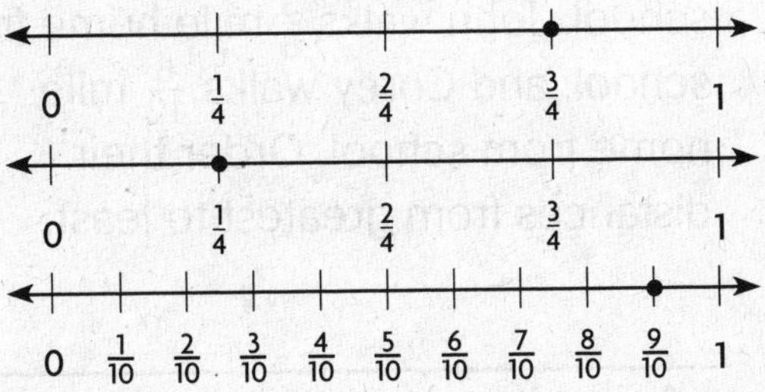

2. $\frac{3}{5}$, $\frac{4}{9}$, $\frac{4}{12}$ ____________

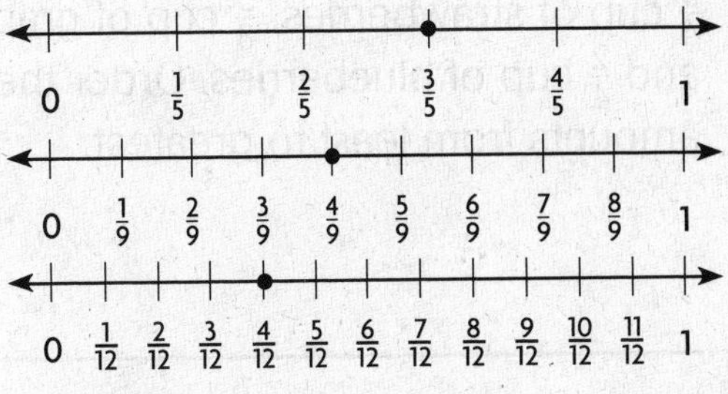

3. $\frac{1}{7}$, $\frac{3}{5}$, $\frac{5}{10}$ ____________

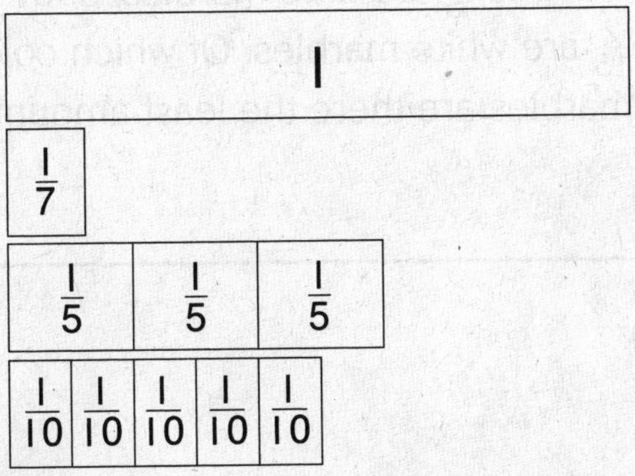

4. $\frac{1}{3}$, $\frac{1}{8}$, $\frac{3}{6}$ ____________

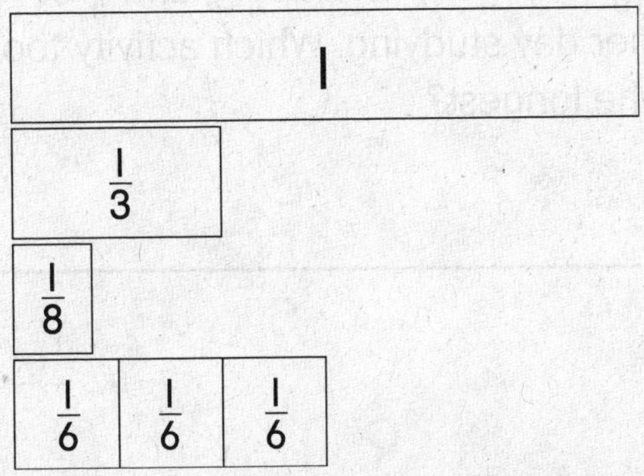

Name________________________________ **Lesson 16.4**

Order Fractions

Order the fractions from least to greater.

1. $\frac{1}{3}, \frac{1}{8}, \frac{1}{6}$ ____

2. $\frac{4}{5}, \frac{3}{5}, \frac{5}{8}$ ____

3. $\frac{4}{10}, \frac{4}{12}, \frac{4}{8}$ ____

4. $\frac{3}{7}, \frac{5}{10}, \frac{5}{8}$ ____

5. $\frac{1}{9}, \frac{4}{5}, \frac{2}{3}$ ____

6. $\frac{5}{6}, \frac{6}{10}, \frac{1}{12}$ ____

7. $\frac{5}{12}, \frac{2}{4}, \frac{4}{6}$ ____

8. $\frac{3}{9}, \frac{2}{10}, \frac{5}{6}$ ____

Order the fractions from greatest to least.

9. $\frac{1}{5}, \frac{1}{4}, \frac{1}{8}$ ____

10. $\frac{4}{9}, \frac{4}{5}, \frac{2}{3}$ ____

11. $\frac{3}{4}, \frac{3}{8}, \frac{3}{5}$ ____

12. $\frac{2}{10}, \frac{2}{5}, \frac{3}{12}$ ____

13. $\frac{5}{12}, \frac{3}{9}, \frac{3}{6}$ ____

14. $\frac{7}{12}, \frac{3}{4}, \frac{2}{4}$ ____

15. $\frac{5}{8}, \frac{4}{6}, \frac{1}{10}$ ____

16. $\frac{3}{5}, \frac{6}{12}, \frac{2}{10}$ ____

Problem Solving and Test Prep

17. Matt made a fruit salad that included $\frac{3}{4}$ cup of strawberries, $\frac{5}{8}$ cup of grapes, and $\frac{2}{4}$ cup of blueberries. Order the amounts from least to greatest.

____

18. Carolyn walks $\frac{4}{6}$ mile home from school. John walks $\frac{3}{8}$ mile home from school, and Corey walks $\frac{6}{12}$ mile home from school. Order their distances from greatest to least.

____

19. Pat spent $\frac{3}{9}$ of her day shopping, $\frac{2}{10}$ of her day exercising, and $\frac{2}{5}$ of her day studying. Which activity took the longest?

____

20. In a jar of marbles, $\frac{3}{10}$ are red marbles, $\frac{1}{5}$ are blue marbles, and $\frac{2}{15}$ are white marbles. Of which color marbles are there the least amount?

____

Name____

Read and Write Mixed Numbers

A **mixed number** is a number made up of a whole number and a fraction.

Write a mixed number for the model.

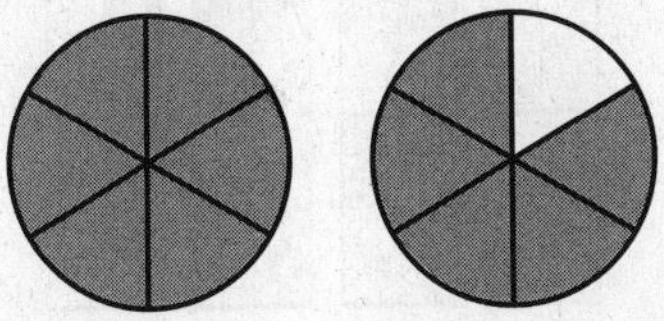

The circle on the left represents 1 whole figure shaded.

The circle on the right represents $\frac{5}{6}$ shaded.

So, $1\frac{5}{6}$ figures are shaded.

Rename $5\frac{5}{6}$ as a fraction.

Model $5\frac{5}{6}$.

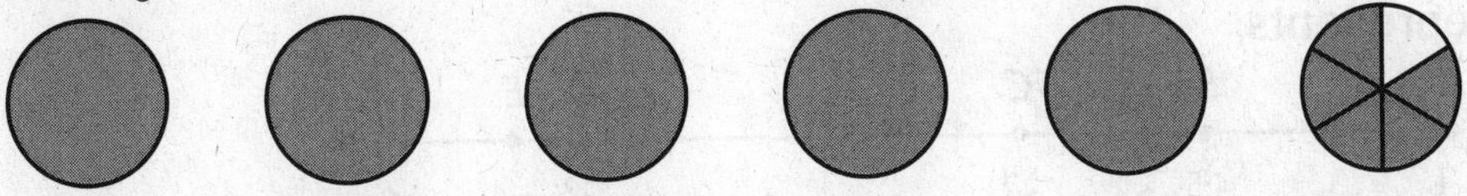

Rename each 1 whole as $\frac{6}{6}$.

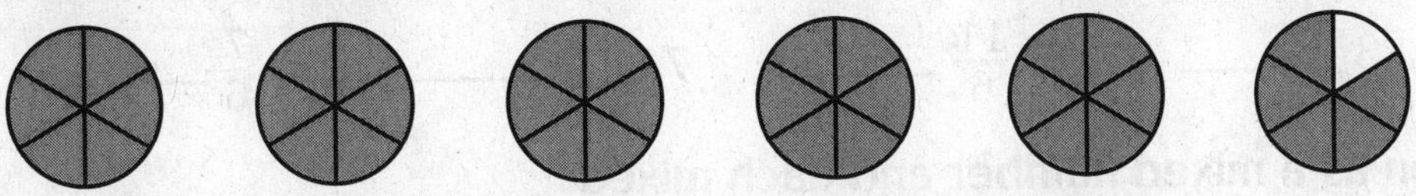

The total number of sixths is the numerator of the fraction. The numerator is 35.

So, $5\frac{5}{6}$ renamed as a fraction is $\frac{35}{6}$.

Write a mixed number for each picture.

1. 

2. 

3. 

Rename the fraction as a mixed number and the mixed number as a fraction.

4. $\frac{5}{3}$ 

5. $3\frac{3}{4}$ 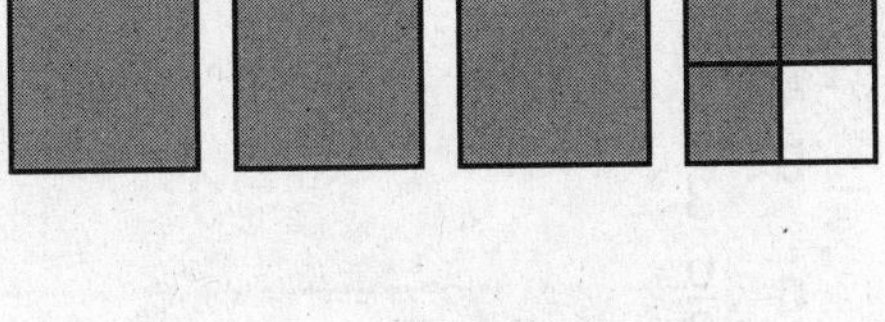

Name____

Lesson 16.5

Read and Write Mixed Numbers

Write a mixed number for each picture.

1. 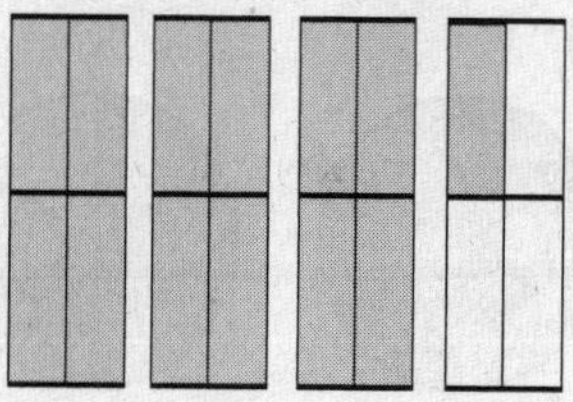

2. 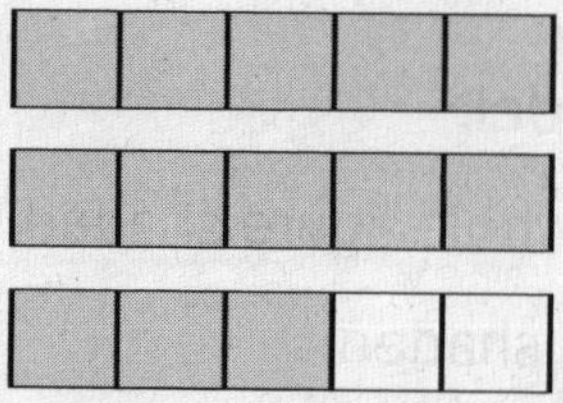

3. 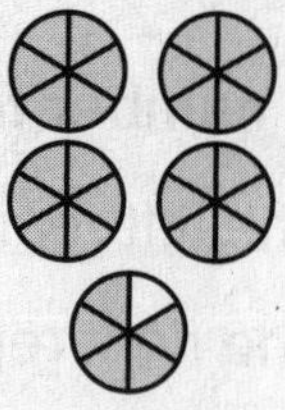

______________ ______________ ______________

For 4–8, use the number line to write the letter each mixed number or fraction represents.

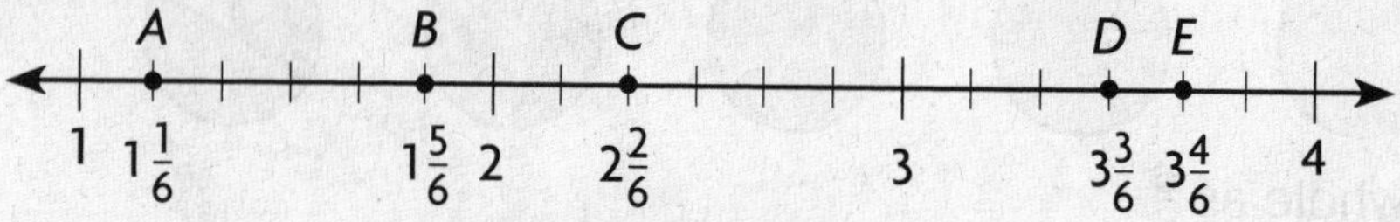

4. $\frac{14}{6}$ ______ 5. $3\frac{4}{6}$ ______ 6. $\frac{11}{6}$ ______ 7. $3\frac{3}{6}$ ______ 8. $\frac{7}{6}$ ______

Rename each fraction as a mixed number and each mixed number as a fraction. You may wish to draw a picture.

9. $5\frac{3}{4}$ 10. $3\frac{2}{10}$ 11. $\frac{38}{6}$ 12. $\frac{23}{3}$ 13. $2\frac{3}{8}$

______ ______ ______ ______ ______

Problem Solving and Test Prep

14. Ned cuts a board that is $5\frac{1}{4}$ inches long. Draw a number line and locate $5\frac{1}{4}$ inches.

15. Julia goes for a bike ride for $1\frac{2}{3}$ hours. Draw a number line to represent the length of time.

16. Denzel makes a cake with $2\frac{2}{3}$ cups of flour. Which shows the mixed number as a fraction?

   A $\frac{4}{3}$

   B $\frac{8}{3}$

   C $\frac{6}{3}$

   D $\frac{10}{3}$

17. Ashley serves $3\frac{5}{8}$ trays of muffins. How many muffins does Ashley serve if each muffin is $\frac{1}{8}$ of a tray?

   A 29

   B 15

   C 24

   D 19

Name______________________________________

Compare and Order Mixed Numbers

You can use models to compare and order mixed numbers.

Compare $6\frac{1}{6}$ and $6\frac{1}{2}$.

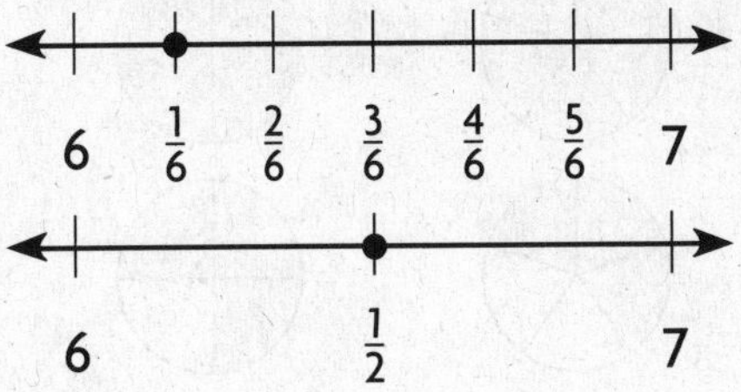

- Draw a number line divided into sixths. Locate $6\frac{1}{6}$.
- Draw a number line divided into halves. Locate $6\frac{1}{2}$.
- Since $6\frac{1}{2}$ is to the right of $6\frac{1}{6}$ on the number line, $6\frac{1}{6} < 6\frac{1}{2}$.

Order the mixed numbers $1\frac{7}{9}$, $1\frac{1}{2}$, and $1\frac{12}{18}$ from greatest to least.

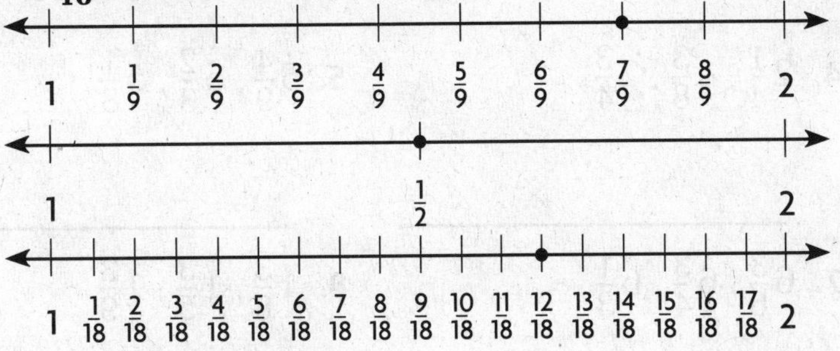

- Draw a number line divided into ninths. Label $1\frac{7}{9}$.
- Draw a number line divided into halves. Label $1\frac{1}{2}$.
- Draw a number line divided into eighteenths. Label $1\frac{12}{18}$.
- Since $1\frac{7}{9}$ is farthest to the right, it is the greatest number.
- Since $1\frac{1}{2}$ is farthest to the left, it is the least number.
- So, the order from greatest to least is $1\frac{7}{9}$, $1\frac{12}{18}$, $1\frac{1}{2}$.

Compare the mixed numbers. Use <, >, or =.

1.

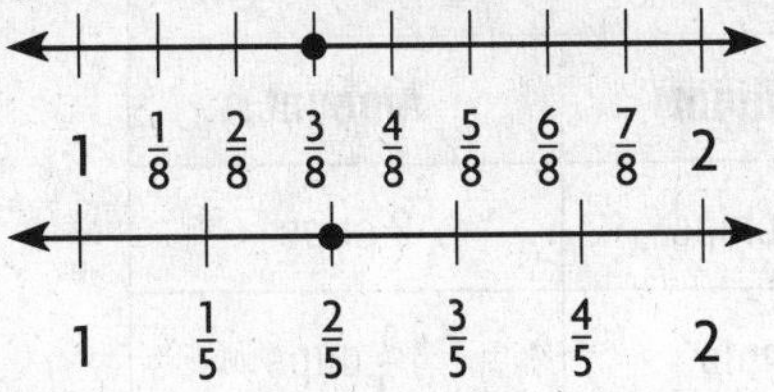

$1\frac{3}{8}$ ● $1\frac{2}{5}$

2.

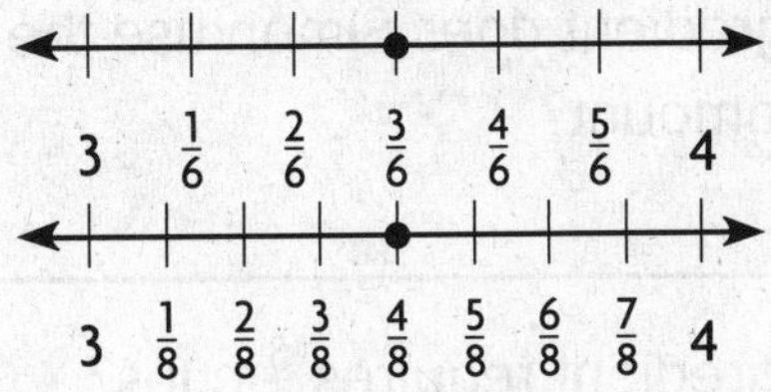

$3\frac{3}{6}$ ● $3\frac{4}{8}$

Order the mixed numbers from greatest to least.

3. $2\frac{7}{9}$, $2\frac{4}{5}$, $3\frac{1}{3}$

4. $4\frac{1}{5}$, $4\frac{10}{11}$, $4\frac{3}{4}$

5. $2\frac{3}{4}$, $3\frac{3}{7}$, $1\frac{2}{8}$

6. $5\frac{1}{5}$, $6\frac{2}{5}$, $6\frac{3}{8}$

__________ __________ __________ __________

Name____

Compare and Order Mixed Numbers

Compare the mixed numbers. Use <, >, or =.

1.

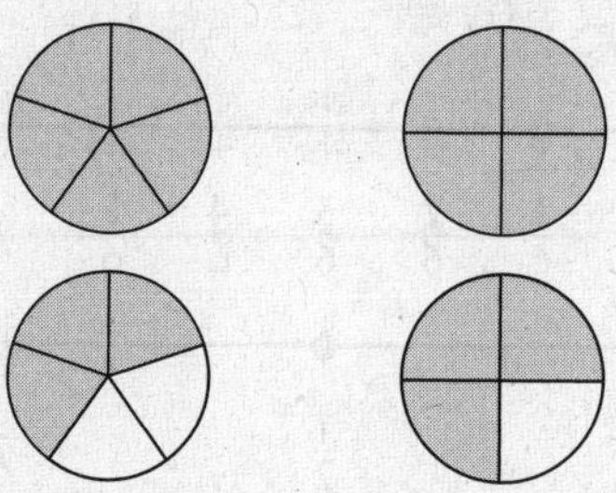

$1\frac{3}{5}$ ● $1\frac{3}{4}$

2.

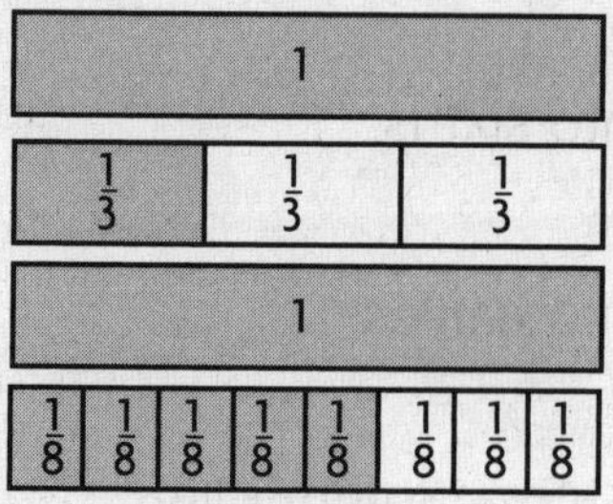

$1\frac{1}{3}$ ● $1\frac{5}{8}$

3.

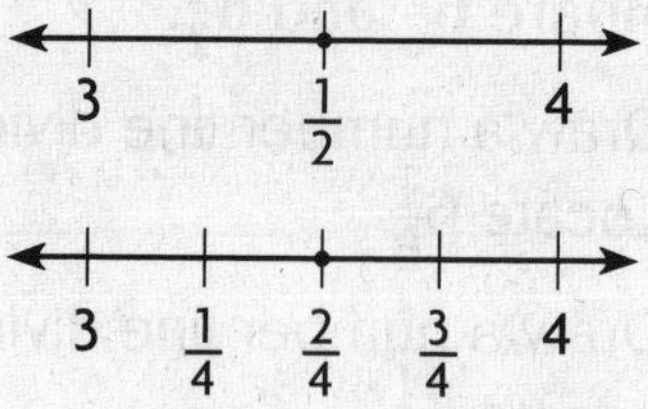

$3\frac{1}{2}$ ● $3\frac{2}{4}$

Order the mixed numbers from least to greatest.

4. $2\frac{1}{4}, 4\frac{3}{8}, 2\frac{3}{4}$ ______________

5. $5\frac{4}{9}, 5\frac{2}{3}, 5\frac{1}{8}$ ______________

6. $3\frac{4}{5}, 3\frac{2}{10}, 3\frac{5}{12}$ ______________

7. $6\frac{3}{6}, 6\frac{3}{4}, 6\frac{1}{3}$ ______________

8. $1\frac{3}{8}, 1\frac{3}{5}, 1\frac{3}{9}$ ______________

9. $7\frac{1}{4}, 7\frac{1}{7}, 7\frac{3}{5}$ ______________

Problem Solving and Test Prep

USE DATA For 10–11, use the table.

Recipe for Trail Mix	
Ingredient	**Amount**
Corn chips	2 cups
Peanuts	$1\frac{1}{3}$ cups
Raisins	$1\frac{2}{3}$ cups

10. Simon makes trail mix for a party. Of which ingredient does Simon use the greatest amount?

____

11. Which ingredient requires $\frac{5}{3}$ cups?

____

12. Jamal plays soccer for $\frac{12}{5}$ hours. Write the amount of time Jamal plays soccer as a mixed number.

____

13. Eddie is at an amusement park and wants to find the ride with the shortest wait. The waits for four rides are shown. Which wait is the shortest?

A $1\frac{4}{5}$ hours

B $1\frac{1}{5}$ hours

C $1\frac{1}{2}$ hours

D $1\frac{2}{3}$ hours

Name______________________________ Lesson 16.7

Problem Solving Workshop Skill: Sequence Information

Gloria gives her youngest cat $1\frac{1}{2}$ cups of food. She gives her oldest cat $\frac{7}{8}$ cup of food and her second-oldest cat $1\frac{3}{4}$ cups of food. Draw number lines to represent the amounts of food. Which cat gets the most food?

1. What are you asked to find?

____

2. What are three amounts of food that you are comparing?

____

3. Label the three number lines below for each amount of food.

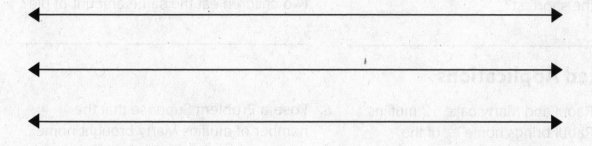

4. According to the number line, which is the greatest amount of cat food?

____

5. Which cat gets the most food?

____

6. How can you check your answer?

____

7. Mike rents three movies. The first movie is $2\frac{1}{3}$ hours long, the second movie is $1\frac{7}{8}$ hours long, and the third movie is $2\frac{2}{5}$ hours long. Order the lengths of the movies from longest to shortest.

____

8. Joe, Lara, and Mike share a bucket of popcorn. Joe eats $2\frac{2}{3}$ cups, Lara eats $2\frac{3}{6}$ cups, and Mike eats $2\frac{1}{5}$ cups. Who eats the most popcorn? Who eats the least? Order their amounts from least to greatest.

____

Name____ **Lesson 16.7**

Problem Solving Workshop Skill: Sequence Information

Problem Solving Skill Practice

Sequence the information to solve.

1. Ben practices piano for $1\frac{2}{3}$ hours. Charlene practices flute for $\frac{3}{4}$ hour. Walter practices drums for $1\frac{1}{2}$ hours. Who practices for the most time? the least?

2. Sara cuts 3 lengths of ribbon. The first piece is $4\frac{5}{8}$ inches long. The second piece is $3\frac{1}{4}$ inches long. The third piece is $4\frac{1}{2}$ inches long. Which ribbon piece is the longest? the shortest?

____

3. Joyce takes a nap for $\frac{8}{5}$ hours. Rex naps for $\frac{5}{3}$ hours. Maya naps for $\frac{5}{4}$ hours. Who takes the longest nap? the shortest?

____

4. Betsy, Latisha, and Ramon eat a pizza pie. Betsy eats $\frac{2}{8}$ of the pie. Latisha eats $\frac{1}{4}$ of the pie. Ramon eats $\frac{3}{8}$ of the pie. Which two children eat the same amount of pie?

____

Mixed Applications

5. Raoul and Marty bake 12 muffins. Raoul brings home $\frac{5}{12}$ of the muffins. Marty brings home $\frac{1}{3}$ of the muffins. Draw number lines to show who brought home more muffins.

6. **Pose a Problem** Suppose that the number of muffins Marty brought home was changed to $\frac{1}{2}$. Write a new problem using this information.

____

____

____

____

____

____

7. Zach makes this number line:

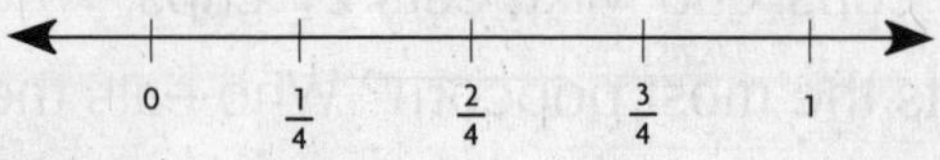

Between which two fractions would you place $\frac{3}{8}$?

____

8. Rina makes this number line:

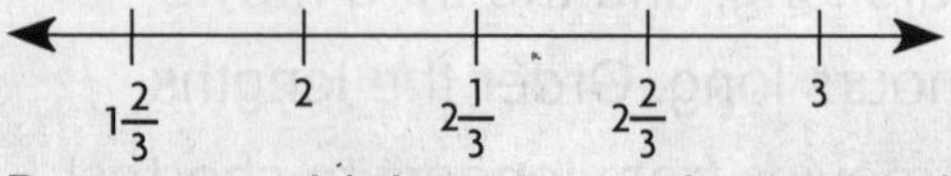

Between which two numbers would you place $2\frac{5}{6}$?

____

Spiral Review

For 1–5, write *prime* or *composite* for each number.

1. 45 ____________
2. 7 ____________
3. 81 ____________
4. 13 ____________
5. 25 ____________

For 8–10, use the spinner below.

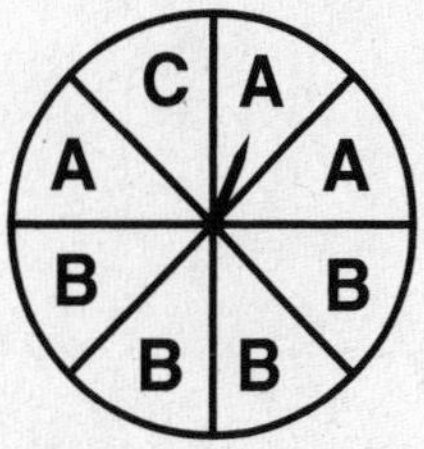

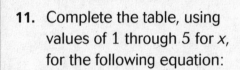

8. Which outcome is most likely?

____

9. Is an outcome of "D" possible?

____

10. Which outcome is least likely?

____

For 6–7, use the diagram below.

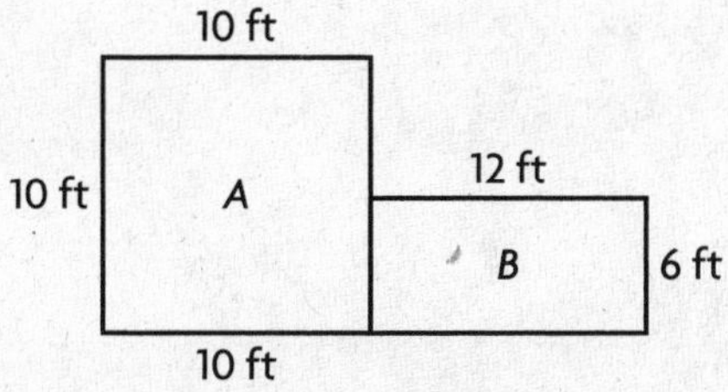

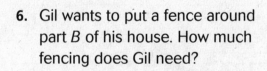

6. Gil wants to put a fence around part *B* of his house. How much fencing does Gil need?

____

7. Now, Gil wants to fill part *A* with tile. First, Gil needs to know the area of part *A*. What is the area of part *A*?

____

11. Complete the table, using values of 1 through 5 for *x*, for the following equation:

$$x + 2 = y$$

Input, *x*					
Output, *y*					

12. Now graph the equation on the coordinate grid below.

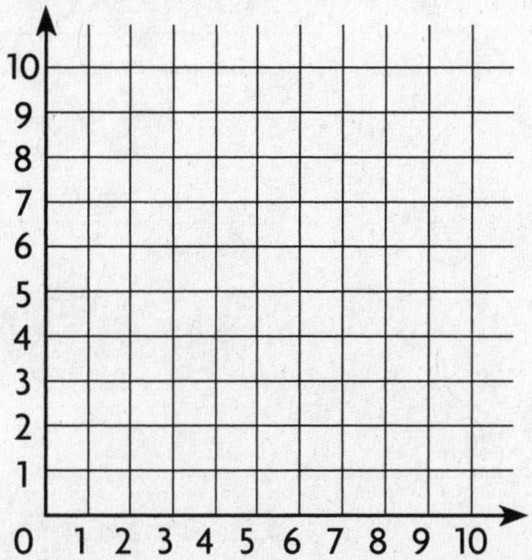

Name____

Model Addition

Number lines can help you add fractions. Fractions that have the same denominator, or bottom number, are called **like fractions**. When fractions have the same denominator, you can add the numerators, or top numbers. The denominator always stays the same.

Find the sum.

$\frac{1}{2} + \frac{1}{2}$

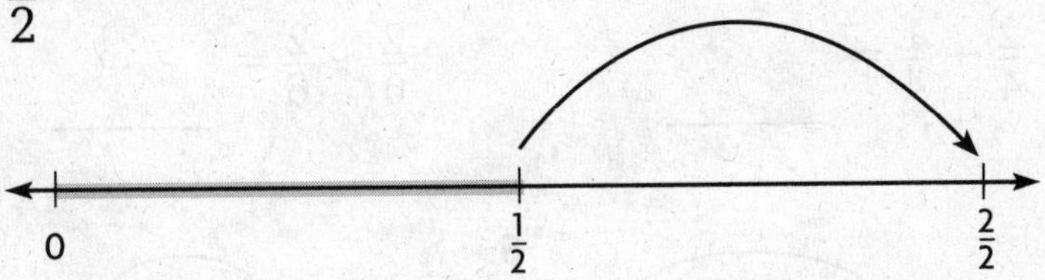

- The denominator tells the number of equal parts the number line is divided into.
- The denominator is 2, so the number line is divided into 2 equal parts.
- Label the number line with 0, $\frac{1}{2}$, and $\frac{2}{2}$.
- Shade the part for the first addend from 0 to $\frac{1}{2}$. To add the second addend, $\frac{1}{2}$, move from $\frac{1}{2}$ to $\frac{2}{2}$.
- Since you moved over 2 equal parts in all, that means there are 2 out of 2 equal parts, or $\frac{2}{2}$.

So, $\frac{1}{2} + \frac{1}{2} = \frac{2}{2}$, or 1.

Find the sum.

1. 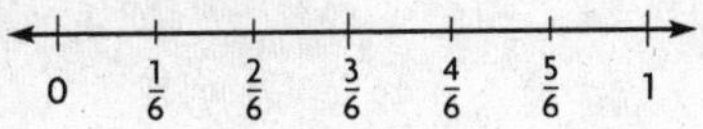

$\frac{5}{6} + \frac{1}{6} =$ ________

2. 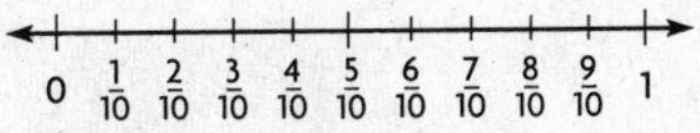

$\frac{5}{10} + \frac{2}{10} =$ ________

3. 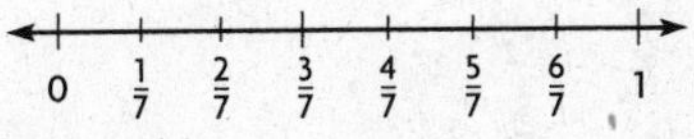

$\frac{3}{7} + \frac{2}{7} =$ ________

Model the sum. Record your answer.

4. $\frac{1}{3} + \frac{1}{3}$

5. $\frac{3}{7} + \frac{1}{7}$

6. $\frac{4}{7} + \frac{2}{7}$

Name____

Model Addition

Find the sum.

1.

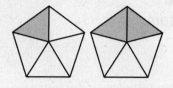

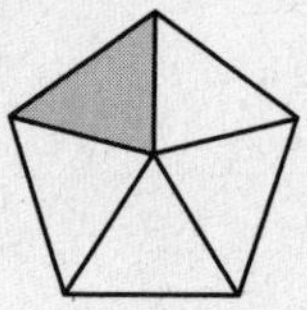

$\frac{1}{5} + \frac{2}{5} =$ ________

2.

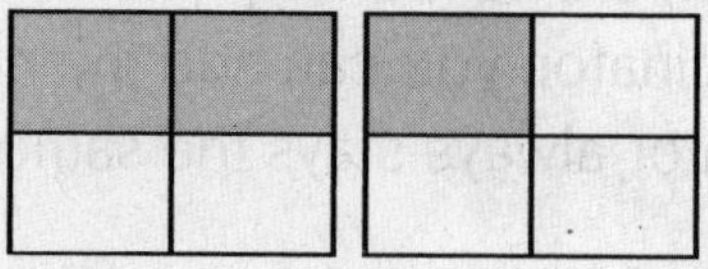

$\frac{2}{4} + \frac{1}{4} =$ ________

3.

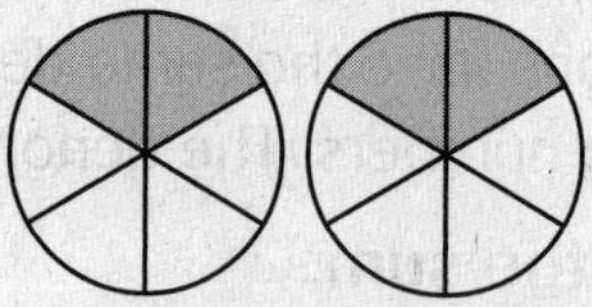

$\frac{2}{6} + \frac{2}{6} =$ ________

4.

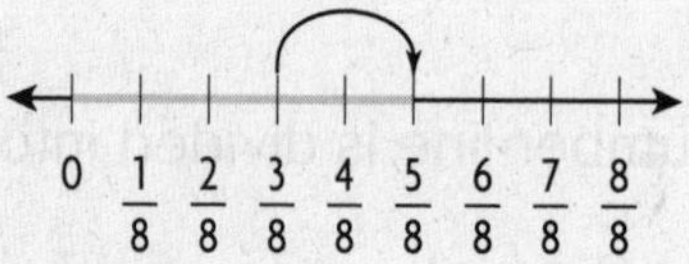

$\frac{3}{8} + \frac{2}{8} =$ ________

5.

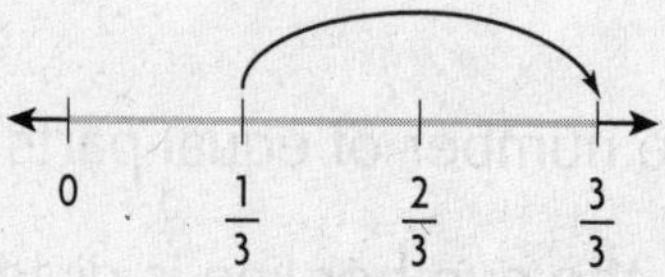

$\frac{1}{3} + \frac{2}{3} =$ ________

6.

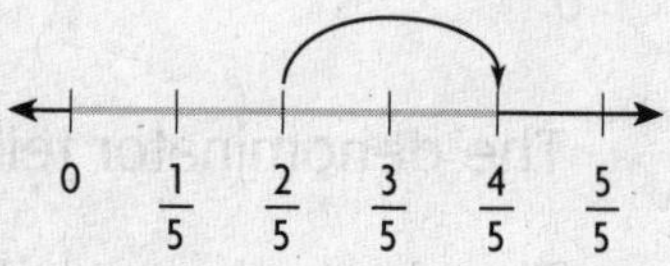

$\frac{2}{5} + \frac{2}{5} =$ ________

Model the sum. Record your answer.

7. $\frac{3}{8} + \frac{1}{8} =$ ________

8. $\frac{4}{9} + \frac{2}{9} =$ ________

9. $\frac{2}{10} + \frac{4}{10} =$ ________

10. $\frac{3}{6} + \frac{1}{6} =$ ________

11. $\frac{4}{12} + \frac{5}{12} =$ ________

12. $\frac{1}{4} + \frac{1}{4} =$ ________

13. $\frac{1}{8} + \frac{5}{8} =$ ________

14. $\frac{3}{6} + \frac{2}{6} =$ ________

15. $\frac{5}{10} + \frac{2}{10} =$ ________

16. $\frac{2}{9} + \frac{3}{9} =$ ________

17. $\frac{6}{12} + \frac{2}{12} =$ ________

18. $\frac{1}{4} + \frac{3}{4} =$ ________

19. $\frac{2}{3} + \frac{1}{3} =$ ________

20. $\frac{6}{9} + \frac{4}{9} =$ ________

21. $\frac{1}{8} + \frac{6}{8} =$ ________

Name____________________________________ **Lesson 17.2**

Model Subtraction

Number lines can be used to subtract fractions. Fractions with common denominators are called **like fractions**. When you have like fractions, you only need to subtract the numerators. The denominator stays the same.

Find the difference.

$\frac{1}{2} - \frac{1}{2}$

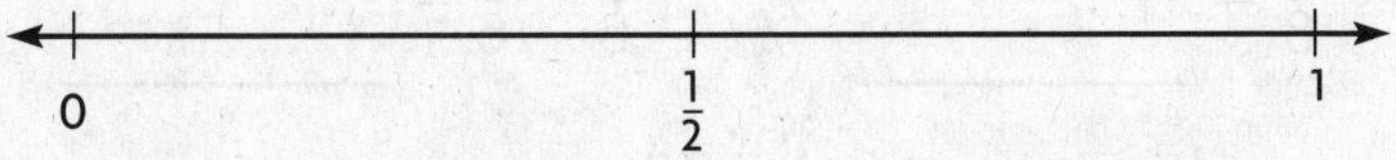

- The denominator is 2, so draw a number line and divide it into 2 equal parts.
- Begin at $\frac{1}{2}$.
- Subtract $\frac{1}{2}$ by counting back 1 part on the number line.
- Since you stopped on 0, $\frac{0}{2}$ of the number line remains. That means $\frac{1}{2} - \frac{1}{2} = \frac{0}{2}$, or 0.

Find the difference.

1.

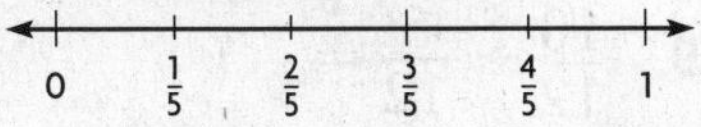

$\frac{4}{5} - \frac{1}{5} =$ ____________

2.

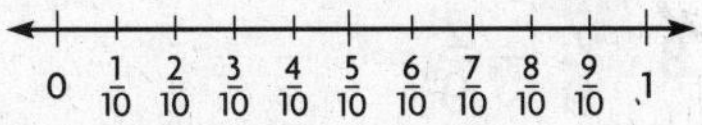

$\frac{6}{10} - \frac{2}{10} =$ ____________

3.

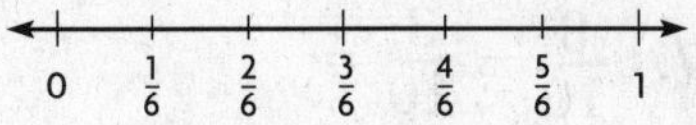

$\frac{2}{6} - \frac{1}{6} =$ ____________

Model the difference. Record your answer.

4. $\frac{9}{14} - \frac{3}{14}$ = ______

5. $\frac{3}{7} - \frac{1}{7}$ = ______

6. $\frac{3}{10} - \frac{1}{10}$ = ______

Name____ **Lesson 17.2**

Model Subtraction

Find the difference.

1.

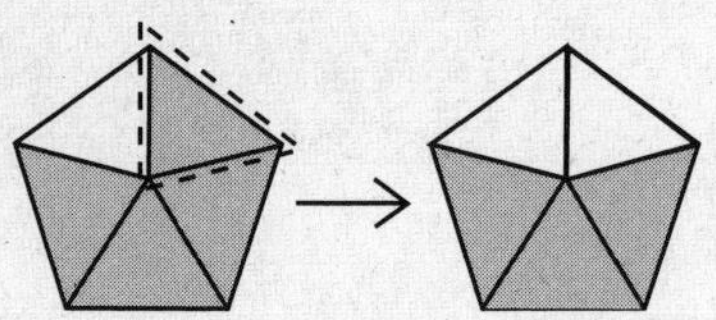

$\frac{4}{5} - \frac{1}{5} =$ ___________

2.

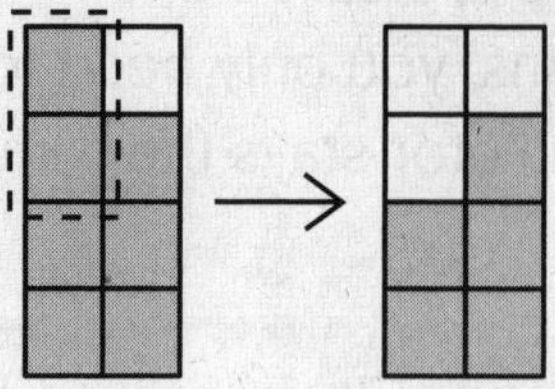

$\frac{7}{8} - \frac{2}{8} =$ ___________

3.

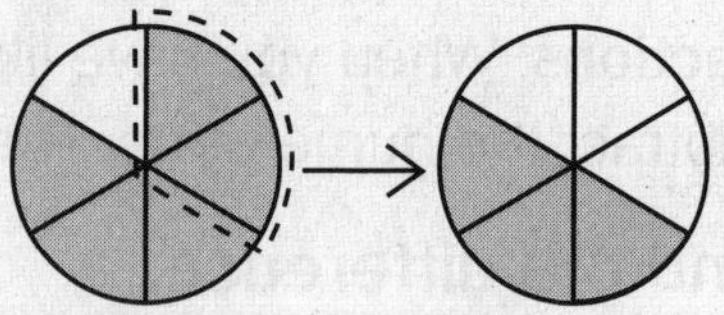

$\frac{5}{6} - \frac{2}{6} =$ ___________

4.

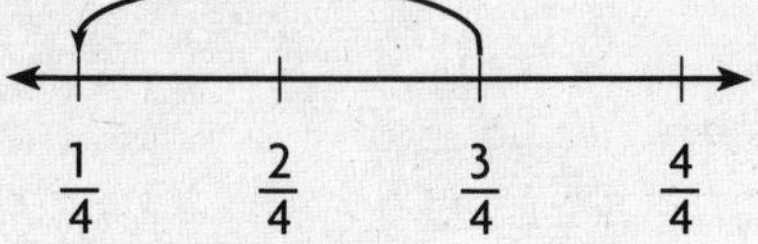

$\frac{3}{4} - \frac{2}{4} =$ ___________

5.

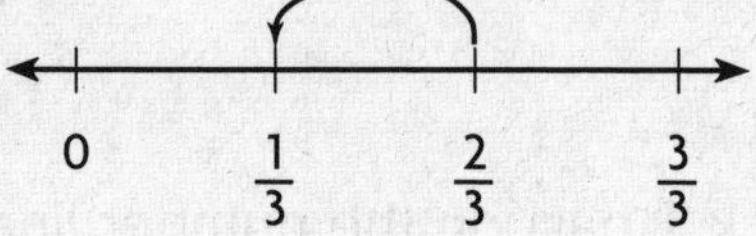

$\frac{2}{3} - \frac{1}{3} =$ ___________

6.

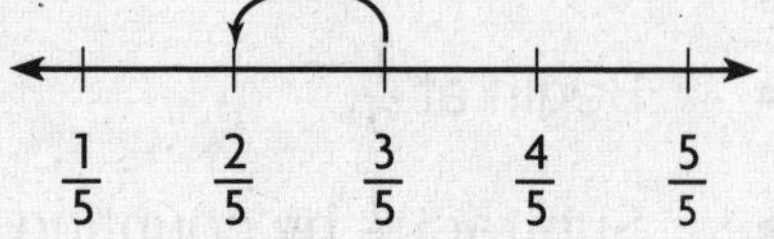

$\frac{3}{5} - \frac{1}{5} =$ ___________

Model the difference. Record your answer.

7. $\frac{8}{10} - \frac{3}{10} =$ ___________

8. $\frac{7}{9} - \frac{2}{9} =$ ___________

9. $\frac{10}{12} - \frac{5}{12} =$ ___________

10. $\frac{5}{6} - \frac{1}{6} =$ ___________

11. $\frac{6}{8} - \frac{1}{8} =$ ___________

12. $\frac{8}{9} - \frac{5}{9} =$ ___________

13. $\frac{7}{8} - \frac{5}{8} =$ ___________

14. $\frac{3}{4} - \frac{1}{4} =$ ___________

15. $\frac{4}{6} - \frac{1}{6} =$ ___________

16. $\frac{8}{9} - \frac{3}{9} =$ ___________

17. $\frac{8}{12} - \frac{2}{12} =$ ___________

18. $\frac{6}{10} - \frac{1}{10} =$ ___________

19. $\frac{2}{3} - \frac{1}{3} =$ ___________

20. $\frac{6}{9} - \frac{4}{9} =$ ___________

21. $\frac{7}{9} - \frac{6}{9} =$ ___________

Name____

Record Addition and Subtraction

When you add or subtract fractions with like denominators, add or subtract only the numerators.

Find and record the sum.

$\frac{4}{6} + \frac{3}{6}$

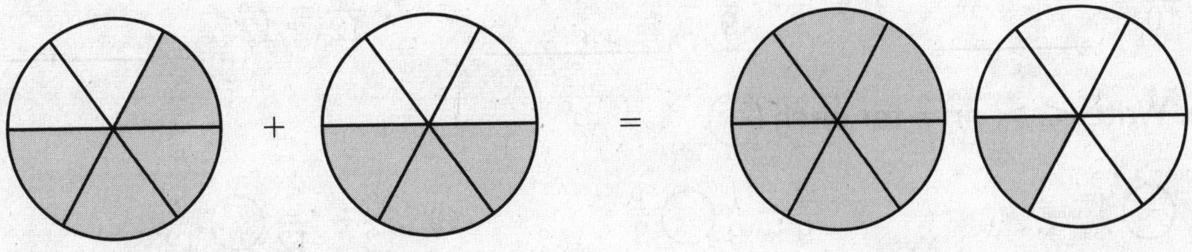

$\frac{\text{4 parts shaded}}{\text{6 equal parts}} + \frac{\text{3 parts shaded}}{\text{6 equal parts}} = \frac{\text{7 parts shaded}}{\text{6 equal parts}}$

$\frac{4}{6} + \frac{3}{6}$ ⟶ Add the numerators. ⟶ Write the denominators. ⟶ $\frac{4 + 3}{6}$ ⟶ $= \frac{7}{6}$, or $1\frac{1}{6}$.

Find and record the sum or difference.

1. $\frac{3}{5} + \frac{4}{5} =$ ______

2. $\frac{3}{4} - \frac{1}{4} =$ ______

3. $\frac{1}{9} + \frac{7}{9} =$ ______

4. $\frac{8}{11} - \frac{3}{11} =$ ______

5. $\frac{10}{12} - \frac{5}{12} =$ ______

6. $\frac{2}{3} + \frac{1}{3} =$ ______

7. $\frac{5}{6} + \frac{4}{6} =$ ______

8. $\frac{6}{10} - \frac{4}{10} =$ ______

Name________________________________

Record Addition and Subtraction

Find and record the sum or difference.

1. $\begin{array}{r} \frac{7}{8} \\ +\frac{3}{8} \\ \hline \end{array}$

2. $\begin{array}{r} \frac{5}{6} \\ -\frac{3}{6} \\ \hline \end{array}$

3. $\begin{array}{r} \frac{3}{12} \\ +\frac{4}{12} \\ \hline \end{array}$

4. $\begin{array}{r} \frac{8}{9} \\ -\frac{1}{9} \\ \hline \end{array}$

5. $\frac{7}{10} + \frac{3}{10} =$ ________

6. $\frac{7}{9} - \frac{4}{9} =$ ________

7. $\frac{4}{12} + \frac{7}{12} =$ ________

Compare. Write <, >, or = for each ○.

8. $\frac{5}{6} - \frac{1}{6}$ ○ 1

9. $\frac{4}{9} - \frac{1}{9}$ ○ $\frac{1}{3}$

10. $\frac{10}{12} - \frac{2}{12}$ ○ $\frac{7}{12}$

ALGEBRA Find the value of *x*.

11. $\frac{2}{7} + \frac{x}{7} = \frac{6}{7}$

$x =$ ______

12. $\frac{x}{3} - \frac{1}{3} = \frac{1}{3}$

$x =$ ______

13. $\frac{4}{5} - \frac{3}{5} = \frac{1}{x}$

$x =$ ______

14. $\frac{4}{x} + \frac{6}{x} = 1$

$x =$ ______

Problem Solving and Test Prep

USE DATA For 15–16, use the bar graph.

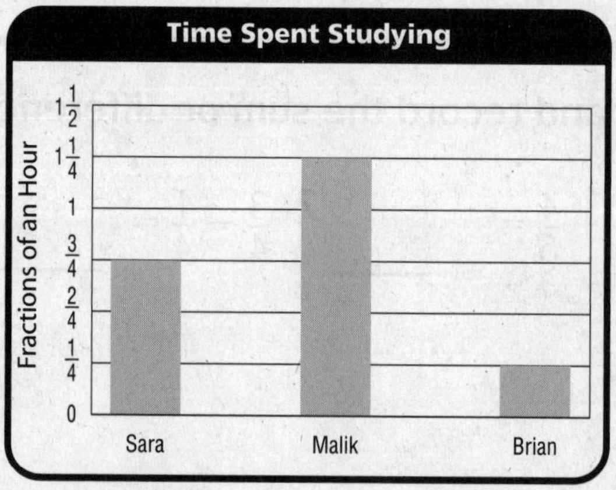

15. How much more time does Sara study than Brian?

____

16. How much more time does Malik study than Sara and Brian together?

____

17. Sue buys $\frac{2}{8}$ pound of walnuts and $\frac{5}{8}$ pound of peanuts. How many pounds of nuts does Sue buy in all?

A $\frac{3}{8}$

B $\frac{2}{16}$

C $\frac{7}{8}$

D $\frac{7}{16}$

18. Juan's garden is divided into eighths. Of the garden, $\frac{3}{8}$ is tomato plants and $\frac{1}{8}$ is cucumber plants. What part of the garden does Juan have left to plant?

____

Name____ Lesson 17.4

Problem Solving Workshop Strategy: Write an Equation

Jen takes a dance class of hip-hop and jazz. The class lasts $\frac{9}{10}$ hour. Hip-hop lasts $\frac{4}{10}$ hour. Which part of the class lasts longer, hip-hop or jazz?

Read to Understand

1. Write the question as a fill-in-the-blank sentence.

____

____

Plan

2. Choose a variable. What does the variable represent?

____

____

Solve

3. Write an equation to find the length of the jazz class. Which part of the class lasts longer?

____

Check

4. How can you check your work?

____

5. Of George's books, $\frac{5}{8}$ are about animals. $\frac{2}{8}$ of the books in his collection are about dogs. What fraction of George's books are about other animals? Write an equation to solve.

____

____

6. Maria spends $\frac{6}{14}$ of her day in school. While at school, she studies math for $\frac{2}{14}$ of her day. What fraction of Maria's day is spent studying other subjects in school? Write an equation to solve.

____

____

AF1.1 Use letters, boxes, or other symbols to stand for any number in simple expressions or equations (e.g., demonstrate an understanding and the use of the concept of a variable).

Name ____________________

Problem Solving Strategy: Write an Equation

Problem Solving Strategy Practice

Write an equation to solve.

1. Jasmine takes a dance class for tap and modern dance. The class lasts for $\frac{5}{6}$ hour. The modern dance part lasts for $\frac{2}{6}$ hour. How long does the tap part last?

2. Martin buys $\frac{5}{8}$ yard of fabric for a project. He has $\frac{2}{8}$ yard of fabric left over after completing the project. How much fabric did Martin use for his project?

3. Robert goes to soccer practice on Saturdays. This week, $\frac{2}{5}$ of the children at practice are girls. What fraction of the children are boys?

4. Harley and Belinda share a bag of crackers. Harley eats $\frac{7}{12}$ of the crackers. Belinda eats $\frac{4}{12}$ of the crackers. What fraction of the crackers do they eat in all?

Mixed Strategy Practice

USE DATA For 5 – 6, use the table.

5. Nicky walks dogs to earn money. How much more time does Nicky spend walking Binky than Pugg?

6. How much time does Nicky spend walking Rusty and Pugg altogether?

Dog Walking Hours

Dog Names	Time Spent
Pugg	$\frac{2}{8}$ hour
Rusty	$\frac{3}{8}$ hour
Binky	$\frac{5}{8}$ hour

7. Michel practices piano for 75 minutes each day. Write a mixed number to show the time Michel practices, in hours.

8. Linda plays 90 minutes of softball a day. Write a mixed number to show the time Linda plays, in hours.

Spiral Review

For 1–4, estimate. Then find the product.

1. $\begin{array}{r} 196 \\ \times\ 10 \\ \hline \end{array}$

2. $\begin{array}{r} 384 \\ \times\ 69 \\ \hline \end{array}$

3. $\begin{array}{r} 877 \\ \times\ 36 \\ \hline \end{array}$

4. $\begin{array}{r} 7{,}200 \\ \times\ 19 \\ \hline \end{array}$

For 5–6, solve a simpler problem.

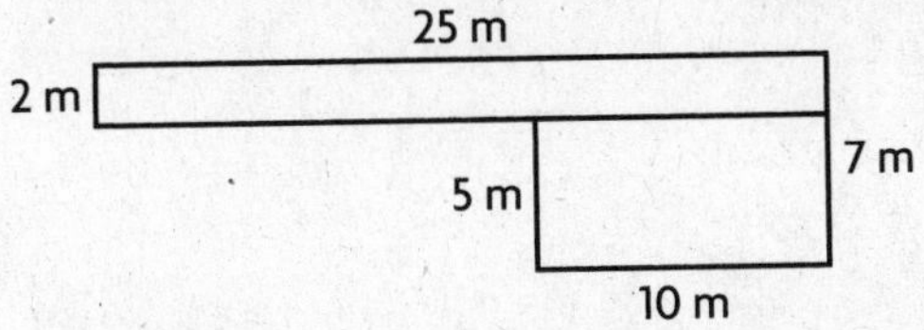

5. What is the total area of the figure?

____

6. What is the perimeter of the figure?

____

7. Sean is given a bag of plastic chips. There is 1 red, 1 black, and 1 green chip. If Sean picks 1 chip out of the bag and puts it back, and then picks another chip out of the bag, what are all of the possible color combinations that Sean can choose? Make an organized list.

8. How many possible outcomes does Sean have?

____

For 9–13, tell what you do first. Then find the value of each expression.

9. $12 + (11 - 3) =$ ____________

10. $(7 + 9) - 14 =$ ____________

11. $25 - (13 + 1) =$ ____________

12. $(11 - 2) + 14 =$ ____________

13. $37 - (19 + 4) =$ ____________

Name____ **Lesson 17.5**

Add and Subtract Mixed Numbers

You can add and subtract mixed numbers with like denominators. Add or subtract the fractions first. Then add or subtract the whole numbers. You can check your answer by making a model of the equation.

Find the sum. $3\frac{1}{6} + 1\frac{3}{6}$

Start with the fractions. Add the numerators first. Keep the denominators the same.	$\frac{1}{6} + \frac{3}{6} = \frac{4}{6}$
Add the whole numbers.	$3 + 1 = 4$
If needed, change any improper fractions in the sum to a mixed number.	$3\frac{1}{6} + 1\frac{3}{6} = 4\frac{4}{6}$
Write the answer in simplest form.	$4\frac{4}{6} = 4\frac{2}{3}$

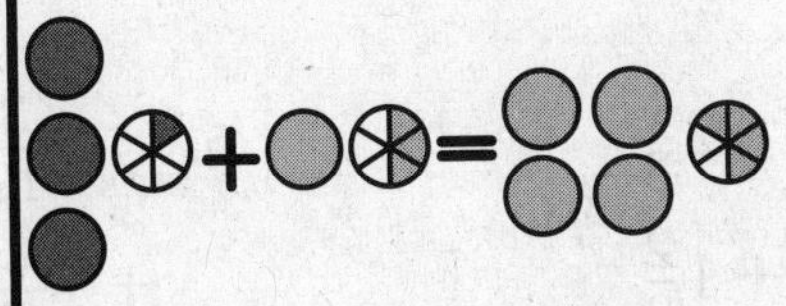

$3\frac{1}{6} + 1\frac{3}{6} = 4\frac{4}{6}$, or $4\frac{2}{3}$

Find the difference.
$1\frac{4}{5} - \frac{2}{5}$

Start with the fractions. Subtract the numerators. Keep the denominators the same.	$\frac{4}{5} - \frac{2}{5} = \frac{2}{5}$
Subtract the whole numbers.	$1 - 0 = 1$
If needed, change any improper fractions in the difference to a mixed number.	$1\frac{4}{5} - \frac{2}{5} = 1\frac{2}{5}$
Write the answer in simplest form.	$1\frac{2}{5}$

Model and record the sum or difference.

1. $1\frac{2}{5} + 1\frac{1}{5}$

2. $4\frac{1}{3} + 4\frac{2}{3}$

3. $6\frac{5}{6} - 2\frac{1}{6}$

4. $3\frac{7}{8} - 2\frac{6}{8}$

5. $2\frac{7}{10} + 3\frac{5}{10}$

6. $3\frac{4}{7} - 2\frac{4}{7}$

7. $2\frac{8}{11} + 2\frac{8}{11}$

8. $4\frac{8}{9} + 3\frac{5}{9}$

Name____

Add and Subtract Mixed Numbers

Model and record the sum or difference.

1. $2\frac{1}{8} + 1\frac{3}{8}$ = ____

2. $1\frac{3}{5} + 3\frac{1}{5}$ = ____

3. $5\frac{7}{10} - 1\frac{2}{10}$ = ____

4. $3\frac{7}{9} - \frac{3}{9}$ = ____

5. $2\frac{1}{3} + 1\frac{2}{3}$ = ____

6. $3\frac{3}{12} + 1\frac{4}{12}$ = ____

7. $2\frac{5}{6} - 1\frac{3}{6}$ = ____

8. $4\frac{8}{12} - 1\frac{3}{12}$ = ____

9. $1\frac{7}{9} - 1\frac{4}{9} =$ ________

10. $2\frac{3}{4} + 2\frac{1}{4} =$ ________

11. $4\frac{4}{10} + \frac{7}{10} =$ ________

ALGEBRA Find the value of *n*.

12. $2\frac{4}{6} + 1\frac{n}{6} = 3\frac{5}{6}$

$n =$ ________

13. $3\frac{n}{8} - 2\frac{3}{8} = 1\frac{2}{8}$

$n =$ ________

14. $5\frac{9}{10} - 5\frac{n}{10} = \frac{2}{10}$

$n =$ ________

Problem Solving and Test Prep

USE DATA For 15–16, use the table.

Ribbons Used for Puppets	
Red	$3\frac{4}{8}$ inches
Blue	$4\frac{1}{4}$ inches
Green	$5\frac{1}{8}$ inches

15. How many more inches are there of green ribbon than blue ribbon?

____

16. Tim adds another $2\frac{2}{8}$ inches of red ribbon. How many inches are there now of both red and green ribbon? ________

17. Sonya drives $2\frac{3}{10}$ miles to the store and $4\frac{4}{10}$ miles to the library. How far does Sonya drive in all?

A $2\frac{7}{10}$
B $6\frac{7}{10}$
C $6\frac{6}{10}$
D $6\frac{7}{20}$

18. Ira uses $4\frac{3}{8}$ cups of almonds, $2\frac{1}{8}$ cups of raisins, and $1\frac{2}{8}$ cups of peanuts to make trail mix. How much more almonds does Ira use than peanuts?

A $4\frac{3}{8}$
B $3\frac{1}{8}$
C $5\frac{5}{8}$
D $2\frac{1}{8}$

Name____ **Lesson 18.1**

Relate Fractions and Decimals

A decimal is a number with one or more digits to the right of the decimal point.

Write the decimal and fraction shown by the model.

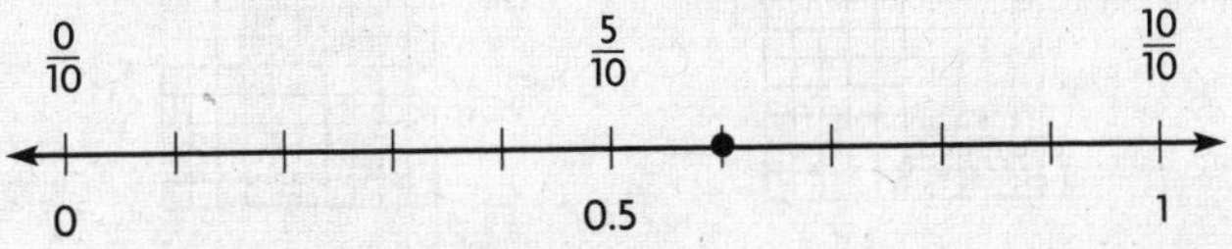

The number line is divided into 10 equal parts.

First, label the parts below the number line using tenths starting at 0 and stopping at 1.

Then, label the parts above the number line using fractions starting at $\frac{0}{10}$ and ending at $\frac{6}{10}$.

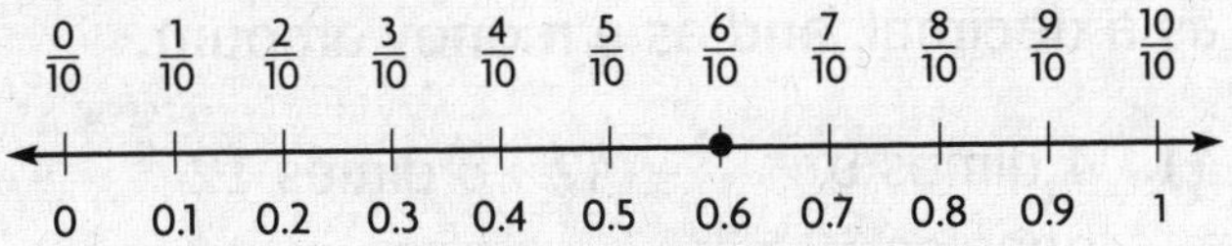

Locate the point on the number line.
The decimal 0.6 and the fraction $\frac{6}{10}$ names the point.

So, the decimal and the fraction shown by the model is 0.6 and $\frac{6}{10}$.

Write the fraction as a decimal.

$\frac{5}{100}$ Read the fraction aloud as five hundredths.

Write the zero in the ones place, and then write the decimal point. 0.■■

Write the zero in the tenths place. 0.0■

Write a 5 in the hundredths place. 0.05

So, $\frac{5}{100}$ written as a decimal is 0.05.

Write the decimal and fraction shown by each model.

1. 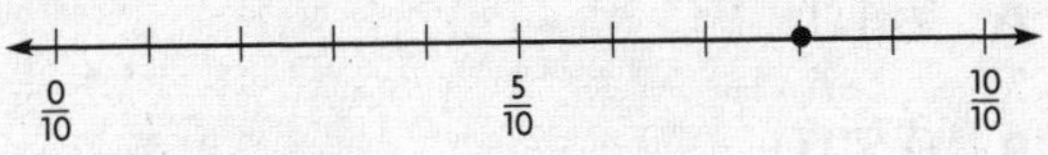

2. 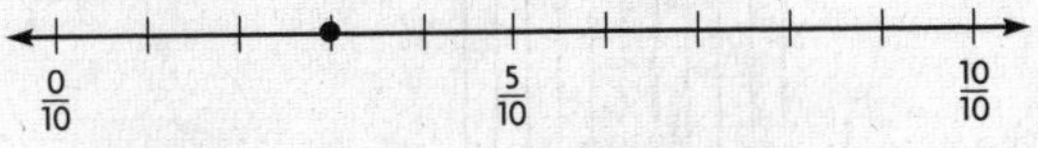

______________________ ______________________

Write each fraction as a decimal. You may draw a picture.

3. $\frac{3}{100}$ ________ **4.** $\frac{4}{8}$ ________ **5.** $\frac{20}{100}$ ________ **6.** $\frac{9}{10}$ ________

Name____________________________________

Lesson 18.1

Relate Fractions and Decimals

Write the decimal and fraction shown by each model

1. 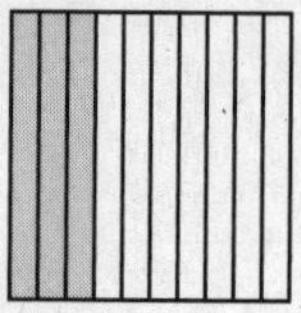 ______

2. 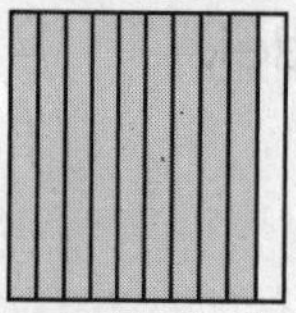 ______

3.  ______

4.  ______

Write each fraction as a decimal. You may draw a picture.

5. $\frac{6}{10}$ ______

6. $\frac{2}{100}$ ______

7. $\frac{1}{10}$ ______

8. $\frac{63}{100}$ ______

Write the amount as a fraction of a dollar, as a decimal, and as a money amount.

9. 6 dimes ______ ______

10. 2 nickels 7 pennies ______ ______

11. 4 dimes 9 pennies ______ ______

12. 8 dimes 12 pennies ______ ______

ALGEBRA Find the missing number.

13. 9 tenths + 7 hundredths = ______

14. 6 tenths + ______ hundredths = 0.66

Problem Solving and Test Prep

15. Write five cents in decimal form.

______

16. Write one and thirty-four hundredths in decimal form.

______

17. Which decimal is shown by the model?

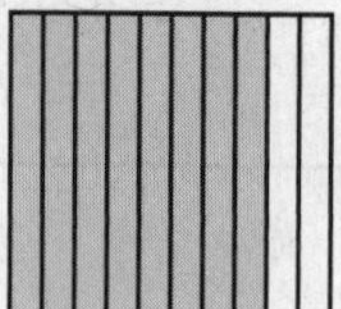

A 0.08

B 0.06

C 0.8

D 0.6

18. Which decimal means the same as $\frac{7}{10}$?

A 7.10

B 0.710

C 0.07

D 0.7

Name____________________________________

Equivalent Decimals

You can use models to show equivalent decimals.

Use a hundredths model. Are the two decimals equivalent?
Write *equivalent* or not *equivalent*.
Model 0.71 and 0.17.

Shade in the appropriate number of squares in each hundredths model.

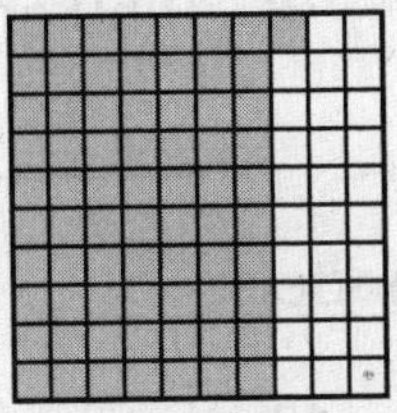

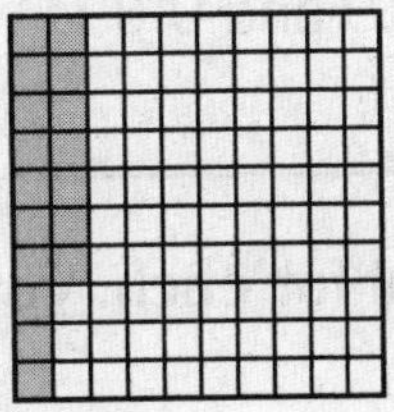

0.71 is 71 hundredths.
Shade in 71 squares.

0.17 is 17 hundredths.
Shade in 17 squares.

See if the models have the same number of squares shaded.
0.71 has more squares shaded than 0.17.

So, 0.71 and 0.17 are *not equivalent*.

Write an equivalent decimal for 0.20. You may use a decimal model.

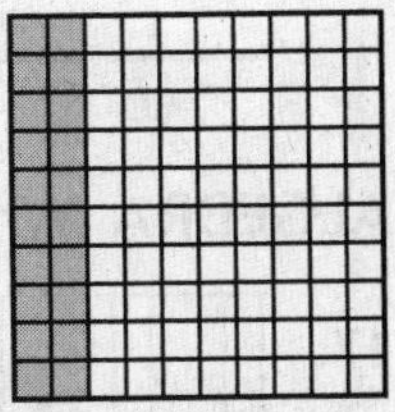

Shade 20 squares of a hundredths model to show 0.20.

There are two columns shaded.

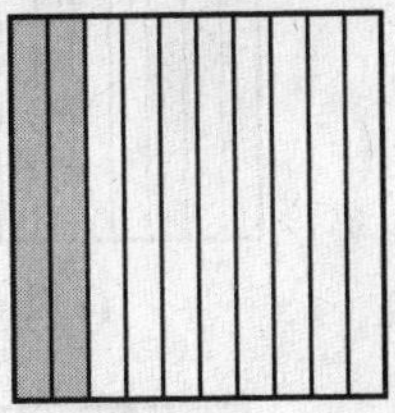

Shade two columns of a tenths model.

The decimal model shows that 0.2 is equivalent to 0.20.
0.2 = 0.20

So, 0.2 is equivalent to 0.20.

Use a tenths model and hundredths model. Are the two decimals equivalent? Write *equivalent* or *not equivalent*.

1. 0.3 and 0.30 ______________

2. 0.72 and 0.27 ______________

3. 0.60 and 0.6 ______________

4. 0.05 and 0.50 ______________

Write an equivalent decimal for each. You may use decimal models.

5. 0.9 ______________

6. 0.2 ______________

7. 0.50 ______________

8. 0.7 ______________

Name____

Lesson 18.2

Equivalent Decimals

Use a tenths model and a hundredths model. Are the two decimals equivalent? Write *equivalent* or *not equivalent*.

1. 0.1 and 0.10 ____________

2. 0.23 and 0.32 ____________

3. 0.65 and 0.56 ____________

4. 0.3 and 0.30 ____________

5. 0.22 and 0.23 ____________

6. 0.9 and 0.09 ____________

7. 0.76 and 0.67 ____________

8. 0.50 and 0.5 ____________

Write an equivalent decimal for each. You may use decimal models.

9. 0.70 ____________

10. $\frac{1}{4}$ ____________

11. 0.2 ____________

12. $\frac{3}{4}$ ____________

13. 0.3 ____________

14. 0.50 ____________

15. $\frac{7}{10}$ ____________

16. 0.90 ____________

ALGEBRA Write an equivalent decimal. Use the models to help.

17.

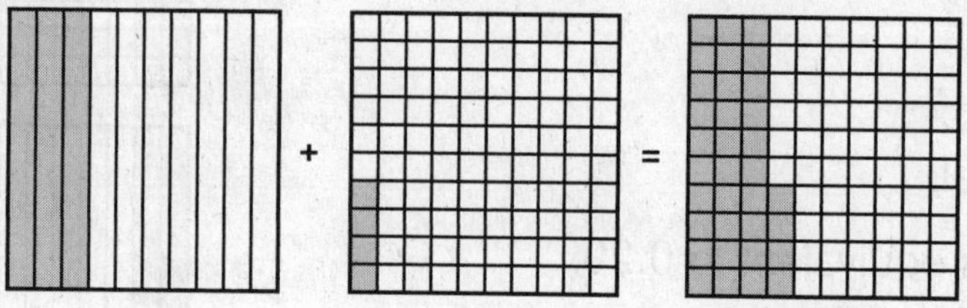

0.3 + 0.04 = ■

18.

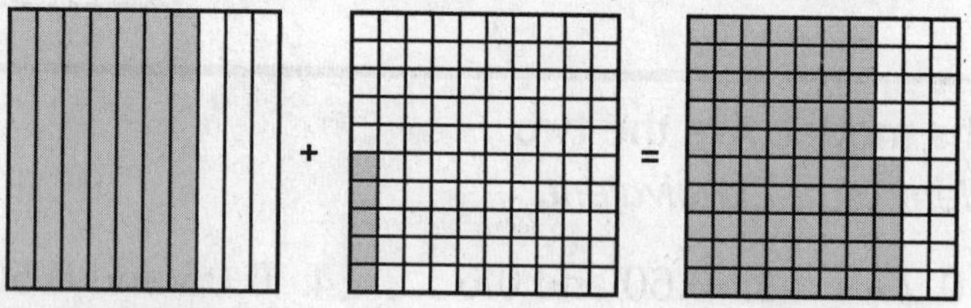

■ + ■ = 0.75

19.

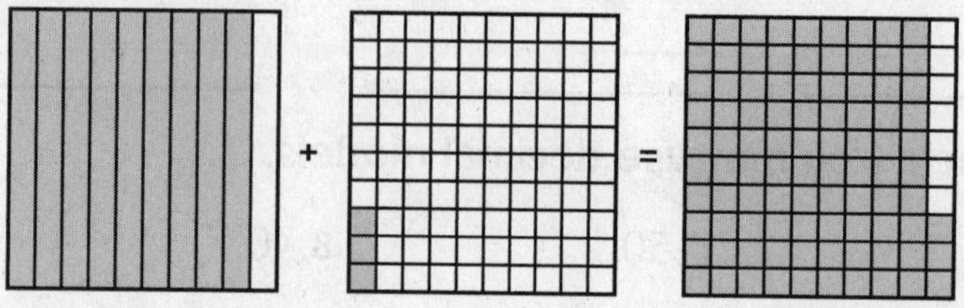

■ + ■ = 0.93

Name____________________ **Lesson 18.3**

Relate Mixed Numbers and Decimals

Write an equivalent decimal and a mixed number for the model.

The number line below is divided into 10 equal parts.

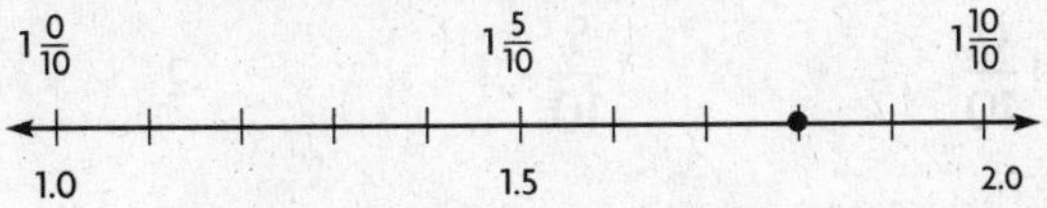

First, label the parts below the number line using tenths starting at 1.0 and stopping at 2.0.

Then, label the parts above the number line using fractions starting at $1\frac{0}{10}$ and stopping at $1\frac{10}{10}$.

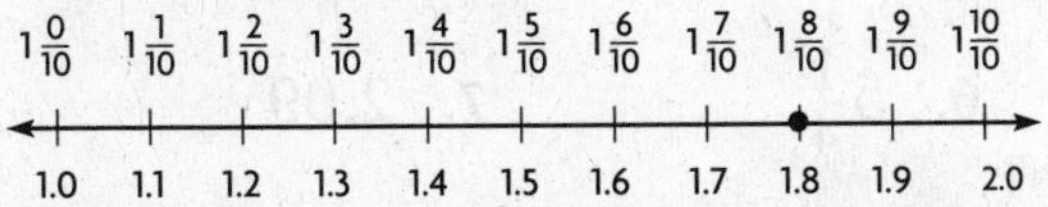

Locate the point on the number line.
The decimal is 1.8 and the mixed number is $1\frac{8}{10}$.

So, the decimal and the fraction shown by the model is 1.8 and $1\frac{8}{10}$.

Write the decimal for the mixed number. Then write the word form.

$2\frac{3}{4}$

- Write the fraction $\frac{3}{4}$ using a denominator of 100.
- Write $\frac{75}{100}$ as a decimal.
- Write the whole number.

So, $2\frac{3}{4}$ written as a decimal is 2.75, or two and seventy-five hundredths.

Write an equivalent decimal and a mixed number. Then write the word form.

1.

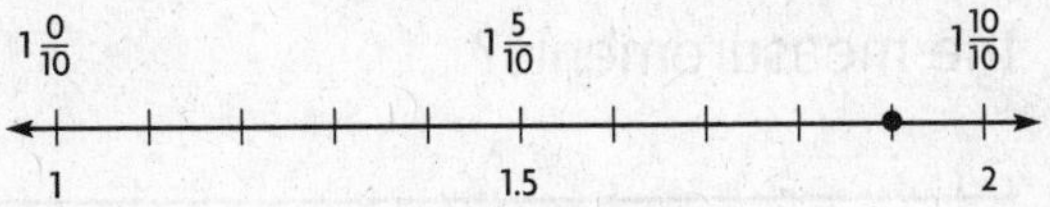

____

2.

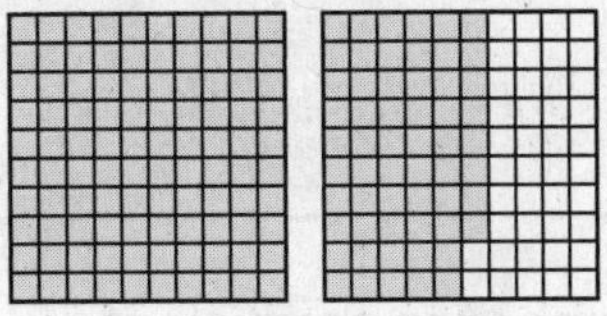

____

Write an equivalent decimal or mixed number. Then write the word form.

3. 5.1

____

4. $2\frac{7}{100}$

____

NS 1.6 Write tenths and hundredths in decimal and fraction notations and know the fraction and decimal equivalents for halves and fourths (e.g., $\frac{1}{2}$ = 0.5 or .50; $\frac{7}{4} = 1\frac{3}{4} = 1.75$).

Name____

Relate Mixed Numbers and Decimals

Write an equivalent decimal and mixed number for each model.

1. 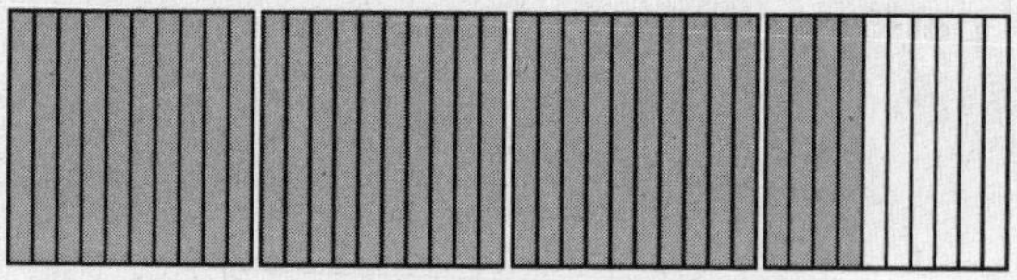

____

2. 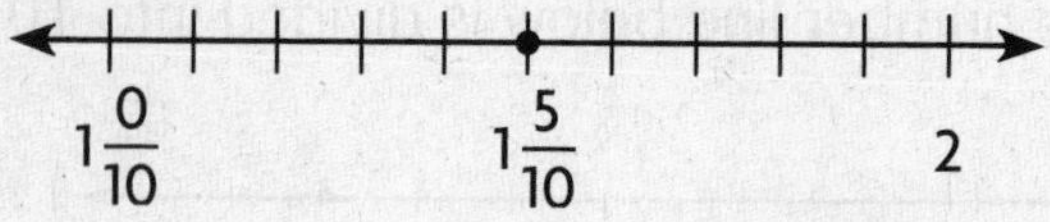

____

Write an equivalent mixed number or a decimal for each. Then write the word form. You may use a model.

3. 6.6 ______ ______ ______

4. $3\frac{90}{100}$ ______ ______ ______

5. 4.75 ______ ______ ______

6. $5\frac{1}{4}$ ______ ______ ______

7. 2.09 ______ ______ ______

ALGEBRA Write the missing number in each ■.

8. $2.4 = 2 + ■$

9. $3.80 = 3 + 0.8 + ■$

10. $5.16 = 5 + ■ + 0.06$

Problem Solving and Test Prep

11. Harriet is thinking of a decimal that is equivalent to eight and one-fifth. What is that decimal?

____

12. A CD case measures four and four-fifths inches by five and three-fifths inches. What is the decimal form of the measurements?

____

13. Which mixed number is equivalent to 3.25?

A $3\frac{1}{4}$

B $3\frac{2}{5}$

C $3\frac{2}{3}$

D $2\frac{9}{100}$

14. In simplest form, what is an equivalent fraction for the decimal 2.36?

A $2\frac{4}{50}$

B $2\frac{3}{10}$

C $2\frac{9}{25}$

D $2\frac{4}{100}$

Spiral Review

For 1–2, tell whether you need an exact answer or an estimate. Then solve.

1. Bianca is writing an article for her school newspaper. So far she has written 362 words. The article cannot be more than 800 words. How many more words can Bianca write in her article?

____

2. Lauren found 212 seashells at a beach in Florida. She found half the amount of seashells at a beach in Maine. About how many fewer seashells did Lauren find at the beach in Maine?

____

For 3–5, tell whether the figure appears to have *line symmetry*, *rotational symmetry*, *both*, or *neither*.

3. 

____

4. 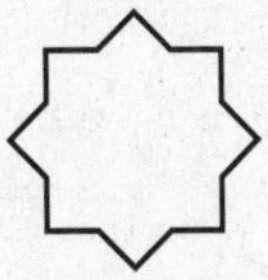

____

5. 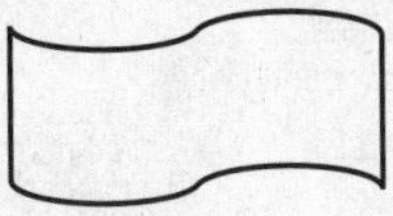

____

For 6–8, use the bar graph.

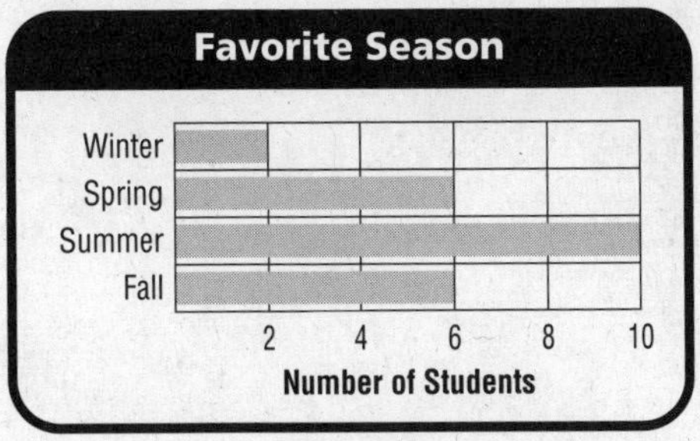

6. Which season was chosen by the fewest students?

____

7. Which two seasons were chosen by the same number of students?

____

8. What interval is used on the scale?

____

For 9–14, solve the equation.

9. $11 + x = 21$

____

10. $12 - y = 4$

____

11. $\square - 13 = 26$

12. $26 + \square = 52$

13. $n - 5 = 50$

____

14. $34 - d = 22$

____

Name____

Lesson 18.4

Compare Decimals

Compare. Write <,>, or = for the ○.

2.2 ○ 2.15

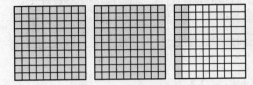

Look at the models.
The model on the left shows 2 whole and 2 tenths shaded.
The model on the right shows 2 whole, 1 tenth and 5 hundredths shaded.

The model on the left shows more shaded.

So, 2.2 > 2.15.

Use the number line to determine whether the number sentence is *true* or *false*.

1.65 > 1.6

Look at the number line.
Locate and label 1.65 and 1.6.

1.65 is to the right of 1.6, so 1.65 is greater.

So, 1.65 > 1.6 is true.

Compare. Write <,>, or = for each ○.

1.

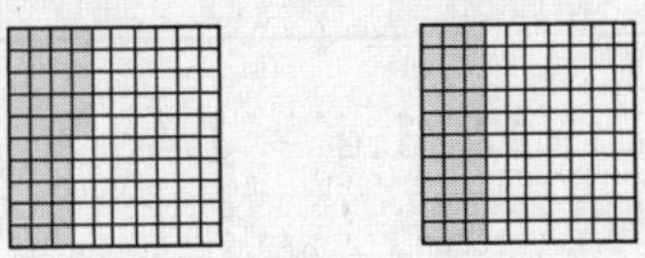

0.35 ○ 0.3

2.

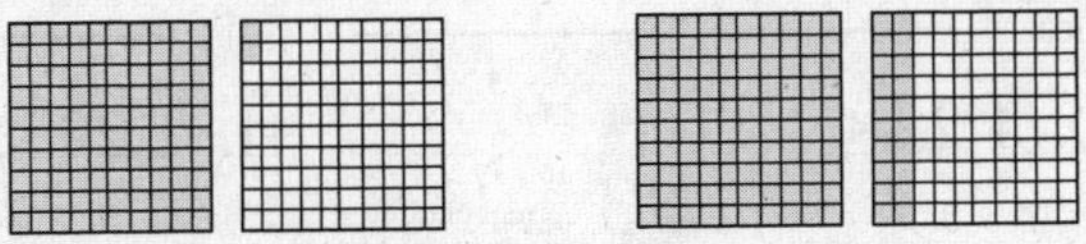

1.02 ○ 1.2

Use the number line above to determine if the following number sentences are *true* or *false*.

3. 1.5 > 1.05 **4.** 1.2 > 1.22 **5.** 1.6 = 1.60

__________ __________ __________

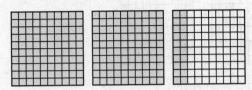

Name____ **Lesson 18.4**

Compare Decimals

Compare. Write <, >, or = for each ◯.

1.

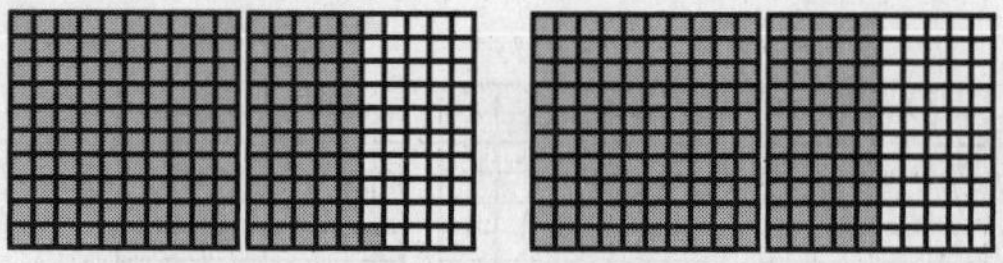

1.51 ◯ 1.5

2.

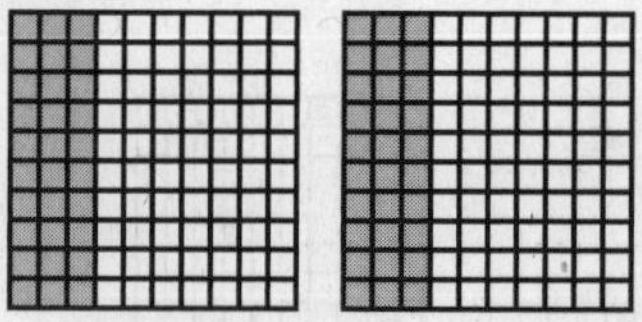

0.30 ◯ 0.3

3.

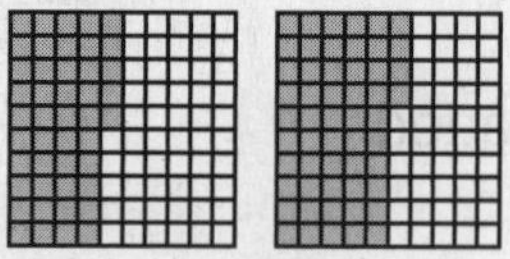

0.45 ◯ 0.54

4.

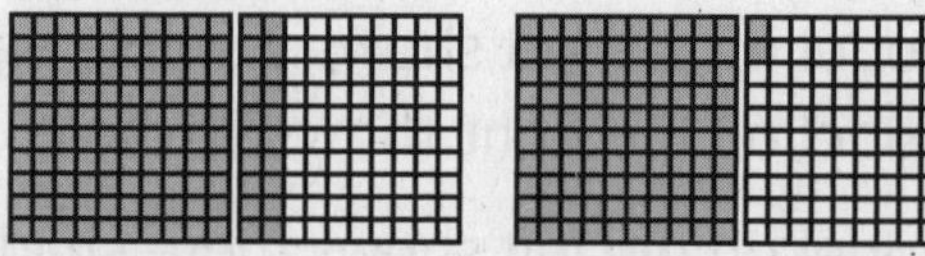

1.20 ◯ 1.02

5.

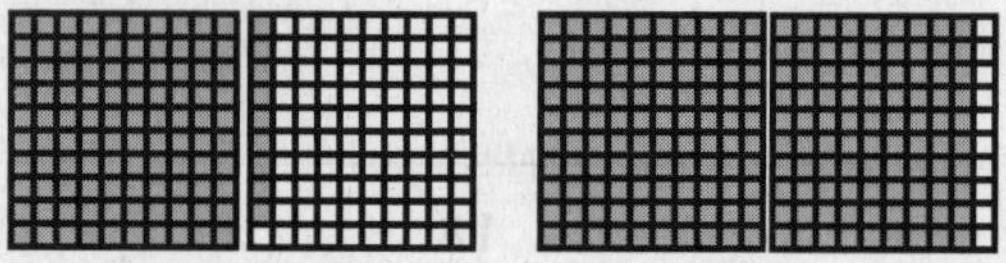

1.09 ◯ 1.90

6.

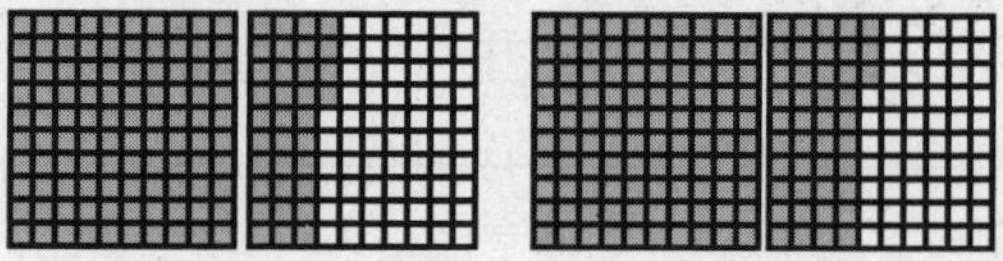

1.34 ◯ 1.43

Use the number line to determine whether the following number sentences are *true* or *false*.

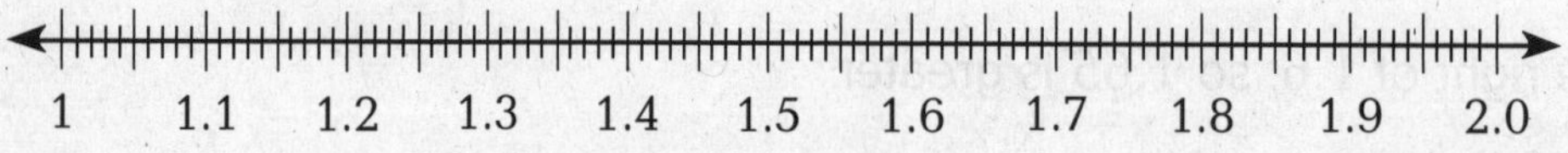

7. $1.25 < 1.52$ ____________

8. $1.70 > 1.7$ ____________

9. $1.21 < 1.2$ ____________

10. $1.22 < 1.11$ ____________

11. $1.29 < 1.92$ ____________

12. $1.4 = 1.40$ ____________

13. $1.09 > 1.08$ ____________

14. $1.66 = 1.67$ ____________

15. $1.37 < 1.35$ ____________

16. $1.55 > 1.45$ ____________

17. $1.0 = 1.00$ ____________

18. $1.9 < 1.99$ ____________

Name ______________________

Order Decimals

Use the number line to order the decimals from least to greatest.
1.7, 1.75, 1.5, 1.05

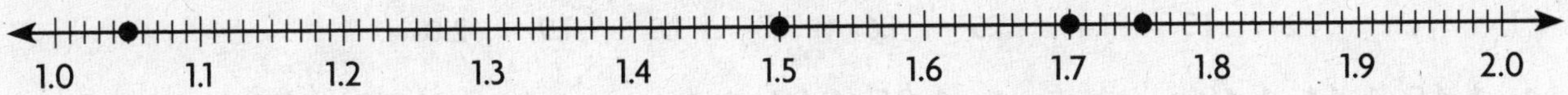

Locate and label 1.7, 1.75, 1.5, and 1.05 on the number line.

1.05 is the farthest to the left on the number line, so it is the least.
1.75 is the farthest to the right on the number line, so it is the greatest.

So, the decimals in order from least to greatest are 1.05, 1.5, 1.7, 1.75.

Order the decimals from greatest to least.
$1.89, $2.15, $1.09

$1.89
$2.15
$1.09

Line up the decimal points.

Then compare the digits beginning with the ones digit or the greatest place.

Since the 2 in $2.15 has the greatest value, it is the largest number.

Check the other two numbers: $1.89 and $1.09.

So, the decimals in order from greatest to least are $2.15, $1.89, and $1.09.

Use the number line above to order the decimals from least to greatest.

1. 1.6, 1.06, 1.61, 1.66

2. 1.2, 1.23, 1.12, 1.21

3. 1.7, 1.77, 1.07, 1.01

4. 1.3, 1.7, 1.04, 1.52

5. 1.63, 1.36, 1.03, 1.6

6. 1.9, 1.91, 1.19, 1.99

Order the decimals from greatest to least.

7. 6.2, 6.02, 6.32, 6.23

8. $2.18, $2.38, $2.08, $2.88

9. 5.5, 5.75, 5.05, 5.65

Name____

Hands On: Order Decimals

Use the number line to order the decimals from least to greatest.

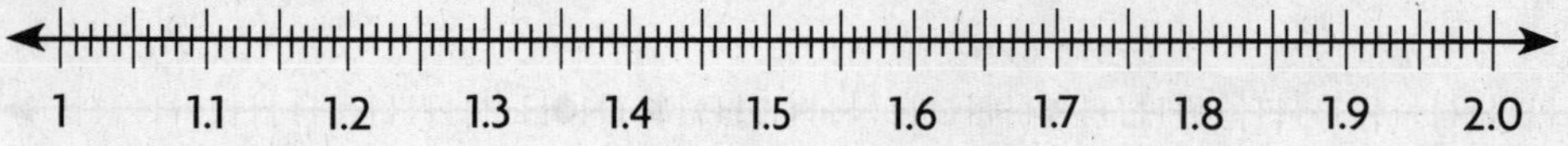

1. 1.45, 1.44, 1.43 ______________

2. 1.05, 1.04, 1.4 ______________

3. 1.78, 1.79, 1.09 ______________

4. 1.33, 1.32, 1.3 ______________

5. 1.2, 1.19, 1.27 ______________

6. 1.05, 1.03, 1.01 ______________

7. 1.02, 1.03, 1.1 ______________

8. 1.84, 1.89, 1.82 ______________

9. 1.66, 1.65, 1.62 ______________

Order the decimals from greatest to least.

10. 1.66, 1.06, 1.6, 1.65 ______________

11. \$5.33, \$5.93, \$5.39, \$3.55 ______________

12. 4.84, 4.48, 4.88, 4.44 ______________

13. \$1.45, \$1.43, \$1.54, \$1.34 ______________

14. 7.32, 7.38, 7.83, 7.23 ______________

15. \$0.98, \$1.99, \$0.89, \$1.89 ______________

16. 0.67, 0.76, 0.98, 1.01 ______________

17. \$1.21, \$1.12, \$1.11, \$1.10 ______________

18. 4.77, 5.07, 5.1, 4.6 ______________

19. 1.21, 1.45, 1.12, 1.44 ______________

20. 2.21, 2.67, 2.66, 2.3 ______________

21. \$9.00, \$9.10, \$9.11, \$9.99 ______________

22. \$5.97, \$5.96, \$6.59, \$5.75 ______________

23. \$3.39, \$3.03, \$3.83, \$3.30 ______________

24. 8.17, 8.05, 8.08, 8.1 ______________

Name____ Lesson 18.6

Problem Solving Workshop Skill: Draw Conclusions

Megan lives 2.4 miles from Bob's Bookstore. Josh lives 2.15 miles away, and Sam lives 2.09 miles away. Who lives closest to Bob's Bookstore?

1. What are you asked to find?

____

2. How can you use information in the problem to draw a conclusion?

____

____

3. What is the answer to the question?

____

4. How can you check your answer?

____

____

Solve the problem.

5. Dane spent $12.50 on school supplies. Peter spent $12.05. Marshall spent $12.25. Which student spent the least on school supplies?

____

6. Sarah leaped 2.1 meters. Kelly leaped 2.23 meters. Linda leaped 2.19 meters. Who leaped the farthest: Sarah, Kelly, or Linda?

____

Name____

Problem Solving Workshop Skill: Draw Conclusions

Problem Solving Skill Practice

Use the information on the chart to draw a conclusion.

1. Jane looks at the ads to the right and wants the best value for her money. If she wants one game, which one should Jane buy, and at which store?

____

____

2. What if Great Games sold playing cards for $3.50? Which store would have the better value?

____

Mixed Applications

USE DATA For 3–4, use the map.

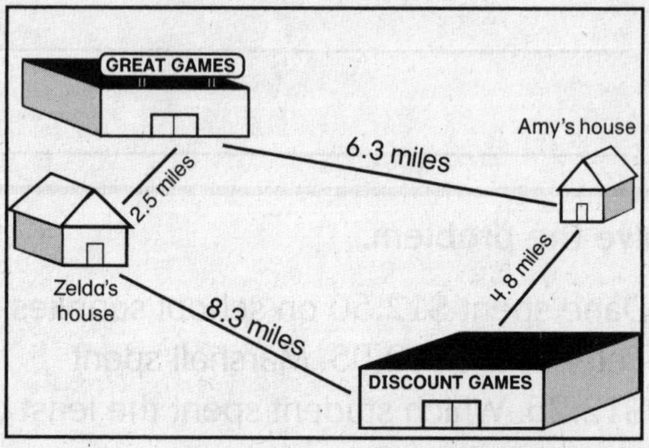

3. Sal lives 4.08 miles from Discount Games. Who lives closer: Amy or Sal?

____

4. Sal lives 6.33 miles from Great Games. List Amy, Sal, and Zelda in the order of least to greatest distance from each store.

____

____

____

5. Patty and Tom went to the mall. Patty bought a T-shirt for $12.35. Tom bought a T-shirt for $13.25. Who paid more money for their T-shirt?

____

Name____________________

Round Decimals

When you round a decimal to the nearest tenth, the rounded number will stop at the tenths place. For example, when 4.73 is rounded to the nearest tenth, it becomes 4.7.

Round 5.55 to the nearest tenth.

- Identify the tenths place.

Think: The 5 to the right of the decimal point is in the tenths place.

- Look at the digit to the right of the tenths place.
  If that number is less than 5, the digit in the tenths place stays the same. If it is 5 or more, the digit is increased by 1.

Think: The digit to the right of the tenths place is 5, so round the number in the tenths place up. Round 5.55 to 5.6.

So, 5.55 rounded to the nearest tenth is 5.6.

Round $48.92 to the nearest dollar.

- When you round to the nearest dollar, you round to the ones place. So, identify the ones place.

Think: The 8 is in the ones place.

- If the number to the right of the ones place is less than 5, the digit in the ones place stays the same. If it is 5 or more, the digit is increased by 1.

Think: 9 is greater than 5, so round the number in the ones place up. Round $48.92 to $49.

So, $48.92 rounded to the nearest dollar is $49.

Round each number to the nearest tenth and each money amount to the nearest dollar.

1. 1.92 ________

2. $21.10 ________

3. 56.16 ________

4. $24.60 ________

5. 6.07 ________

6. 28.34 ________

7. 6.37 ________

8. 55.92 ________

9. $81.99 ________

10. 8.35 ________

11. 48.48 ________

12. 8.77 ________

13. 39.94 ________

14. $6.03 ________

15. 18.26 ________

16. 77.12 ________

Name____ **Lesson 19.1**

Round Decimals

Round each number to the nearest tenth and each money amount to the nearest dollar.

1. 7.38 ______ **2.** 43.56 ______ **3.** 199.62 ______ **4.** 76.04 ______

5. $22.51 ______ **6.** $8.87 ______ **7.** $255.02 ______ **8.** $655.78 ______

Round each number to the nearest whole number.

9. 7.23 ______ **10.** 5.49 ______ **11.** 51.51 ______ **12.** 388.90 ______

13. 299.45 ______ **14.** 49.99 ______ **15.** 87.46 ______ **16.** 6.66 ______

Problem Solving and Test Prep

USE DATA For 17–18, use the map.

Weight per Fleece for Selected States (2004)

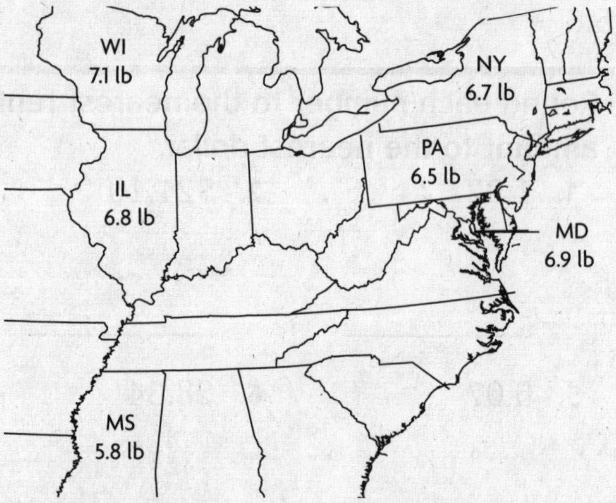

17. Round weight per fleece to the nearest whole number. Which states now show equal weight per fleece?

____

18. Order the states shown according to the heaviest weight per fleece to the lightest weight.

____

19. Mr. Scott gets 29.57 miles per gallon in his van. Which shows the distance rounded to the nearest tenth?

A 29.5 miles
B 29.4 miles
C 29.6 miles
D 29.7 miles

20. Paul is 5.89 feet tall. Which shows Paul's height to the nearest tenth?

A 5.2 feet
B 5.9 feet
C 5.6 feet
D 5.8 feet

Name ______________________________ **Week 29**

Spiral Review

For 1–6, write the numbers in order from least to greatest.

1. 1,904; 1,494; 1,600

____

2. 1,900,451; 11,825,000; 1,900,541

____

3. 6,991; 68,114; 6,000,348

____

4. 73,458; 73,485; 73,084

____

5. 996,000; 969,001; 9,900,000

____

6. 83,001,758; 83,100; 82,100,758

____

For 9, use the data to make a double-bar graph.

Favorite Cookie Type		
	Boys	Girls
Oatmeal	4	6
Chocolate Chip	12	8
Ginger Snap	2	4

9.

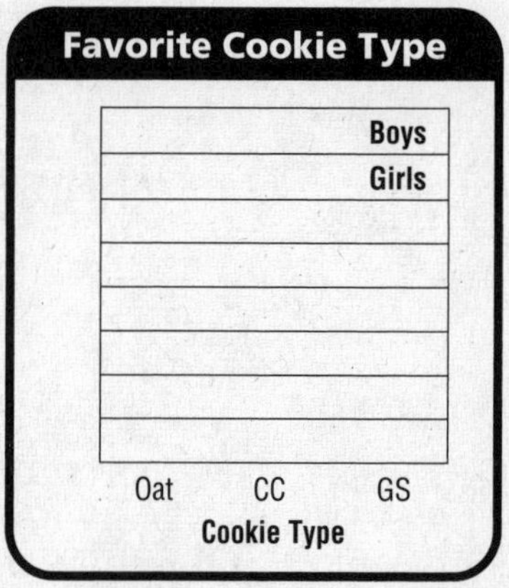

For 7–8, solve a simpler problem.

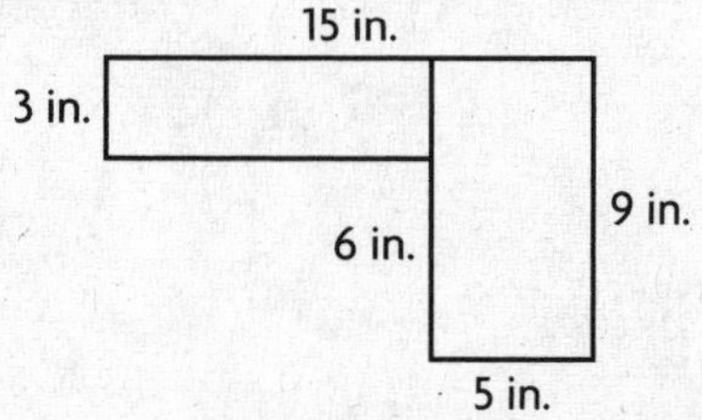

7. What is the total area of the figure?

____

8. What is the total perimeter of the figure?

____

For 10–13, write the value of each expression.

10. $32 + 9 - (9 \times 4)$

____

11. $2 \times (5 + 2) \times 2$

____

12. $(12 \div 3) + (7 \times 2)$

____

13. $(21 + 4) \div 5$

____

Name____

Estimate Decimal Sums and Differences

You can use **rounding** to estimate decimal sums and differences.

Estimate the difference.

```
  $91.24
– $39.87
```

Round each number to the nearest dollar.

```
  $91.24 →   $91
– $39.87 → – $40
```

Estimate by subtracting the rounded numbers.
So, a good estimate for $91.24 – $39.87 is $51.

```
  $91
– $40
  $51
```

Estimate the sum.

```
  4.09
  8.98
+ 1.43
```

```
  4.09 →   4
  8.98 →   9
+ 1.43 → + 1
```

Round each number to the nearest whole number.
Estimate by adding the rounded numbers.

```
  4
  9
+ 1
 14
```

So, a good estimate for 4.09 + 8.98 + 1.43 is 14.

Estimate the sum or difference.

1. 3.9 + 1.8

2. $52.35 – 40.22

3. 5.02 + 4.96 + 1.78

4. $31.99 + 12.12

5. 81.25 – 29.50

6. 23.92 – 11.03

7. 3.95 + 7.11 + 2.18

8. 58.75 – 45.11

9. $1.87 + 1.09

10. 6.67 + 2.25

11. 8.67 + 5.89 + 6.92

12. $72.99 – 32.87

Name______________________________

Estimate Decimal Sums and Differences

Estimate the sum or difference.

1. 6.42 + 8.55 ____________
2. 12.88 + 9.52 ____________
3. \$12.24 − \$8.27 ____________
4. 53.51 − 48.66 ____________
5. \$44.03 − \$15.97 ____________
6. 3.03 + 5.80 ____________
7. 502.22 − 497.53 ____________
8. \$71.04 + \$8.49 ____________

Estimate to compare. Write <, >, or = for each ◯.

9. 43.22 + 15.67 ◯ 81.77 − 32.54
10. 62.48 − 12.02 ◯ 15.65 + 23.99
11. 86.99 − 47.22 ◯ 15.42 + 12.60
12. 31.88 + 16.02 ◯ 75.61 − 40.65

Problem Solving and Test Prep

USE DATA For 13–14, use the table.

California Elementary School Enrollment

Grade Span	Students Enrolled (In Thousands)
K–2	20.9
K–3	41.8
K–4	46.6

13. About how many more students are enrolled in K–4 than K–2 schools?

____

14. About how many students are enrolled in K–3 and K–4 schools in all?

____

15. David drove 99.15 miles in January and 88.98 miles in February. About how many more miles did David drive in January than in February?

A 10 miles

B 20 miles

C 30 miles

D 100 miles

16. Mr. Frances drives 35.62 miles to work. His wife drives 27.25 miles to work. About how many more miles does Mr. Frances drive than Mrs. Frances?

A 8 miles

B 10 miles

C 9 miles

D 3 miles

Name____

Model Addition

You can use decimal models to help you add decimals.

Use a model to find the sum.

$$\begin{array}{r} 1.66 \\ +\ 1.07 \\ \hline \end{array}$$

Step 1: Shade squares to represent 1.66.

Think: Shade all the squares in one model and 66 squares in another model.

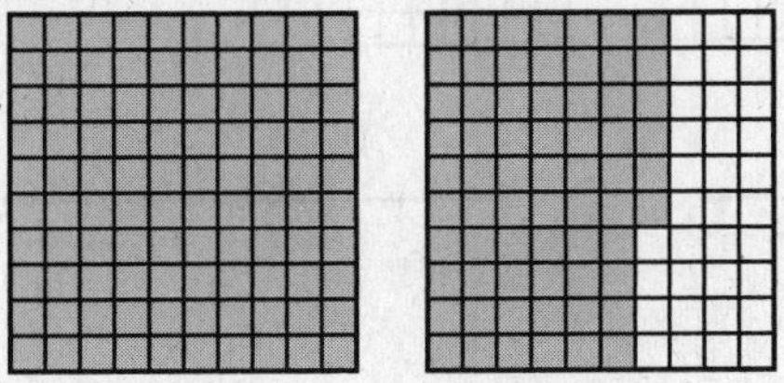

Step 2: Shade in squares to represent 1.07.

Think: Add one whole model and add 7 squares to the model of 1.66.

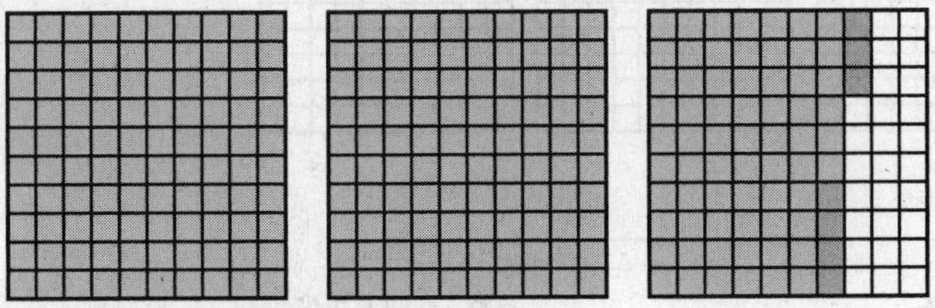

Step 3: Count all the squares.

Think: There are 2 whole models shaded and 73 squares of a hundredths model, or 2.73.

So, $1.66 + 1.07 = 2.73$.

Use models to find the sum.

1. $2.3 + 0.59$

____

2. $1.4 + 0.22$

____

3. $\begin{array}{r} 1.27 \\ +\ 1.15 \\ \hline \end{array}$

____

4. $\begin{array}{r} 0.81 \\ +\ 0.43 \\ \hline \end{array}$

____

Name____

Lesson 19.3

Model Addition

Use models to find the sum.

1. $\begin{array}{r} 0.56 \\ +0.45 \\ \hline \end{array}$

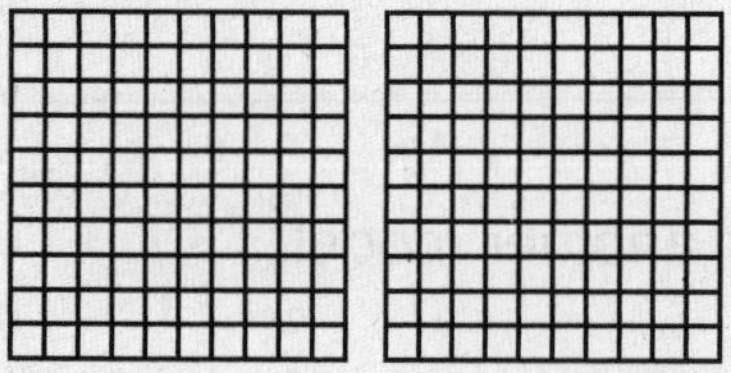

____

2. $\begin{array}{r} 0.4 \\ +0.7 \\ \hline \end{array}$

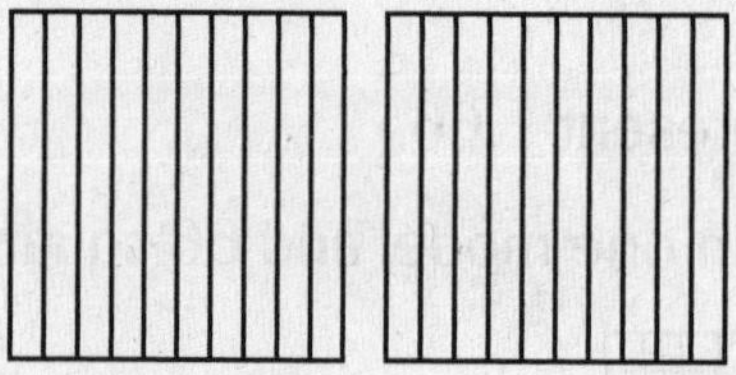

____

3. $\begin{array}{r} 0.25 \\ +0.07 \\ \hline \end{array}$

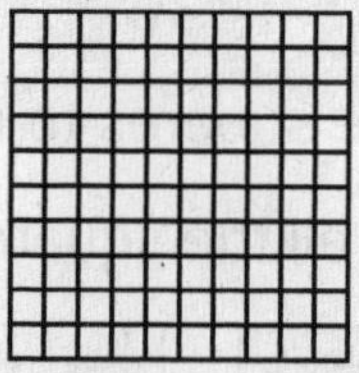

____

4. $\begin{array}{r} 1.05 \\ +0.78 \\ \hline \end{array}$

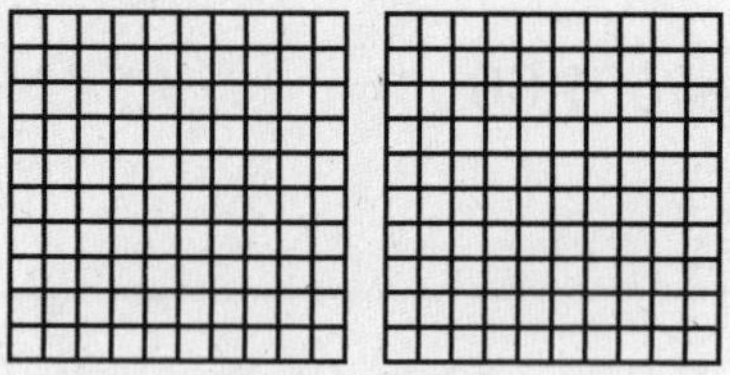

____

5. $\begin{array}{r} 0.38 \\ +1.93 \\ \hline \end{array}$

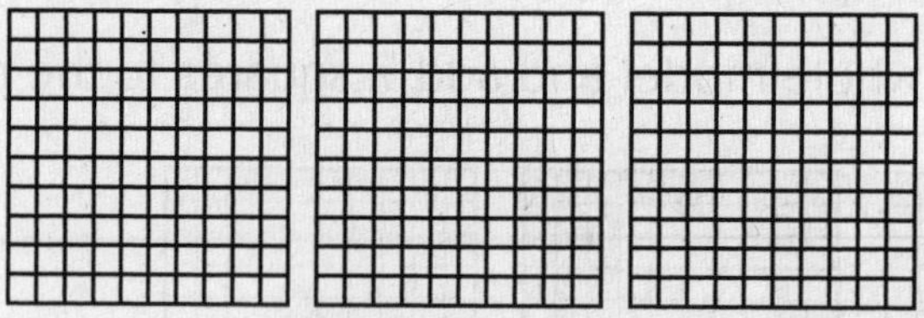

____

6. $\begin{array}{r} 0.44 \\ +1.08 \\ \hline \end{array}$

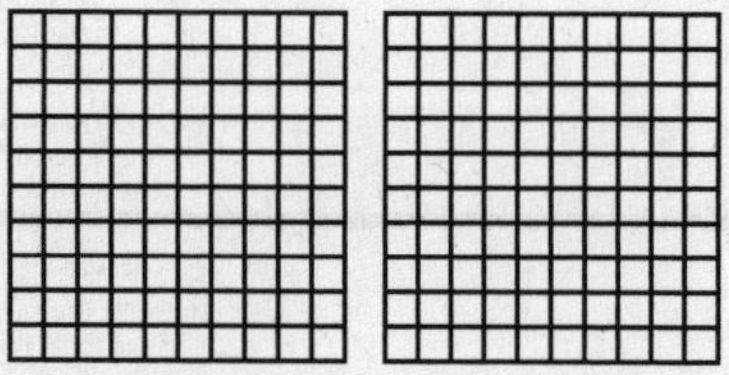

____

7. $\begin{array}{r} 1.06 \\ +0.67 \\ \hline \end{array}$

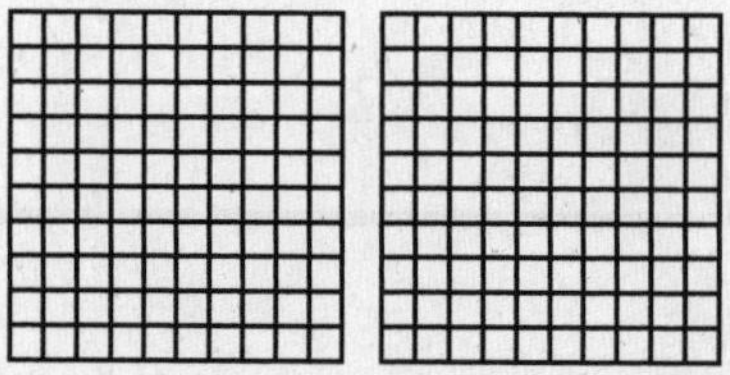

____

8. $\begin{array}{r} 0.16 \\ +1.55 \\ \hline \end{array}$

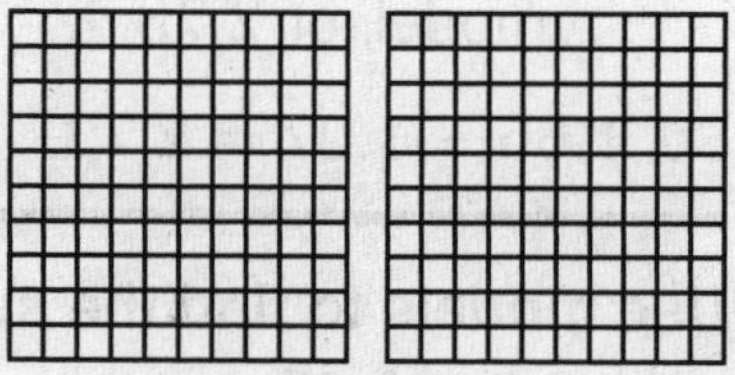

____

ALGEBRA Use the models to find the missing addend.

9.

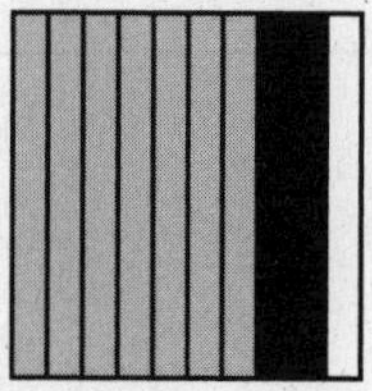

$0.7 + \blacksquare = 0.9$

10.

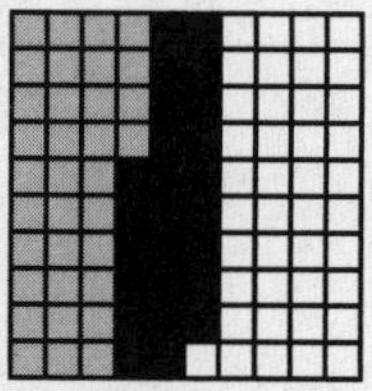

$0.34 + \blacksquare = 0.59$

Model Subtraction

You can use decimal models to help you subtract decimals.

Use a model to find the difference.

$$\begin{array}{r} 1.12 \\ -\ 0.45 \\ \hline \end{array}$$

Step 1: Shade squares to represent 1.12.

Think: Shade in all the squares in one model and 12 squares in another model.

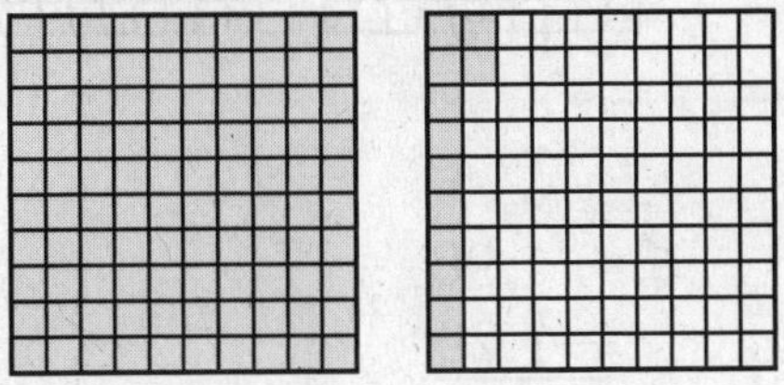

Step 2: Draw Xs to represent 0.45.

Think: Draw Xs on 45 squares of the shaded part.

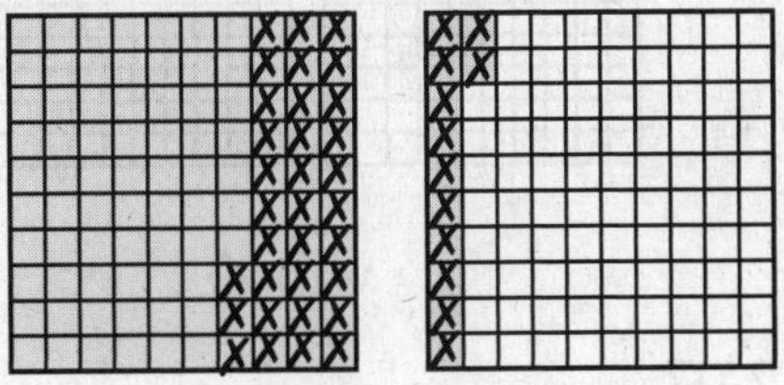

Step 3: Count all of the shaded squares that do not have Xs on them.

Think: There are 67 shaded squares that do not have Xs.

So, 1.12 − 0.45 = 0.67.

Use models to find the difference.

1. 1.4 − 0.61 ____________

2. 1.6 − 1.08 ____________

3.
$$\begin{array}{r} 0.84 \\ -\ 0.17 \\ \hline \end{array}$$
____

4.
$$\begin{array}{r} 1.39 \\ -\ 1.14 \\ \hline \end{array}$$
____

Name______________________________________ Lesson 19.4

Model Subtraction

Use models to find the difference.

1. 0.57 − 0.18

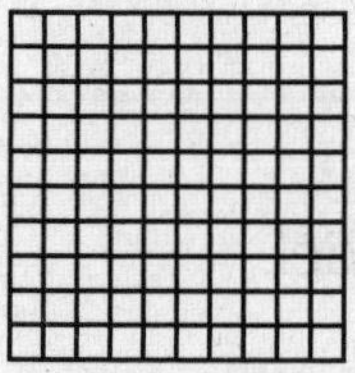

2. 0.7 − 0.3

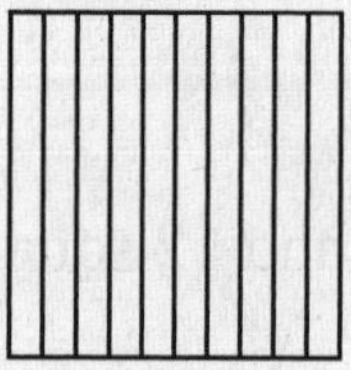

3. 1.07 − 0.42

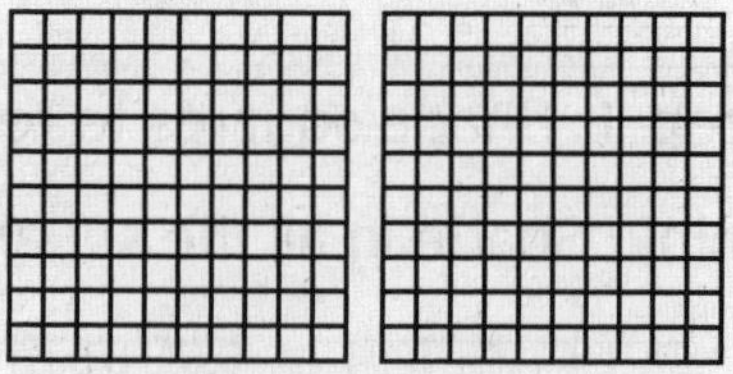

4. 1.09 − 0.90

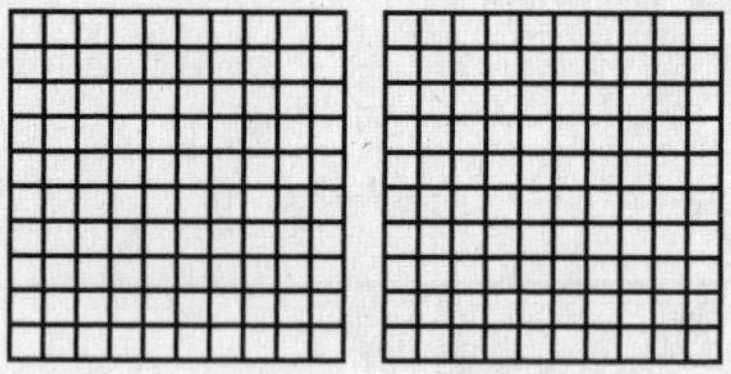

5. 1.00 − 0.63

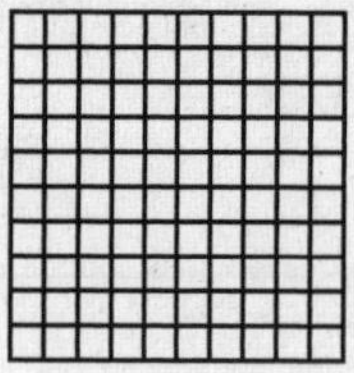

6. 1.98 − 1.29

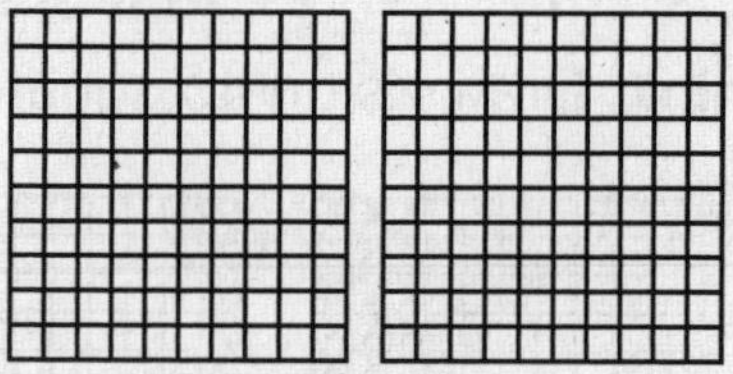

7. 2.73 − 1.79

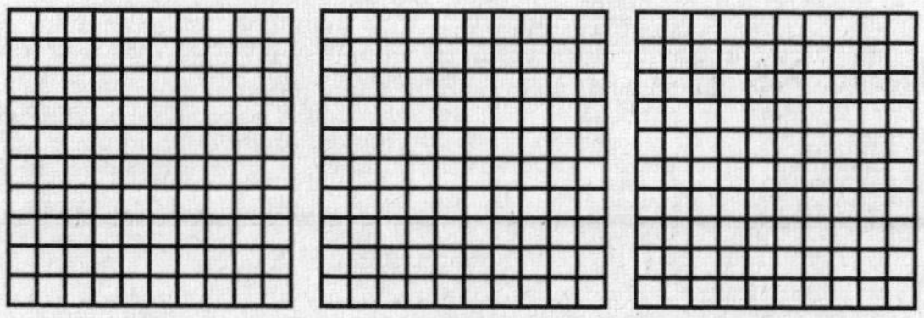

____

8. 2.92 − 2.07

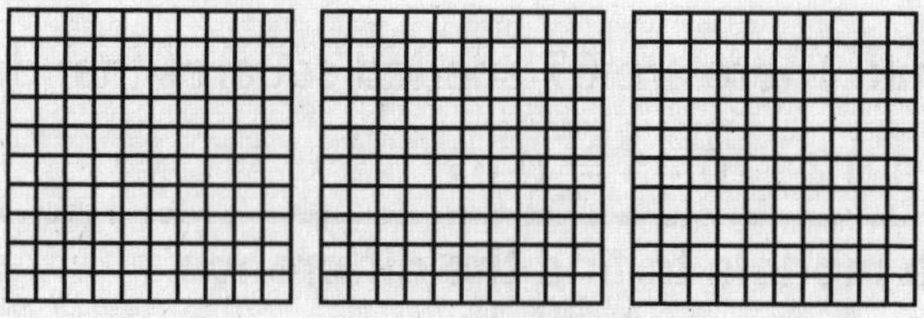

____

ALGEBRA **Use the models to find the missing number.**

9.

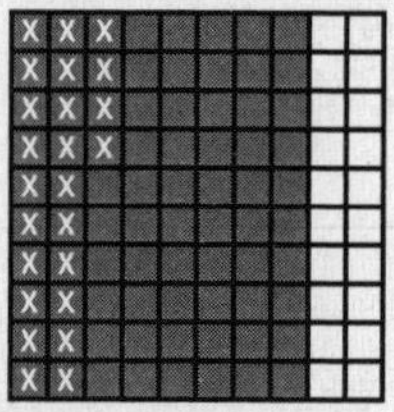

0.80 − ■ = 0.56

____

10.

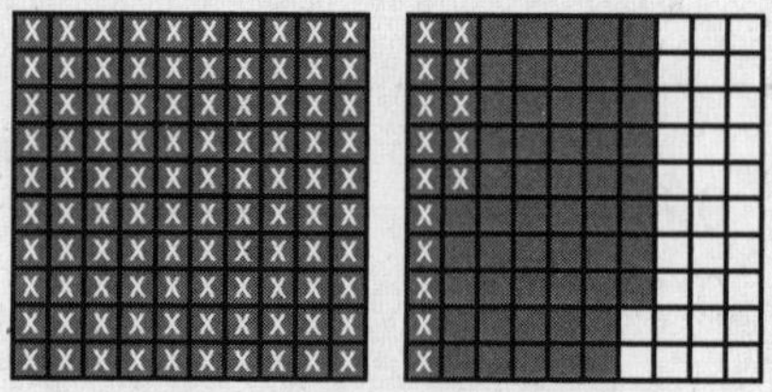

■ − 1.15 = 0.53

____

Name____ **Lesson 19.5**

Record Addition and Subtraction

Equivalent decimals can help add or subtract numbers that do not have the same number of digits after the decimal point.

Estimate. Then record the sum.

$23.44
+ $19.85

Estimate the sum by rounding to the nearest dollar.

$23.44 → $23
+ $19.85 → + $20
$43

Line up the decimal points and add to find the sum.

1 1
$23.44
+ $19.85
$43.29

So, $23.44 + $19.85 is $43.29.
Since $43.29 is close to the estimate of $43, it is reasonable.

Estimate. Then record the difference.

67.10
− 9.98

Estimate the difference by rounding to the nearest whole number.

67.1 → 67
− 9.98 → − 10
57

Line up the decimal points. Place zeros to the right of the decimal point so each number has the same number of digits after the decimal point.

67.10
− 9.98

Subtract as you would with whole number. Place the decimal point in the difference.

5 16 10 10
~~6~~~~7~~.~~1~~~~0~~
− 9.98
57.12

So, 67.1 − 9.98 = 57.12.
Since 57.12 is close to the estimate of 57, it is reasonable.

Estimate. Then record the difference.

1. 7.5 − 2.3	**2.** $4.55 + 1.97	**3.** 24.3 − 18.18	**4.** 71.7 + 2.34
5. 8.2 + 3.5	**6.** $45.36 − 19.55	**7.** 61.7 − 12.84	**8.** 52.99 + 65.20

Name ______________________________

Record Addition and Subtraction

Estimate. Then record the sum or difference.

1. $\begin{array}{r} 5.43 \\ -2.54 \\ \hline \end{array}$
2. $\begin{array}{r} 2.89 \\ +1.22 \\ \hline \end{array}$
3. $\begin{array}{r} \$41.32 \\ -\$37.44 \\ \hline \end{array}$
4. $\begin{array}{r} 2.29 \\ +1.53 \\ \hline \end{array}$

5. \$21.82 + \$13.09 ______
6. 42.14 + 24.36 ______
7. \$94.23 − \$65.44 ______
8. 57.22 − 53.88 ______

Compare. Write <, >, or = for each ◯.

9. \$5.15 + \$0.10 ◯ \$4.84 + \$0.35
10. 3.78 + 2.51 ◯ 9.54 − 3.30

ALGEBRA Find the missing decimals. The sums are given at the end of each row and the bottom of each column.

11.	13.06	4.12	22.77	☐
12.	67.77	☐	15.14	83.64
13.	0.98	73.22	☐	80.78
14.	☐	78.07	44.49	204.37

Problem Solving and Test Prep

15. Sierra had 8.25 feet of plastic wrap. Then she used 3.75 feet. How much does Sierra have left?

____

16. Lyle spent \$2.47 on peanut butter, \$3.56 on jelly, and \$2.37 on a loaf of bread. How much did Lyle spend in all?

____

17. Lauren saved \$9.25 of her allowance. Her best friend saved \$2.45 less than she did. How much did Lauren's best friend save?

A \$6.80
B \$11.70
C \$7.20
D \$5.90

18. Jason bought pants on sale for \$25.89. The original price was \$33.98. How much did Lyle save?

A \$8.25
B \$8.19
C \$8.11
D \$8.09

Name ________________________

Spiral Review

For 1–4, use the thermometer to find the temperature shown by each letter.

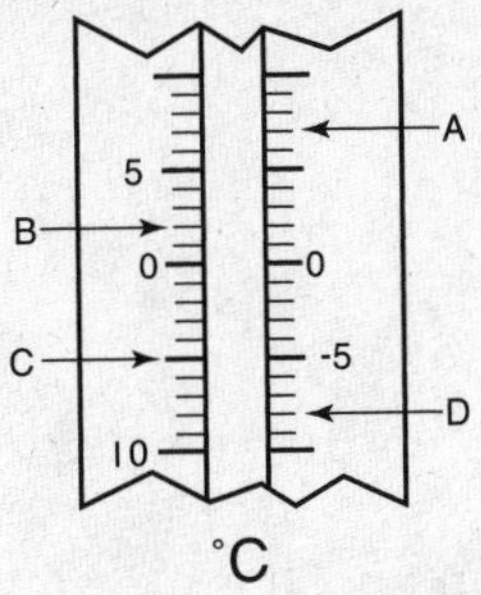

1. A ________

2. B ________

3. C ________

4. D ________

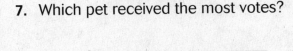

For 5–6, draw two examples of each quadrilateral in the box below.

5. It has no parallel sides.

6. It has 4 equal sides.

For 7–9, use the Favorite Pet graph.

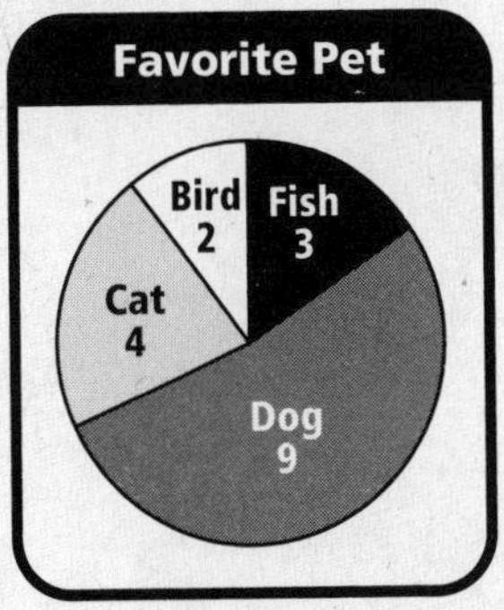

7. Which pet received the most votes?

____

8. Which pet received 4 votes?

____

9. How many people voted in all?

____

For 10–15, find the value of the variable. Then write a related sentence.

10. $9 \times c = 45$

____

11. $36 \div x = 6$

____

12. $a \div 3 = 7$

____

13. $6 \times 6 = n$

____

14. $t \div 9 = 9$

____

15. $4 \times r = 24$

____

Name ____________________

Problem Solving Workshop Strategy: Compare Strategies

The vending machine has 10 drinks left. Some are juice and some are water. There are more cans of juice left than bottles of water. What possible combinations of water and juice could be left in the machine?

Read to Understand

1. What are you asked to find?

____

Plan

2. What strategy could you use to solve this problem?

____

Solve

3. Show how you use the strategy from exercise 2 to solve the problem. What is your answer?

____

Check

4. How can you check to see if your answer is reasonable?

____

Choose a strategy to solve.

5. Jen has a red shirt, a blue shirt, and a green shirt. She has black pants and blue pants. how many different shirt-pant combinations can she make? Solve and explain the strategy you used.

____

6. The mail comes at 1:00 P.M. on Monday, 2:00 P.M. on Tuesday, 1:00 P.M. on Wednesday, and 2:00 P.M. on Thursday. What time will it come on Friday? Solve and explain the strategy you used.

____

Name____

Problem Solving Workshop Strategy: Compare Strategies

Problem Solving Strategy Practice

Predict and test or make a table to solve.

1. Dana will buy chips from a vending machine. The chips cost $2.45. Dana has 2 dollar bills, 3 quarters, 3 dimes, and 4 nickels. What are two different ways Dana can pay for the chips?

____

2. Victor has a $1 bill, 4 quarters, and 2 dimes. He will borrow some money from a friend to buy a bag of chips for $2.45. What coin or coins must he borrow in order to pay for the chips?

____

3. A sandwich costs $1.00 in a vending machine. How many different ways can you pay the exact amount in coins if you only have nickels and quarters?

____

4. Sugar-free gum costs $0.85 in a vending machine. If you have one quarter, how many dimes would you need to buy a pack of sugar-free gum?

____

Mixed Applications

USE DATA For 5–6, use the table.

5. Tanya spent $9.80 at the pool. What did Tanya pay for?

____

6. Libby paid for herself and two sisters to go to the pool. She also bought 3 towels and a bathing cap. How much did Libby spend?

____

Community Center Pool	
Item	**Prices**
Entrance Fee	$1.50
Bathing Cap	$2.75
Towel	$5.55

7. Henry had the exact change to pay for a $0.50 pencil. He paid with 6 coins. What could those coins be?

____

8. In Exercise 1, how much money will Dana have left over after she buys the chips?

____

Name____

Points, Lines, and Rays

Term	Definition	Example
point	names an exact location in space	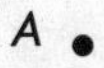
line segment	part of a line; has two endpoints.	
line	straight path of points that continues without end in both directions	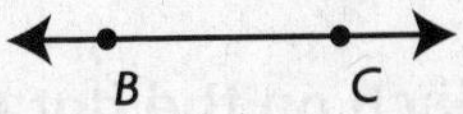
ray	part of a line that has one endpoint and continues without end in one direction	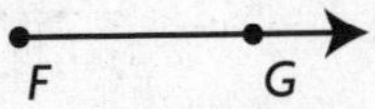
plane	flat surface that continues without end in all directions	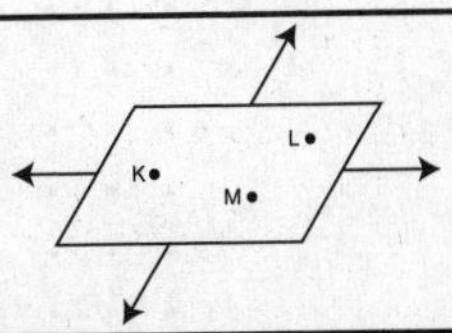

Name a geometric term that best represents the object.

flagpole

- A flagpole does not continue without end in both directions. The tops and bottom of the pole are like endpoints.
- So, a flagpole is like a line segment.

laser beam

- A laser beam comes out of a box or pointer and shines without end in one directlon.
- So, a laser beam is like a ray.

Name a geometric term that best represents the object.

1. highway ____________

2. tip of a marker ____________

3. screwdriver ____________

4. football field ____________

5. grain of salt ____________

6. two way arrow ____________

7. meadow ____________

8. piece of rope ____________

9. flashlight beam ____________

10. extension cord ____________

11. farm field ____________

12. railroad track ____________

Name____

Points, Lines, and Rays

Name the geometric term that best represents the object.

1. top of a desk ____________

2. chalk tray ____________

3. a point from Earth into space ____________

4. NNE on a compass ____________

Name an everyday object that represents the term.

5. point ____________

6. ray ____________

7. line segment ____________

8. plane ____________

Draw and label an example of each on the dot paper.

9. plane *ABC*

10. line segment *DE*

11. ray *FG*

12. point *H*

Problem Solving and Test Prep

USE DATA For 13–16, use the photograph.

13. What geometric term describes the place where the ceiling meets a wall?

____

14. What features in the hallway show planes?

____

15. What geometric term best describes the arrow?

A line
B line segment
C point
D ray

16. Which geometric term best describes the black dot on the window?

A line
B line segment
C point
D ray

Name____

Classify Angles

An **angle** is formed when two rays share the same endpoint, or **vertex**. Angles are measured in degrees (°). A **right angle** measures exactly 90°. An **acute angle** always measures *less* than 90°. An **obtuse angle** always measures *more* than 90°.

Classify the angle as *acute, right,* or *obtuse*.

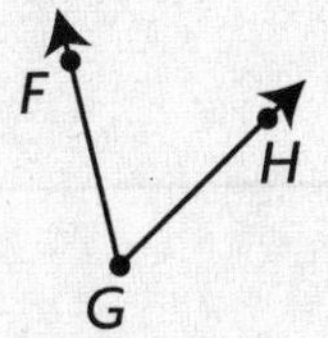

- ∠*FGH* appears to be less than 90°. Any angle between 0° and 89° is classified as acute.
- So, ∠*FGH* is an acute angle.

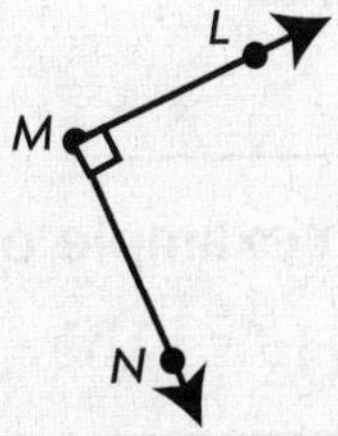

- ∠*LMN* measures 90°. An angle that measures exactly 90° is a right angle.
- So, ∠*LMN* is a right angle.

Classify each angle as *acute, right,* or *obtuse*.

1. 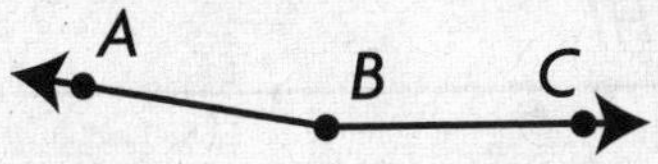

____

2. 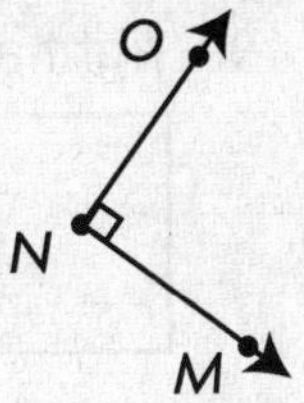

____

3. 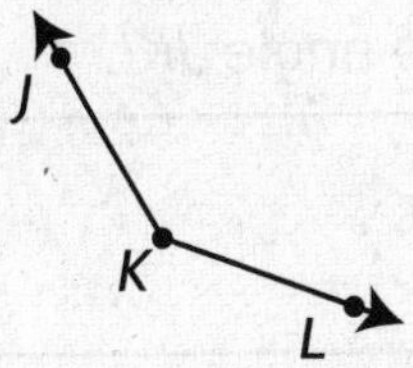

____

4. 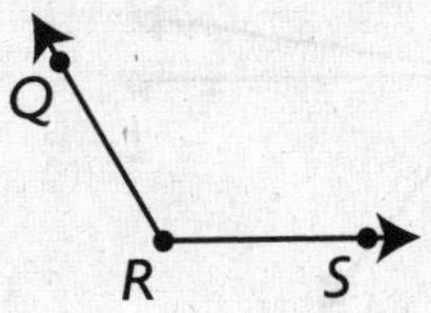

____

5. 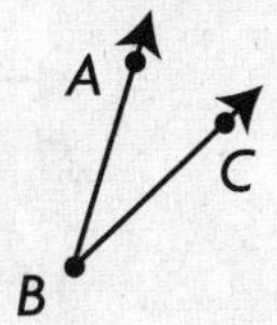

____

6. 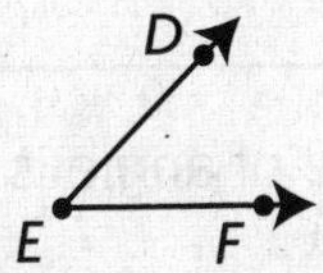

____

MG 3.5 Know the definitions of a right angle, an acute angle, and an obtuse angle. Understand that 90°, 180°, 270°, and 360° are associated, respectively, with $\frac{1}{4}$, $\frac{1}{2}$, $\frac{3}{4}$, and full turns.

Name ____________________

Classify Angles

Classify each angle as *acute, right,* or *obtuse.*

1.

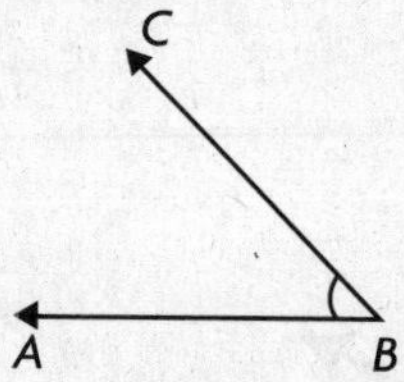

____

2.

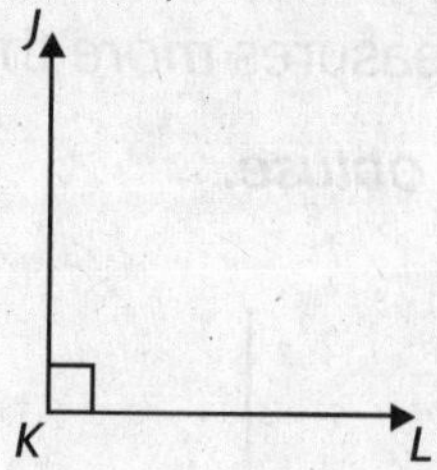

____

3.

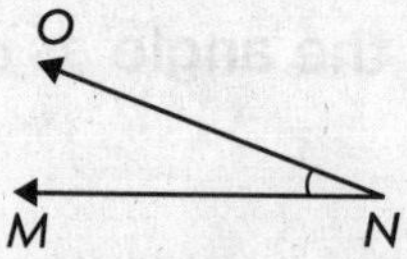

____

Draw and label an example of each.

4. acute angle *PQR*

5. obtuse angle *STU*

6. right angle *DEF*

7. acute angle *XYZ*

8. obtuse angle *JKL*

9. right angle *GHI*

Problem Solving and Test Prep

USE DATA For 10–11, use the angles shown.

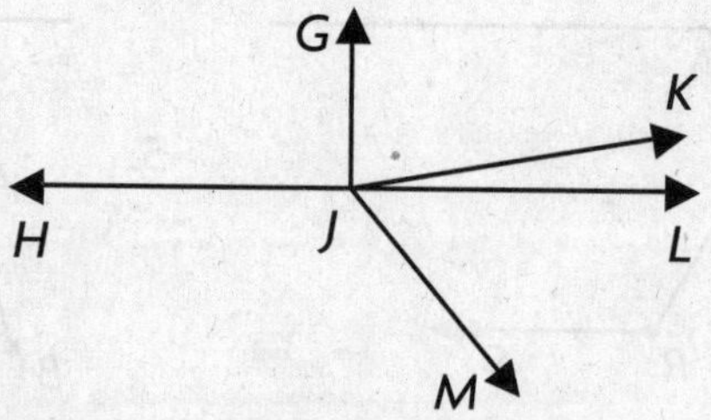

10. Which angles appear to be acute?

____

11. What type of angle is angle *HJM*?

____

12. At what time do the hands on a clock represent a right angle?

A 9:15 **C** 9:00

B 11:30 **D** 6:00

13. Which is the measure of a right angle?

A 45° **C** 110°

B 90° **D** 180°

Name ______________________________________ **Lesson 20.3**

Line Relationships

Pairs of lines may be **intersecting, parallel**, or **perpendicular**.

Type of Line	Definition	Example
intersecting	lines that cross each other at one point and form four angles	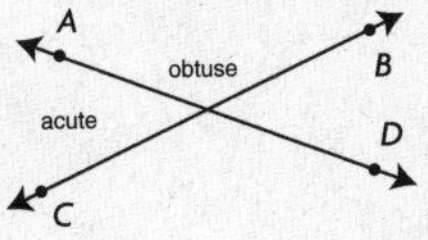
parallel	lines that never intersect and are always the same distance apart	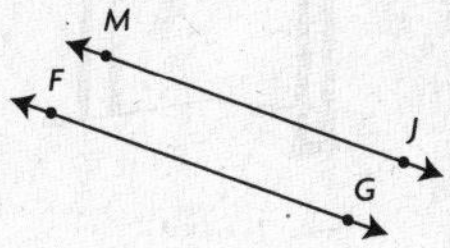
perpendicular	lines that intersect and form four right angles	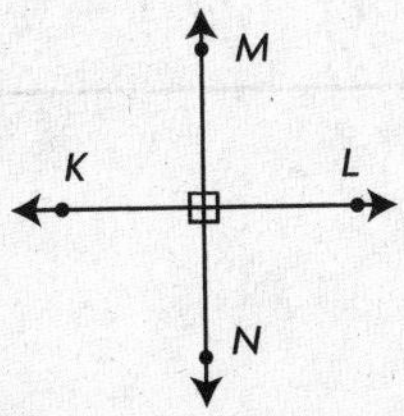

Name any line relationships you see in the figure below. Write *intersecting, parallel*, or *perpendicular*.

- The lines dividing the colored stripes are parallel lines.
- The lines that form the top of the kite are intersecting lines.
- The lines where the kite poles intersect the stripes are perpendicular lines.

Name any line relationships you see in each figure. Write *intersecting, parallel*, or *perpendicular*.

1. 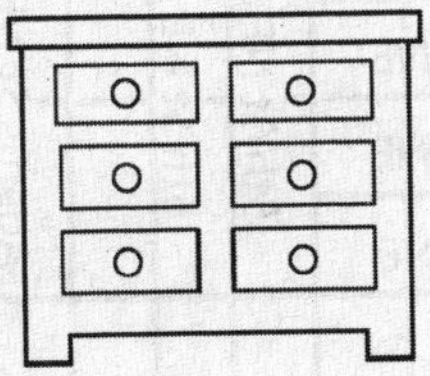

____

2. 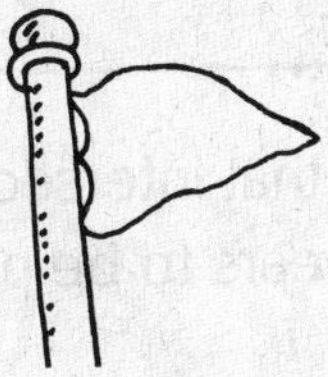

____

3. 

____

4. 

____

Name ______________________________

Line Relationships

Name any line relationships you see in each figure. Write *intersecting, parallel,* or *perpendicular.*

1. 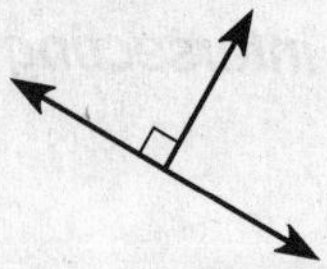

____

2. 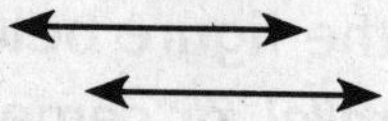

____

3. 

____

4. 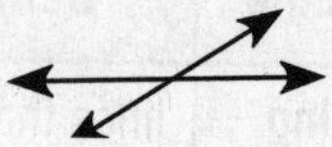

____

5. 

____

6. 

____

7. 

____

8. 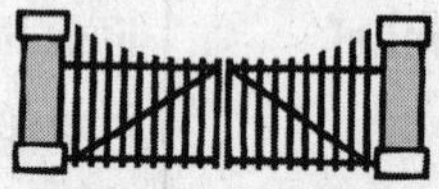

____

Problem Solving and Test Prep

USE DATA For 9–10, use the map.

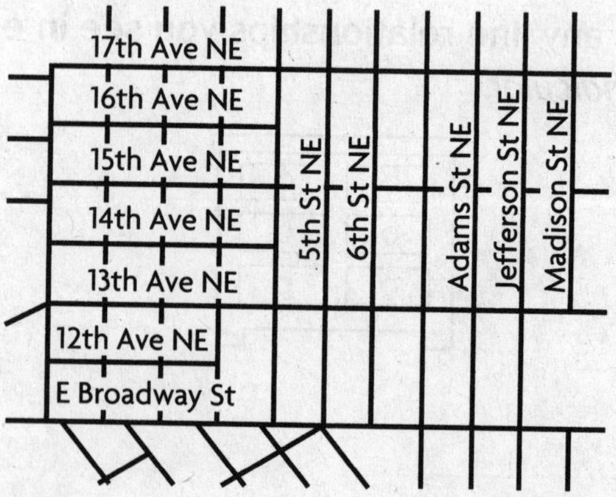

9. Name a street that appears to be parallel to E Broadway St.

____

10. Name a street that intersects Madison St. NE and appears to be parallel to 15th Ave. NE.

____

11. Which best describes intersecting lines?

A They never meet.

B They form four angles.

C They form only obtuse angles.

D They form only acute angles.

12. Which best describes parallel lines?

A They never meet.

B They form four angles.

C They form only obtuse angles.

D They form only acute angles.

Name ______________________________ **Week 31**

Spiral Review

For 1–3, write two equivalent fractions for each point.

1.

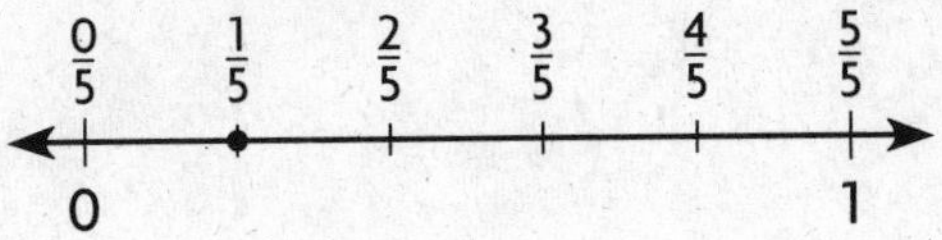

____

2.

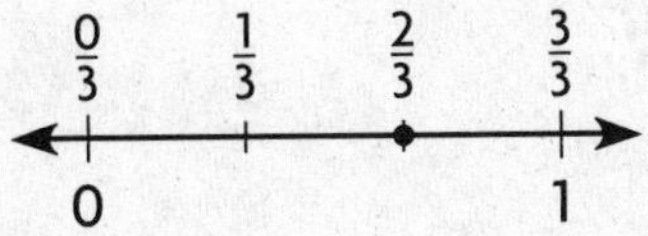

____

3.

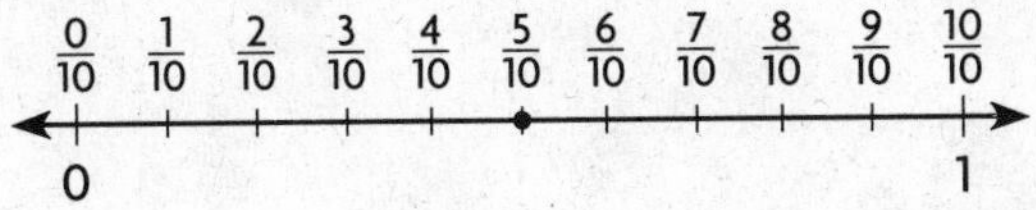

____

For 5–7, use the graph.

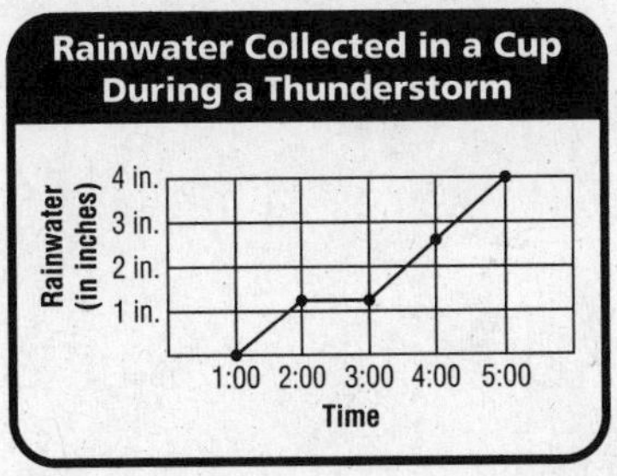

5. How many inches of rain were in the cup at 5:00?

____

6. At what time did the cup contain 3 inches of rainwater?

____

7. What can you conclude about the rainfall between 2:00 and 3:00?

____

For 4, make a table using values of 1 through 5 for *x*. Then graph the equation.

4. $y = x + 1$

x	1	2	3	4	5
y					

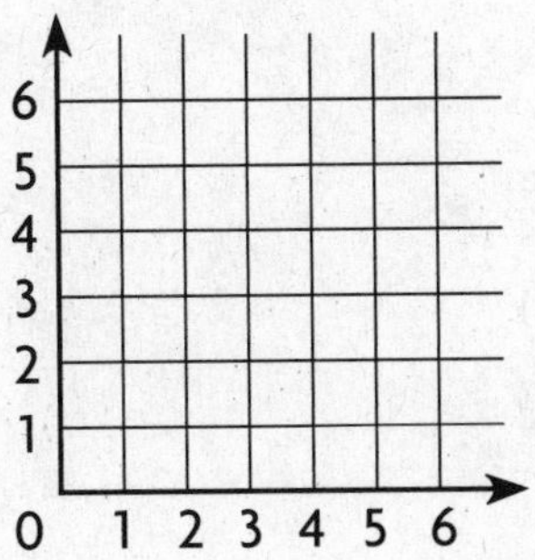

For 8–11, use the grid below. Write the ordered pair for each point.

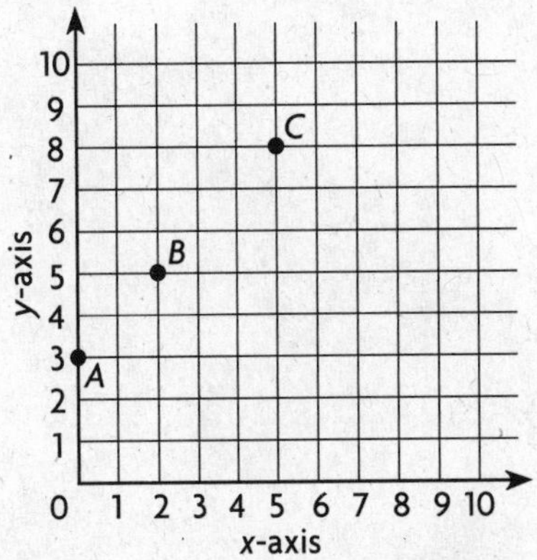

8. *A* ______________

9. *B* ______________

10. *C* ______________

11. What is the rule? ______________

Name ________________________________ **Lesson 20.4**

Polygons

Polygons are closed figures that are formed by 3 or more straight sides and connected by a line segments. A **regular polygon's** sides are all the same length, and its angles are all the same measure.

Polygon	Sides and Angles	Example
triangle	3 sides 3 angles	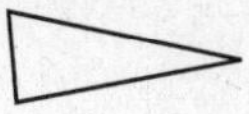
quadrilateral	4 sides 4 angles	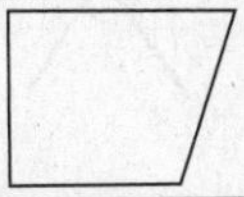
pentagon	5 sides 5 angles	
hexagon	6 sides 6 angles	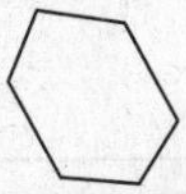
octagon	8 sides 8 angles	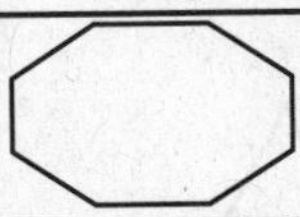
decagon	10 sides 10 angles	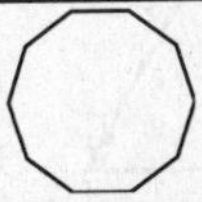

Name the polygon. Tell whether it appears to be *regular* or *not regular*.

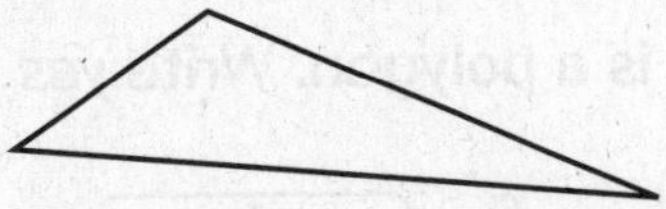

- The polygon has 3 sides, so it is a triangle.
- The sides of the triangle are not all the same length, so the triangle is not regular.

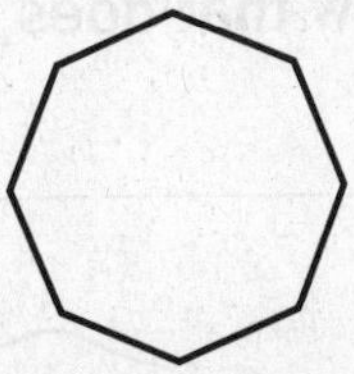

- The polygon has 8 sides, so it is an octagon.
- The angles of the octagon are all the same measure, so the octagon is regular.

Name the polygon. Tell whether it appears to be *regular* or *not regular*.

1. 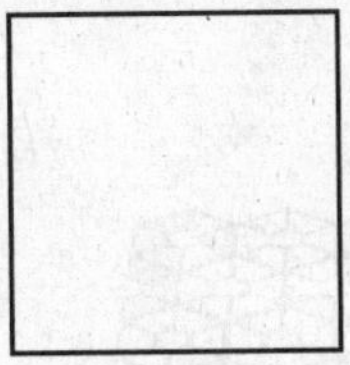

____

2. 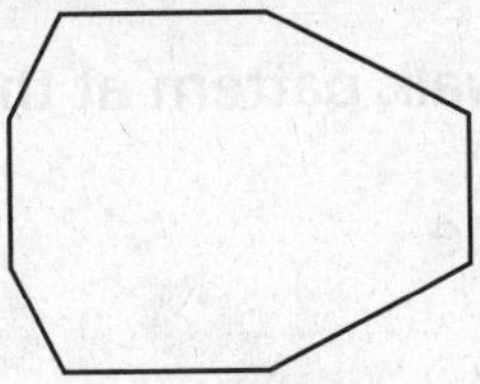

____

3. 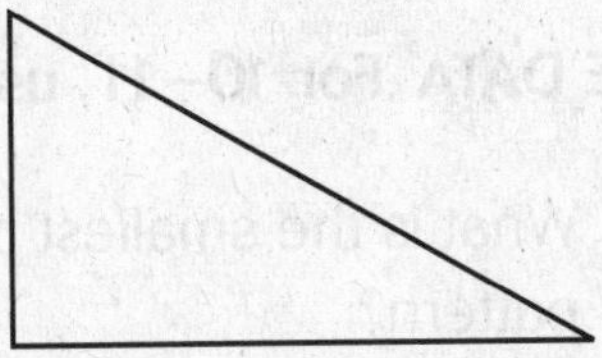

____

4. 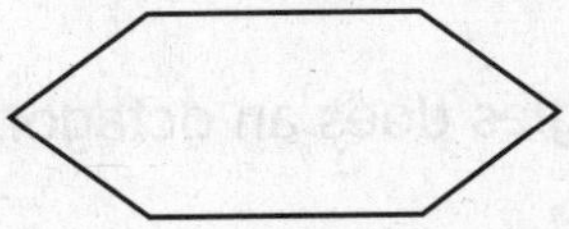

____

5. 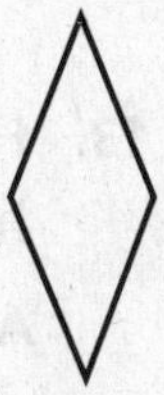

____

6. 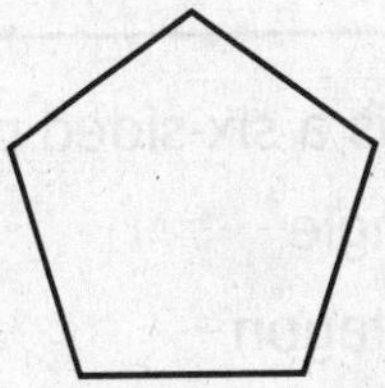

____

Name______________________________

Polygons

Name the polygon. Tell if it appears *regular* or *not regular*.

1. 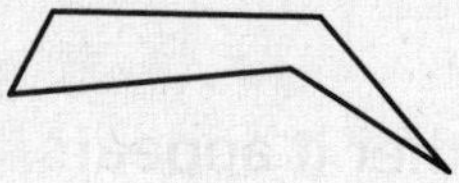 ____________

2. 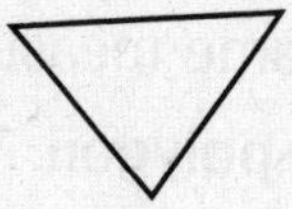 ____________

3.  ____________

4. 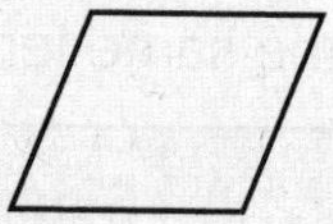 ____________

Tell if each figure is a polygon. Write *yes* or *no*.

5. 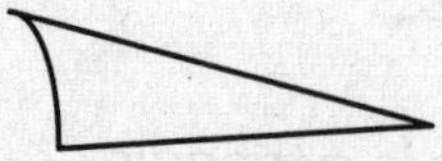 ____________

6.  ____________

7.  ____________

8. 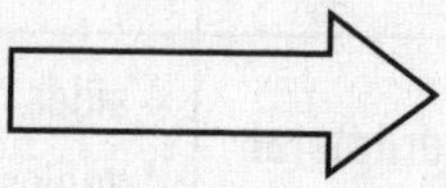 ____________

Choose the figure below that does not belong. Explain.

____

9.

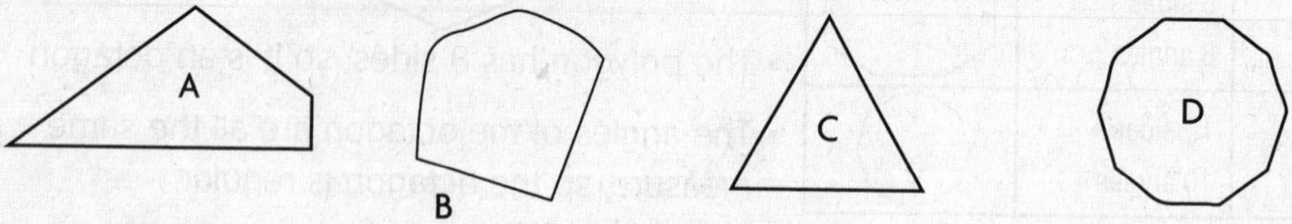

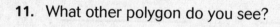

Problem Solving and Test Prep

USE DATA For 10–11, use the sidewalk pattern at the right.

10. What is the smallest polygon in the pattern?

____

11. What other polygon do you see?

____

12. Which is a six-sided polygon?

A triangle
B pentagon
C octagon
D hexagon

13. How many angles does an octagon have?

A 8
B 9
C 10
D 7

Name____ **Lesson 20.5**

Classify Triangles

Triangle	Description	Example
equilateral	3 equal sides	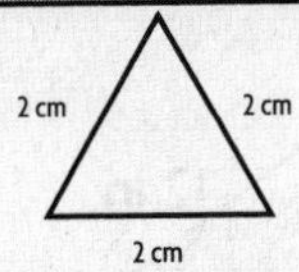
isosceles	two equal sides	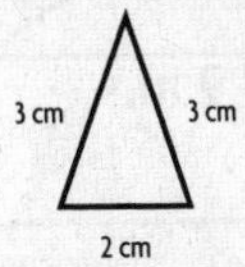
scalene	no equal sides	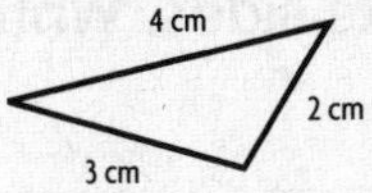
right	1 right angle	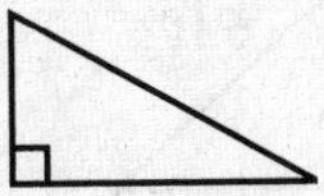
acute	3 acute angles	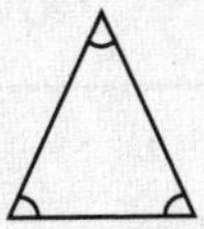
obtuse	1 obtuse angle	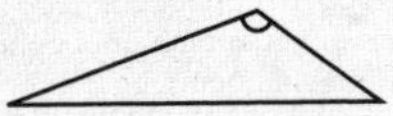

Classify each triangle. Write *isosceles, scalene,* or *equilateral.* Then write *right, acute,* or *obtuse.*

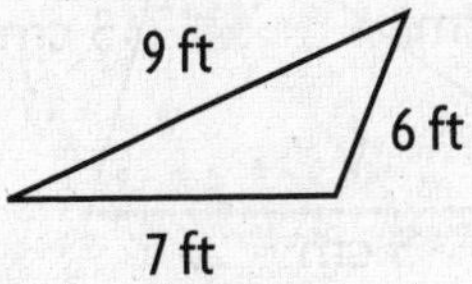

- Each side is a different length, so the triangle is scalene. There is one obtuse angle, so the triangle is obtuse.

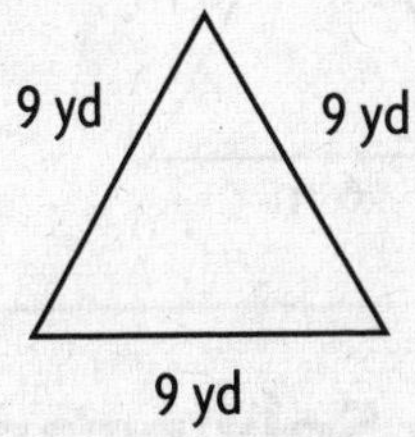

- All three sides are equal, so the triangle is equilateral. There are 3 acute angles, so the triangle is acute.

Classify each triangle. Write *isosceles, scalene,* or *equilateral.* Then write *right, acute,* or *obtuse.*

1.

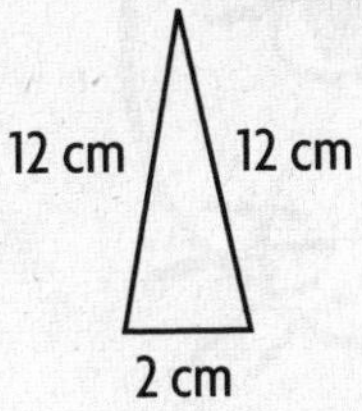

____

2.

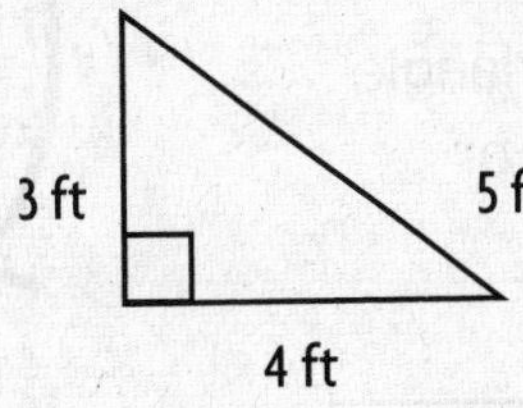

____

3.

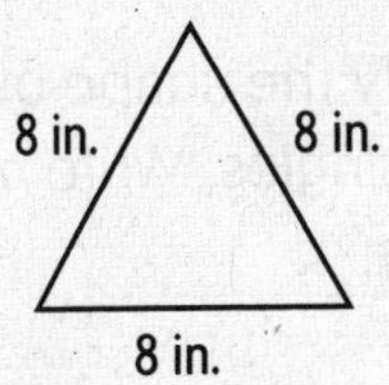

____

4.

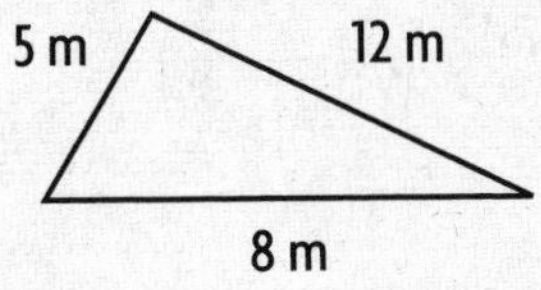

____

5.

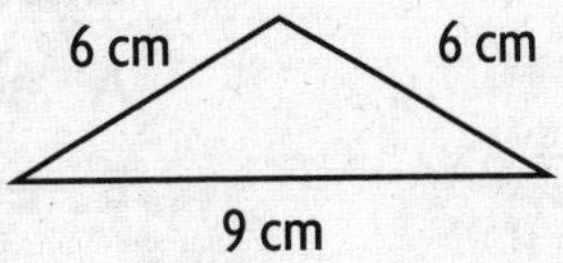

____

6.

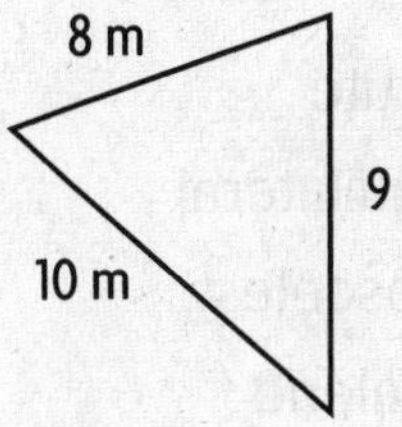

____

Name____

Classify Triangles

Classify each triangle. Write *isosceles, scalene,* or *equilateral.* Then write *right, acute,* or *obtuse.*

1.

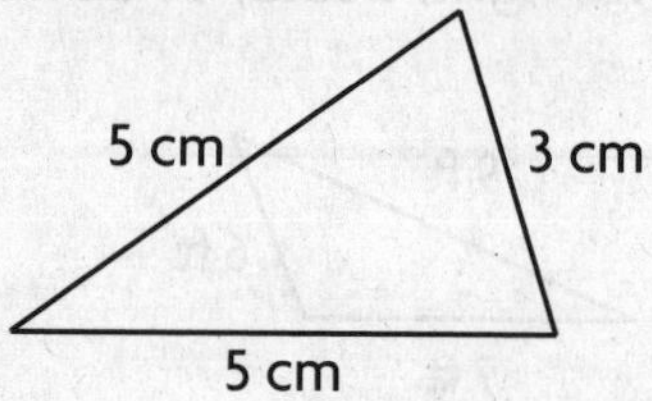

____

2.

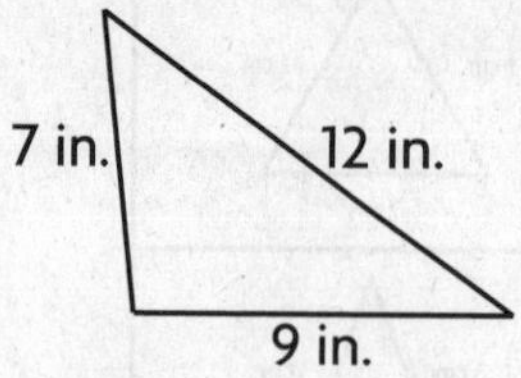

____

3.

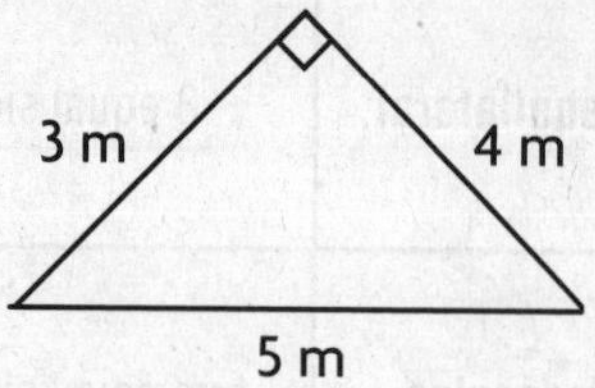

____

Classify each triangle by the lengths of its sides. Write *isosceles, scalene,* or *equilateral.*

4.

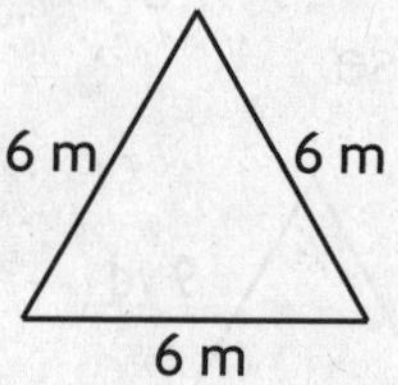

____

5.

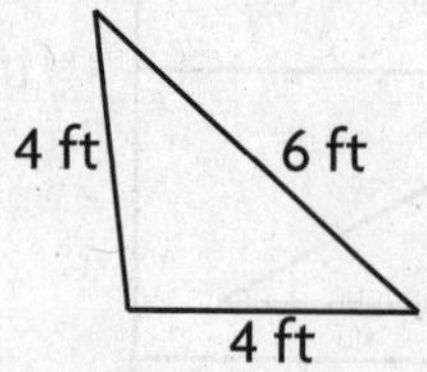

____

6.

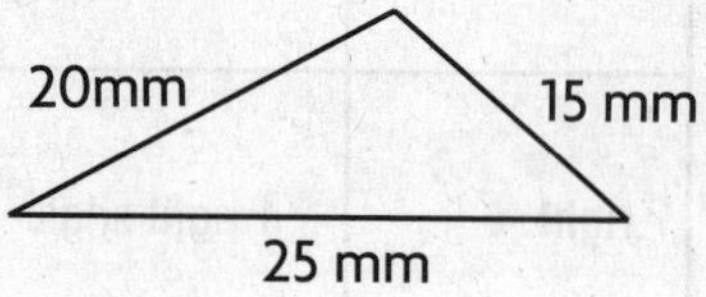

____

Problem Solving and Test Prep

USE DATA For 7–8, use the drawing.

7. Classify the shape of the gray triangle by the length of its sides. Write *isosceles, scalene,* or *equilateral.*

   ____

8. Classify the shape of the gray triangle by its angles. Write *right, acute,* or *obtuse.*

   ____

9. What kind of triangle has 2 equal sides?

   A acute

   B equilateral

   C isosceles

   D scalene

10. What kind of triangle has no equal sides?

    A acute

    B equilateral

    C isosceles

    D scalene

Name____ **Lesson 20.6**

Classify Quadrilaterals

A **quadrilateral** is a polygon with 4 sides and 4 angles.

Quadrilateral	Description
parallelogram	- 2 pairs of parallel sides - opposite sides equal
square	- 2 pairs of parallel sides - 4 equal sides - 4 right angles
rectangle	- 2 pairs of parallel sides - opposite sides equal - 4 right angles
rhombus	- 2 pairs of parallel sides - 4 equal sides
trapezoid	- exactly 1 pair of parallel sides

Classify the figure in as many of the following ways as possible. Write *quadrilateral, parallelogram, rhombus, rectangle, square,* or *trapezoid.*

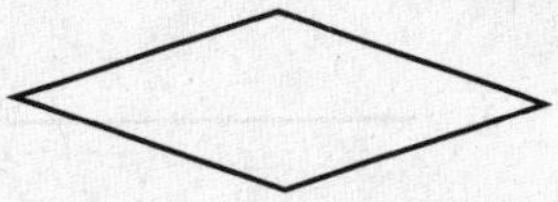

- The figure has 4 sides, so it is a quadrilateral. It has 2 pairs of parallel sides, so it is a parallelogram. All 4 sides are equal, so it is a rhombus.

Draw an example of a quadrilateral that has no parallel sides

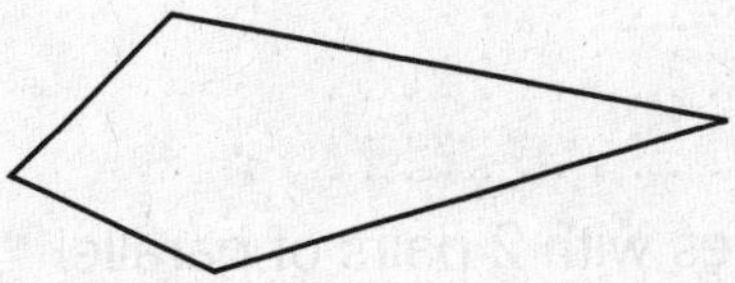

- The figure has 4 sides but none of the sides are parallel.

Classify the figure in as many of the following ways as possible. Write *quadrilateral, parallelogram, rhombus, rectangle, square,* or *trapezoid.*

1. 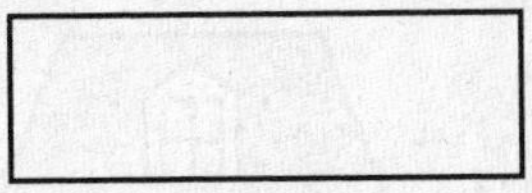

____

2. 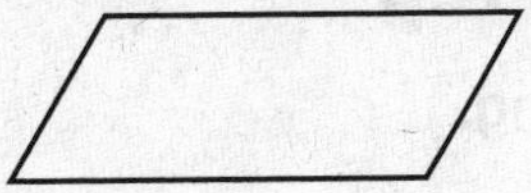

____

3. 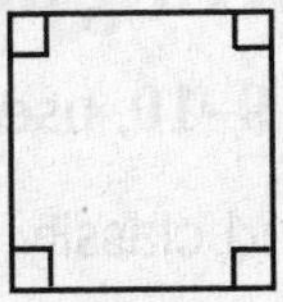

____

Draw an example of each quadrilateral.

4. It has 2 pairs of parallel sides.

5. It has 2 equal sides.

6. It has only 1 pair of parallel sides.

Name____ Lesson 20.6

Classify Quadrilaterals

Classify each figure in as many of the following ways as possible. Write *quadrilateral, parallelogram, rhombus, rectangle, square,* or *trapezoid.*

1.  **2.**  **3.** 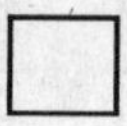 **4.** 

________________ ________________ ________________ ________________

________________ ________________ ________________ ________________

Draw and label an example of each quadrilateral.

5. 2 pairs of parallel sides and opposite sides equal

6. 4 equal sides with 4 right angles

7. 4 equal sides with 2 pairs of parallel sides

8. no pairs of parallel sides

Problem Solving and Test Prep

USE DATA For 9–10, use the drawing.

9. Describe and classify the roof of the Victorian dollhouse.

____

10. What are the different ways to classify the windows?

____

11. Which is the best description of the figures shown below?

A parallelograms **C** rectangles
B quadrilaterals **D** trapezoids

12. Which is the best description of the figures?

A parallelograms **C** rectangles
B quadrilaterals **D** trapezoids

Name____ Lesson 20.7

Circles

A **circle** is a closed, round plane figure. Circles have a **center, diameters, radii,** and **chords.**

Part of Circle	Description	Example
chord	a line segment whose two endpoints are on the circle	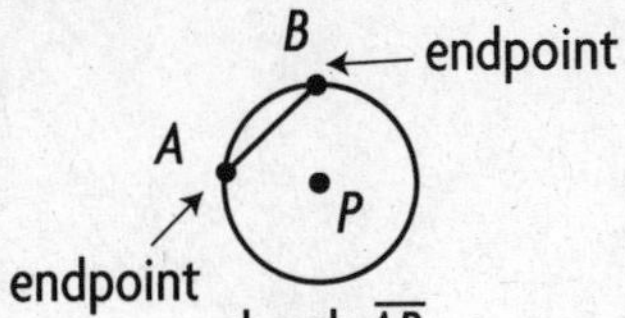 chord: $\overline{AB}$
diameter	a chord that passes directly through the center of the circle	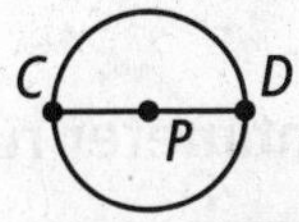 diameter: $\overline{CD}$
radius	a line segment with one endpoint in the center and one on the circle	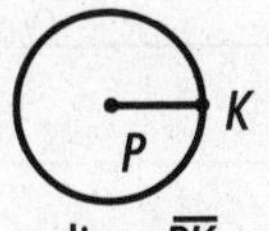 radius: $\overline{PK}$

Construct circle *J* with a 2-centimeter radius. Label each of the following.

a. chord *EF* **b. diameter *BC***

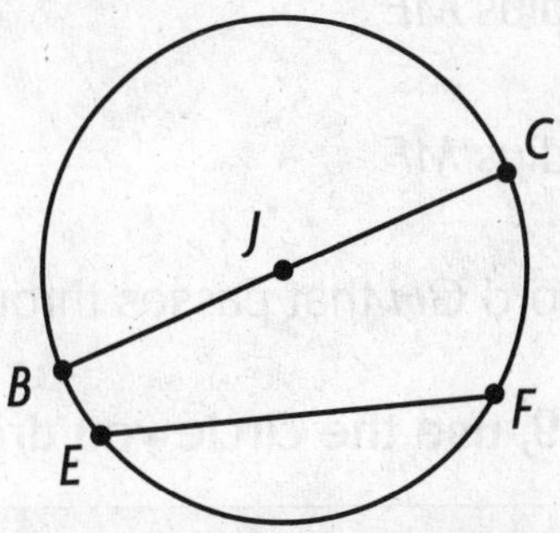

• Circle *J* has chord *EF* on it. Chord *EF* is a line segment whose two endpoints are on the circle. Diameter *BC* passes directly through the center of the circle. Radius *JC* is 2 cm.

Draw circle *P* with a 3-centimeter radius. Label each of the following.

1. chord *MK*
2. radius *PB*
3. diameter *AB*

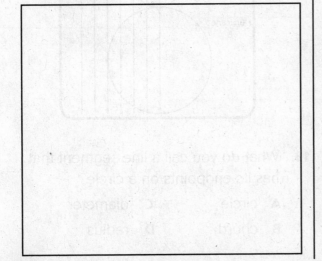

Draw circle *L* with a 4-centimeter radius. Label each of the following.

4. chord *XY*
5. radius *LA*
6. diameter *CN*

Name____

Circles

In the space provided, construct circle *M* with a radius of 2 cm. Label each of the following.

1. chord *AB*

2. diameter *CD*

3. radius *ME*

4. radius *MF*

5. chord *GH* that passes through the center

For 6–9, use the circle you drew and a centimeter ruler to complete the table.

	Name	Part of Circle	Length in cm
6.	*ME*		
7.	*CD*		
8.	*AB*		
9.	*GH*		

Problem Solving and Test Prep

USE DATA For 10–11, use the diagram.

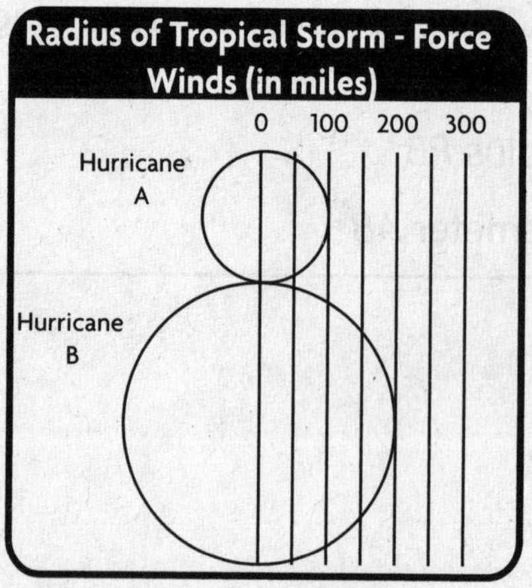

10. What is the diameter of hurricane A in miles?

____

11. What is the radius of hurricane B in miles?

____

12. What is the length of the diameter of a circle with a radius of 6 cm?

 A 3 cm C 9 cm
 B 6 cm D 12 cm

13. What do you call a line segment that has its endpoints on a circle?

 A circle C diameter
 B chord D radius

Name ______________________________ **Week 32**

Spiral Review

For 1–6, find the sum or difference.

1. 16,733 + 67,001

2. 370,400 + 466,989

3. 98,532 − 77,226

4. 200,000 − 100,060

5. 900,040 − 200,020

6. 890,000 + 267,600

For 7–8, find the length of each line segment.

7.

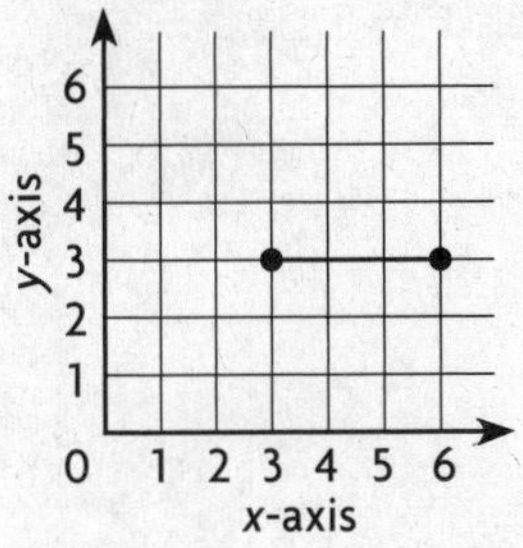

____

8.

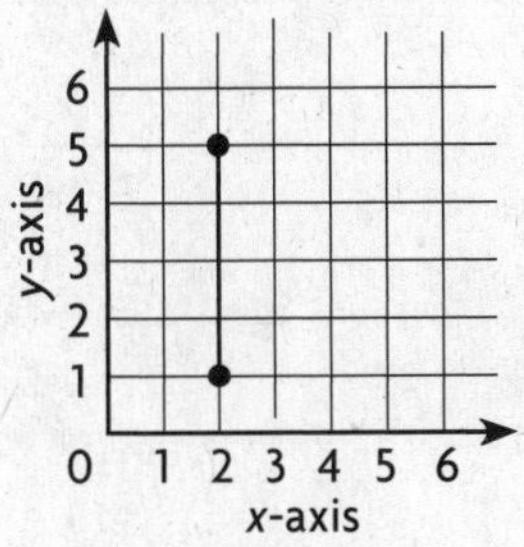

____

For 9, use the data to make a line graph.

Filling Up Mario's Swimming Pool				
Time	**1:00**	**4:00**	**6:00**	**8:00**
Water in feet	1	3	5	7

9.

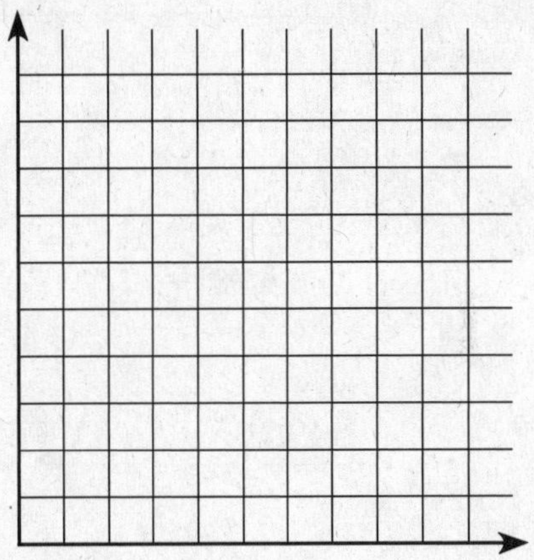

For 10–14, tell whether each equation is true. If not, explain why.

10. $(11 - 6) \times 3 = 5 \times 3$

____

11. $(9 - 4) \times 5 = (3 + 2) \times 5$

____

12. $(1 \times 6) \times 6 = (2 + 3) \times 6$

____

13. $(36 \div 6) \times 2 = (49 \div 7) \times 2$

____

14. $(64 \div 8) \times 2 = (56 \div 7) \times 2$

____

Name____

Lesson 20.8

Problem Solving Workshop Strategy: Use Logical Reasoning

George wants to design a swimming pool that has at least two sets of parallel sides and at least two sets of equal sides. Identify which figures shown below appear to be like George's design?

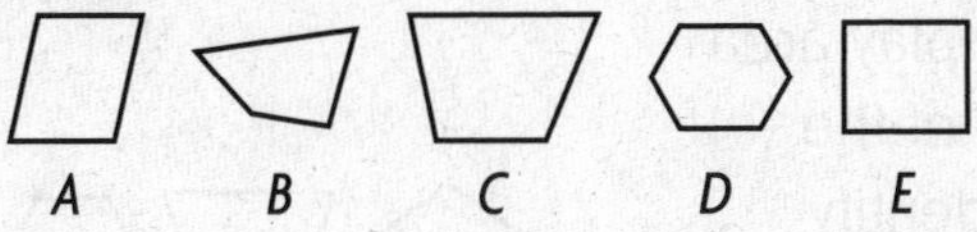

Read to Understand

1. What are you trying to find out in this problem?

____

Plan

2. How can logical reasoning strategy help solve this problem?

____

Solve

3. Complete the Venn diagram. Write the letter of the figures that fit in each category. Figure *A* has been done for you.

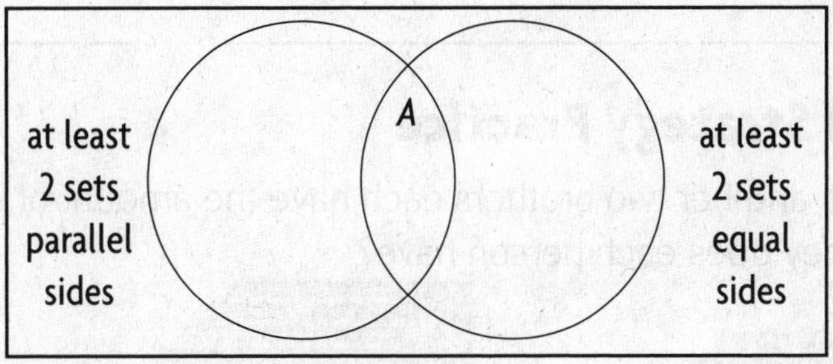

4. What is the solution to this problem?

____

Check

5. How can you check that your answer is correct?

____

Solve.

6. Pete's sandbox has at least 1 right angle. Which figure pictured above looks like the sandbox?

____

7. A piece of fabric has at least 1 set of parallel sides and at least 2 obtuse angles. Which figures pictured above would look like the fabric?

____

Name____________________ **Lesson 20.8**

Problem Solving Workshop Strategy: Use Logical Reasoning

Problem Solving Strategy Practice

For 1–3, use the figures at the right.

1. Lenny's parents put a play area in their backyard. All the sides of the play area are of equal length and none of the angles are acute or square. Identify the figure shown that appears to be like Lenny's play area?

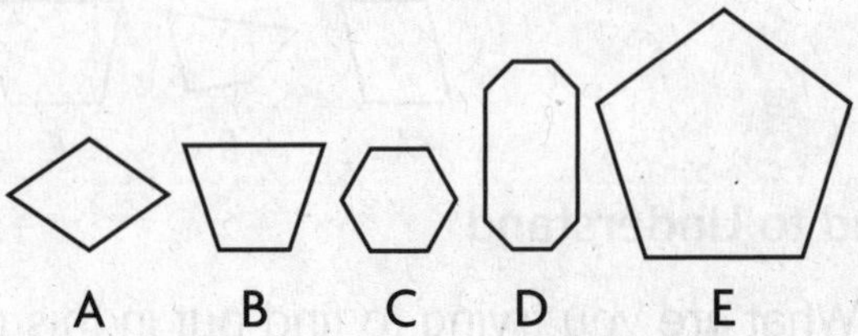

A B C D E

____

2. Cyd is designing a garden that has no parallel sides and all obtuse angles. Identify the figure shown that appears to be like Cyd's design.

____

3. The shape of Holly's backyard has two parallel sides and two acute angles. Identify the figure shown that appears to be like Holly's backyard.

____

Mixed Strategy Practice

4. Willa and her two brothers each have the amount of money shown below. How much money does each person have?

Willa

Bob

Jon

____

5. After Della tossed coins into a pool, James dove in to pick up the quarter. Then Della dove in to pick up her remaining 30 cents. How much money did Della toss into the pool? ____________________

6. Han's backyard was shaped like a square with all right angles. Classify the shape in as many ways as possible.

____

Name____ **Lesson 21.1**

Congruent Figures

Figures that have the exact same size and shape are **congruent**. Figures are not congruent if they have the same shape but are a different size, or if they have the same size but are a different shape.

Tell whether the two figures are ***congruent*** **or** ***not congruent.***

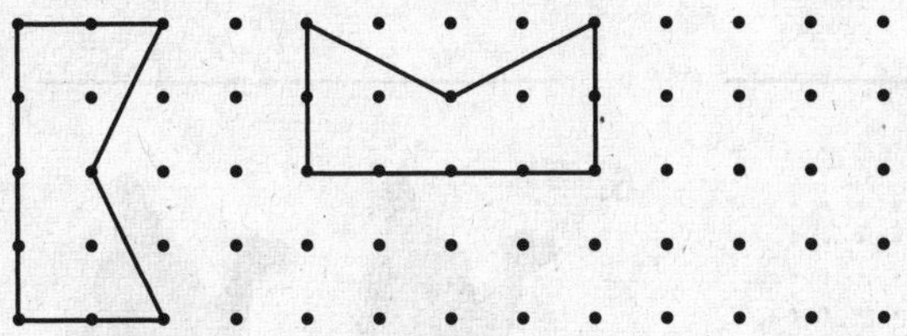

- The figures have the same shape.
- The figures also have the exact same size. You can count the dots to check.
- Figures that have the exact same shape and size are congruent.
- So, the two figures are *congruent*.

Tell whether the two figures are ***congruent*** **or** ***not congruent.***

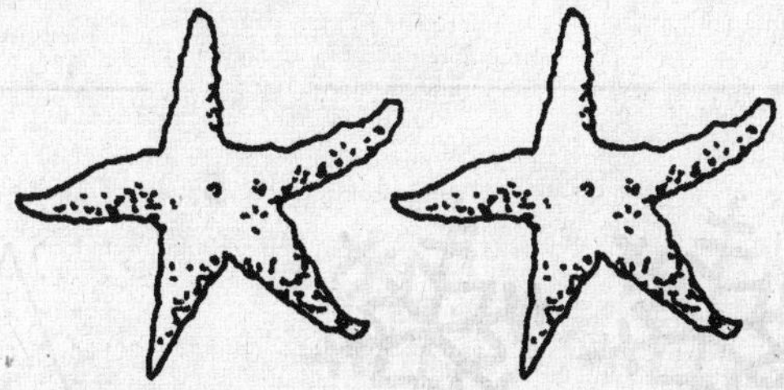

- Both of the figures have the same shape.
- Both of the figures have the same size.
- Figures that have the exact same shape and size are congruent.
- So, the two figures are congruent.

Tell whether the two figures are ***congruent*** **or** ***not congruent.***

1. 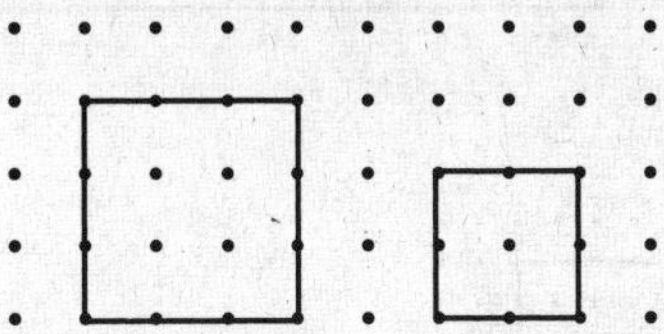
____

2. 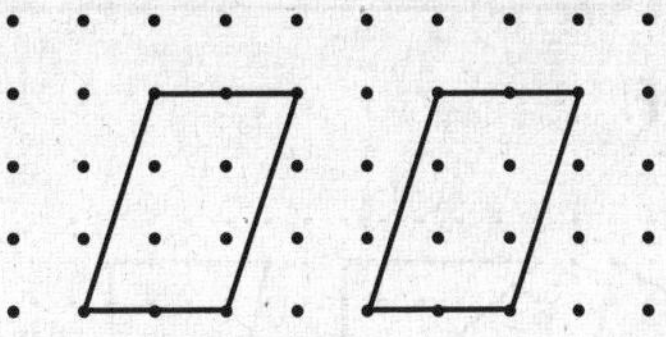
____

3. 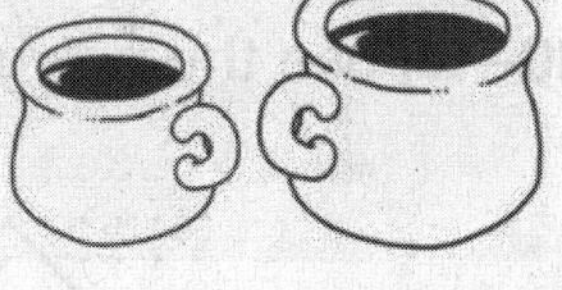
____

4. 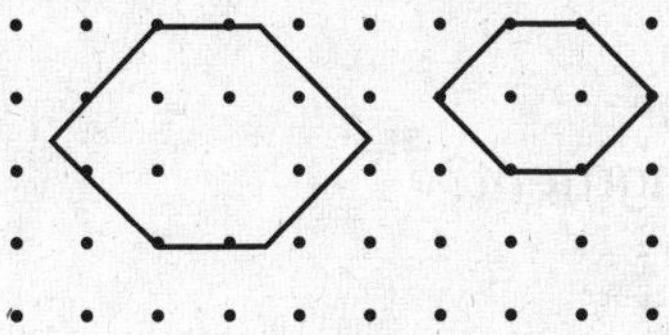
____

5. 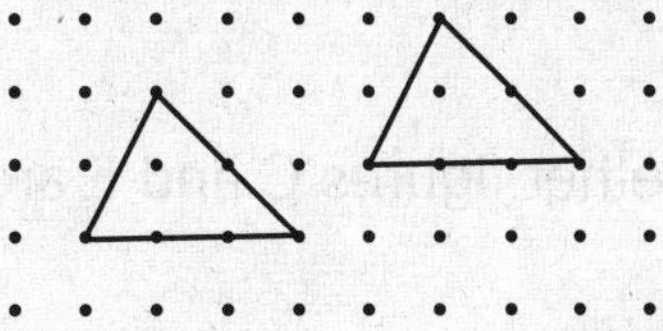
____

6. 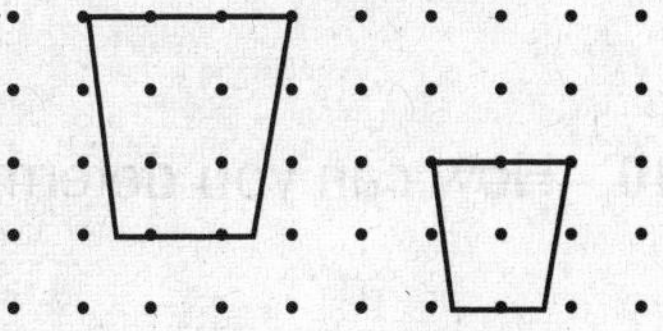
____

7. 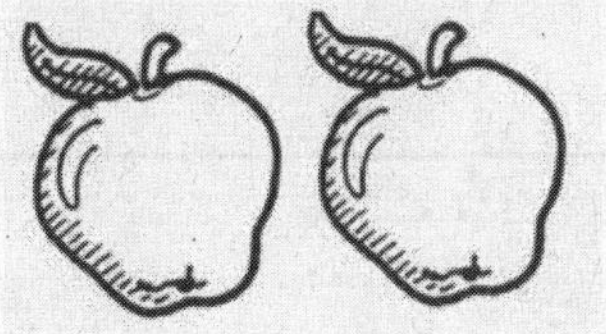
____

8. 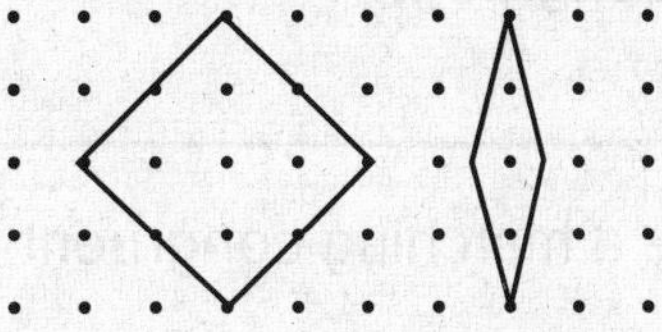
____

9. 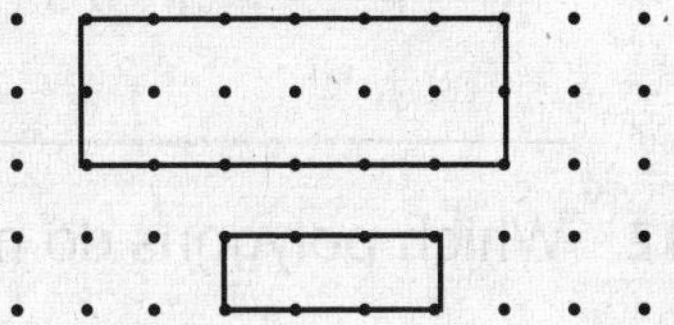
____

Name______________________________ **Lesson 21.1**

Congruent Figures

Tell whether the two figures are *congruent* or *not congruent*.

1. 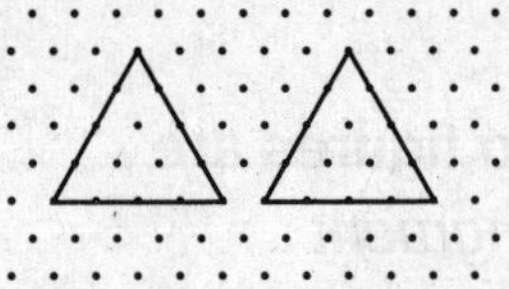

____

2. 

____

3. 

____

4. 

____

5. 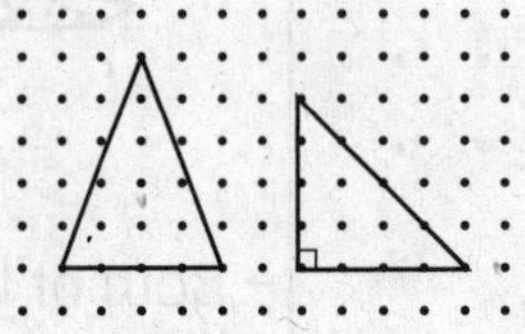

____

6. 

____

7. 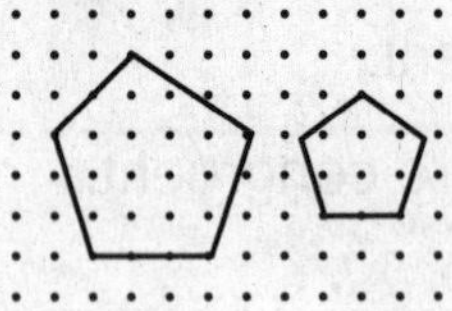

____

8. 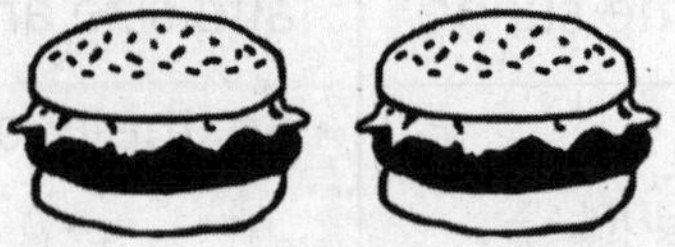

____

9. 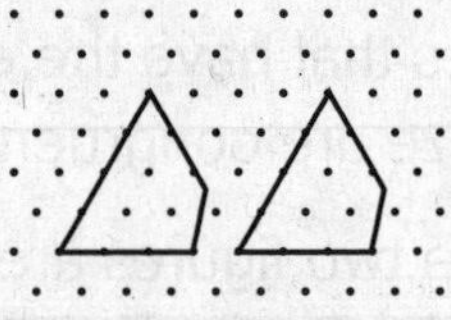

____

For 10–12, use the polygons A–F.

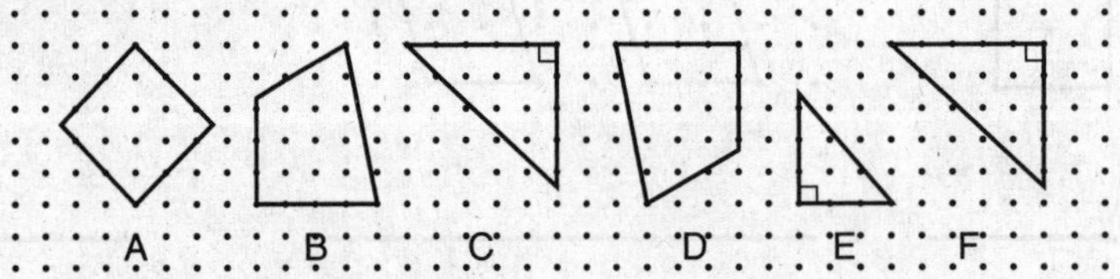

10. How can you determine whether figures C and E are congruent?

____

11. What pairs of polygons are congruent?

____

12. Which polygons do not have a matching congruent figure?

____

Name ____

Turns

A $\frac{3}{4}$ turn covers 270°. A $\frac{1}{2}$ turn covers 180°. A $\frac{1}{4}$ turn covers 90°.

Tell whether the rays on the circle show a $\frac{1}{4}$, $\frac{1}{2}$, $\frac{3}{4}$, or full turn. Then identify the number of degrees the rays have been turned clockwise or counterclockwise.

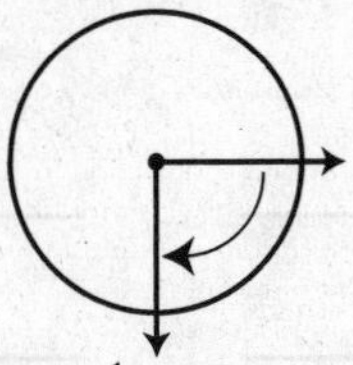

- The rays show a $\frac{1}{4}$ turn, because the distance between the rays covers only $\frac{1}{4}$ of the circle.
- The rays have been moved 90°, since 360° ÷ 4 = 90°.
- The ray is going in the same direction a clock moves, so it is clockwise.

Tell whether the figure has been turned 90°, 180°, 270°, or 360° clockwise or counterclockwise..

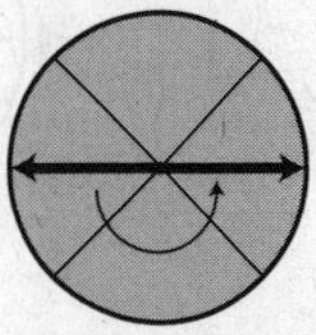

- The arrow moved halfway around the spinner. The arrow was facing directly to the left. Now, it is facing directly to the right.
- When a figure turns all the way around, it moves 360°. Since the figure only turned halfway around, it moved half of 360° or 180°
- The ray is going in the opposite direction a clock moves, so it is counterclockwise.

Tell whether the rays on the circle show a $\frac{1}{4}$, $\frac{1}{2}$, $\frac{3}{4}$, or full turn. Then identify the number of degrees the rays have been turned clockwise or counterclockwise.

1.

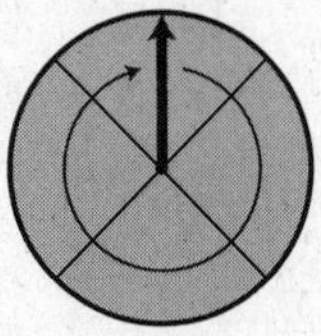

____

2.

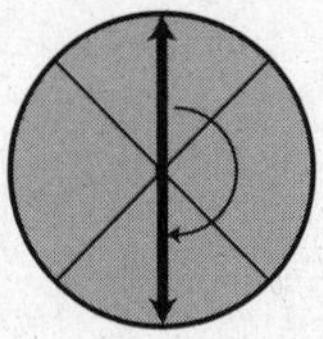

____

3.

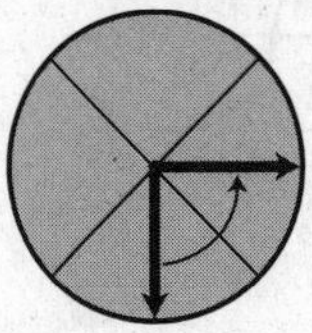

____

Tell whether the figure has been turned 90°, 180°, 270°, or 360° clockwise or counterclockwise.

4.

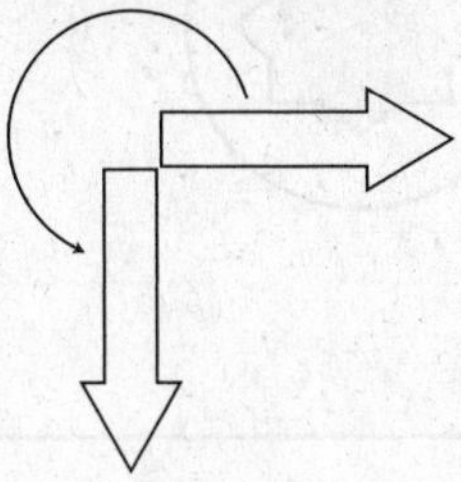

____

5.

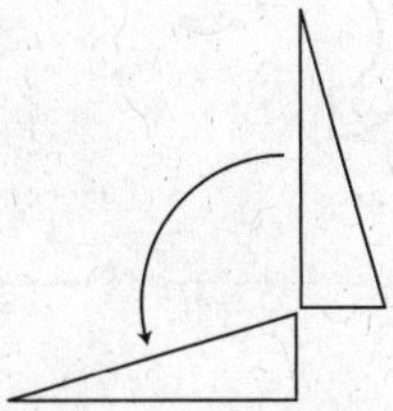

____

6.

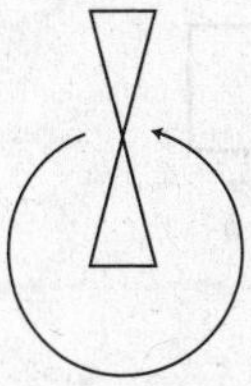

____

Name____

Lesson 21.2

Turns

Tell whether the rays on the circle show a $\frac{1}{4}$, $\frac{1}{2}$, $\frac{3}{4}$, or full turn. Then identify the number of degrees the rays have been turned clockwise or counterclockwise.

1.
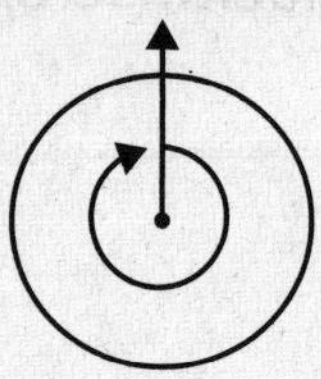
____
____

2.
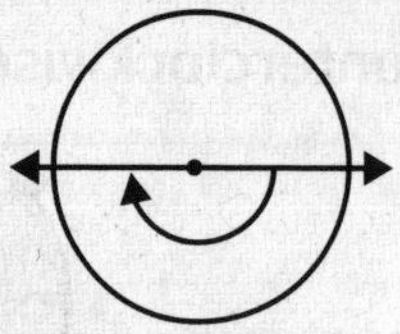
____
____

3.
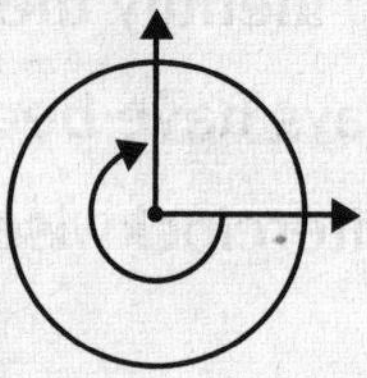
____
____

4.
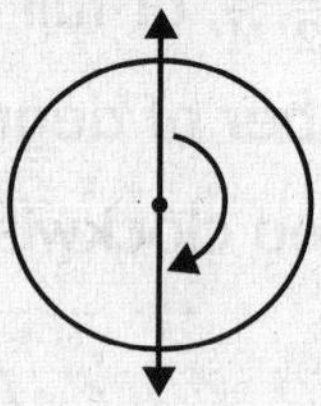
____
____

5.
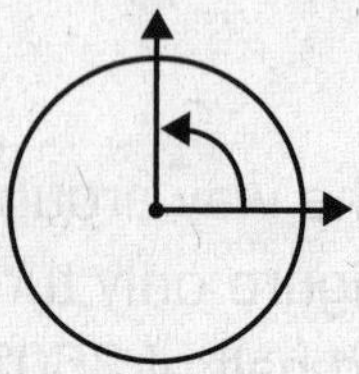
____
____

6.
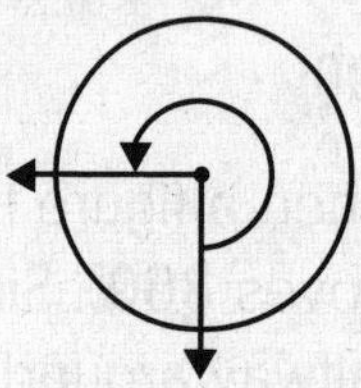
____
____

7.
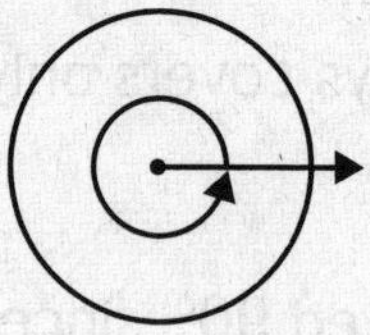
____
____

8.
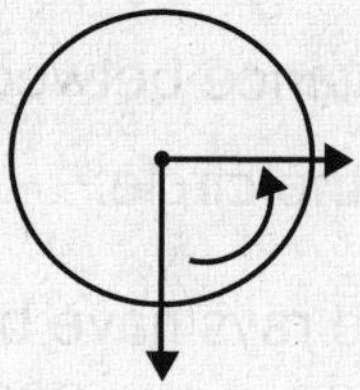
____
____

Tell whether the figure has been turned 90°, 180°, 270°, or 360° clockwise or counterclockwise.

9.
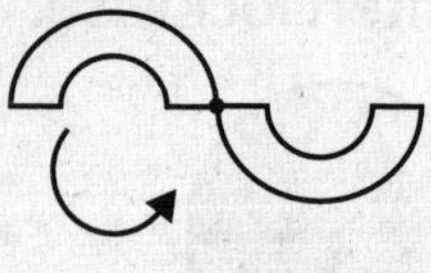
____

10.
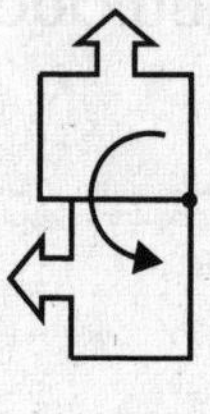
____

11.
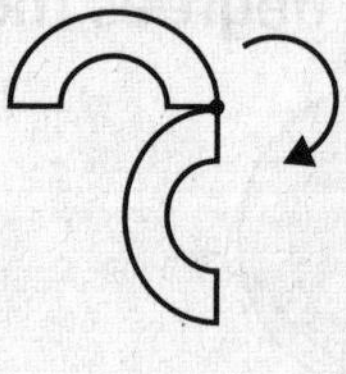
____

12.
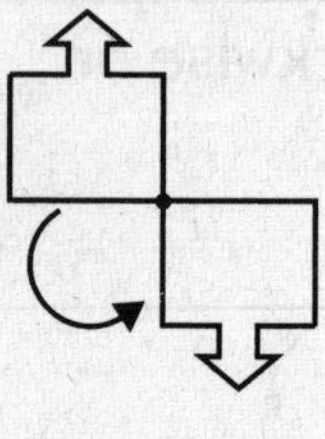
____

13.
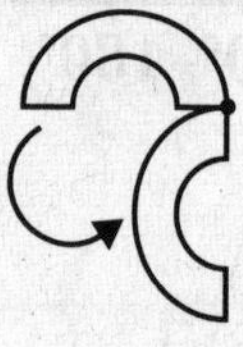
____

14.
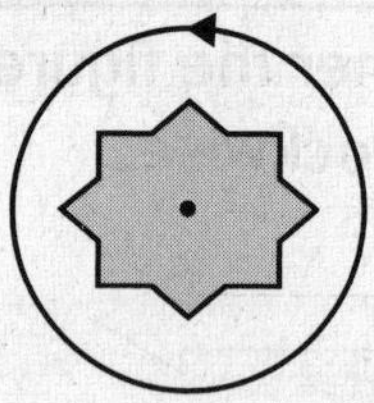
____

Name____

Symmetry

If you draw a line to divide a figure in half, and the two halves are identical to one another, then the figure has **line symmetry**. If a figure can be rotated around a center point and still look the same in at least two positions, it has **rotational symmetry**.

Tell whether the figure has *line symmetry*, *rotational symmetry*, *both*, or *neither*.

A B

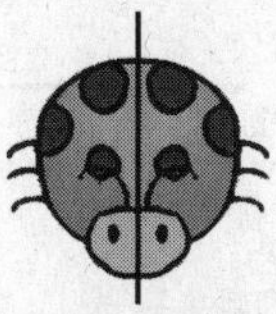

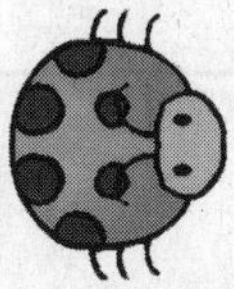

- In figure A, the bug is exactly the same on each side. So, the bug has line symmetry.
- In figure B, the bug has been rotated. It looks different from figure A. So, the bug does not have rotational symmetry.
- So, the bug has line symmetry.

Tell whether the figure has *line symmetry*, *rotational symmetry*, *both*, or *neither*.

A B

- In figure A, the penny does not look exactly the same on both sides of the line. So, the penny does not have line symmetry.
- In figure B, the penny has been rotated. It now looks different from figure A. So, the penny does not have rotational symmetry.
- So, the answer is neither.

Tell whether the figure has *line symmetry, rotational symmetry, both,* or *neither.*

1. 

____

2. 

____

3. 

____

4. 

____

5. 

____

6. 

____

7. 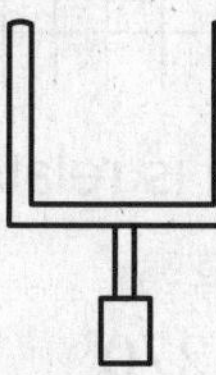

____

8. 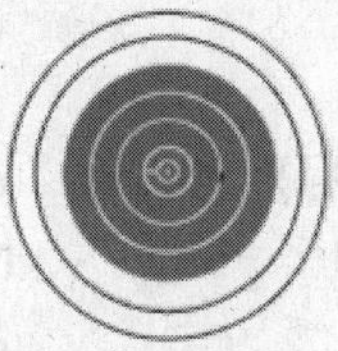

____

9. 

____

Name____

Symmetry

Tell whether the figure appears to have *line symmetry, rotational symmetry, both*, or *neither*.

1.  ____________

2.  ____________

3.  ____________

4.  ____________

5.  ____________

6.  ____________

7.  ____________

8.  ____________

Draw the line or lines of symmetry.

9. 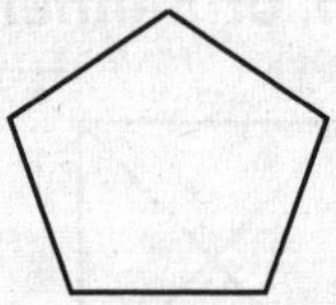

10. 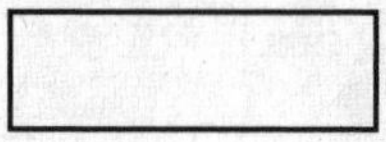

11. 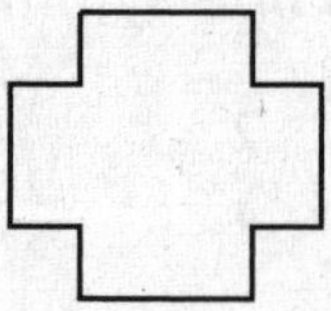

12. 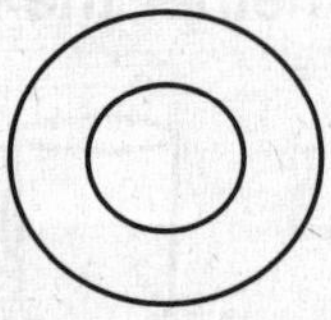

Problem Solving and Test Prep

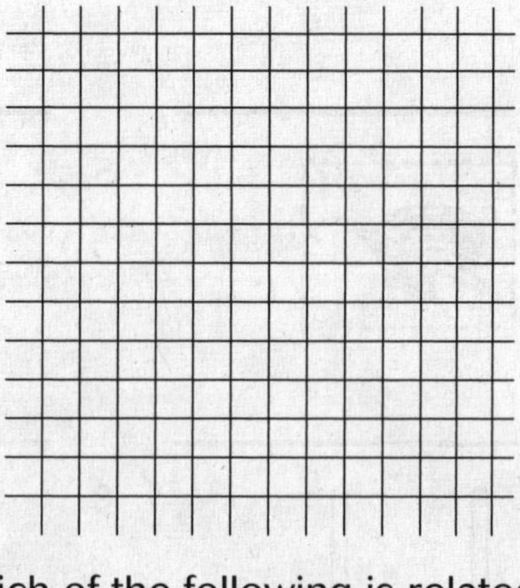

13. On the grid paper at the right, draw and label a figure that has 3 lines of symmetry.

14. On the grid paper at the right, draw and label a figure that has both line and rotational symmetry.

15. Which best describes the symmetry of the letter A?

A line
B rotational
C both
D none

16. Which of the following is related to a $\frac{3}{4}$ turn?

A 90°
B 180°
C 270°
D 360°

Name ____________________

Problem Solving Workshop Strategy: Compare Strategies

Buddy's Rabbit Design

Buddy made the two rabbits shown at the right from pattern blocks. Are the rabbits congruent? Explain.

Read to Understand

1. What are you asked to find out in this problem?

____

Plan

2. What are some strategies you can use to solve this problem?

____

Solve

3. Use the strategies below to solve the problem.

Act it Out

a. Take the blocks from one rabbit and put them over the ________ of the other rabbit.

b. See if the blocks are the exact same ______ and ________.

Draw a Diagram

a. Draw a diagram by tracing the blocks of the rabbit on the left of paper. Then place the blocks from the other rabbit onto your diagram.

b. See if the blocks are the exact same ______ and. ________ .

4. Are the rabbits congruent? ____________________

Check

5. Which strategy was more helpful, Act it Out or Draw a Diagram? Why?

____

____

____

Act it out to solve.

6. Tim made the sun below using pattern blocks. Does Tim's sun have line symmetry?

____

7. Mara made the leaf below using pattern blocks. Does Mara's leaf have rotational symmetry?

____

Name________________________________ Lesson 21.4

Problem Solving Workshop Strategy: Compare Strategies

Problem Solving Strategy Practice

1. Dustin made the dog at the right from pattern blocks. Does Dustin's dog have line symmetry?

____

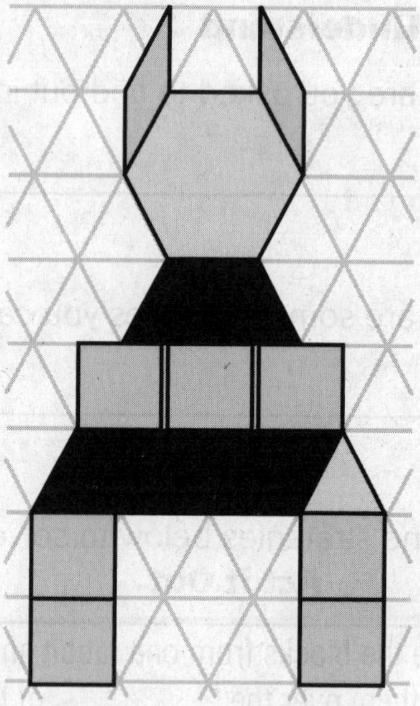

2. What individual blocks in Dustin's dog have rotational symmetry?

____

3. What individual blocks in Dustin's dog do not have rotational symmetry?

____

Problem Solving and Test Prep

USE DATA For 4–5, use Dustin's pattern block dog above.

4. How many pattern-block dogs does Dustin need to make if he wants to place the dogs in an arrangement that has rotational symmetry?

____

5. List the name of each figure Dustin used along with the number of blocks used of each in order from the least number of used blocks to the greatest number of used blocks. Use < or =.

____

6. Sara made a bird using 20 pattern blocks. If she used 4 blocks for each wing, how many pattern blocks did Sara use for the body?

____

7. Sara made 25 copies of her bird for a border around her sister's room. How many pattern pieces did Sara need in all?

____

Spiral Review

For 1–5, choose a method. Then find the product.

1. $80 \times 23 =$ ________

2. $67 \times 30 =$ ________

3. $33 \times 90 =$ ________

4. $45 \times 50 =$ ________

5. $11 \times 20 =$ ________

Favorite Fruit

Fruit	Number of Votes
Apples	16
Bananas	16
Oranges	12

10. Make a bar graph using the information in the table above.

11. What is another type of graph you could use to represent this data?

____

____

For 6–9, name a geometric term that best represents the object. Use the terms *line*, *plane*, *point*, and *ray*.

6. highway ________

7. center of a clock ________

8. shooting arrow ________

9. flag ________

For 12–16, find the missing number. Tell which addition property you used.

12. $31 + \square = 31$

____

13. $35 + \square = 23 + 35$

____

14. $12 + (19 + 9) = (12 + 19) + \square$

____

15. $\square + 191 = 191$

____

16. $\square + (19 + 100) = (11 + 19) + 100$

____

Name____

Geometric Patterns

Geometric patterns are patterns based on color, size, shape, position, and number of figures. A geometric pattern repeats over and over again.

Write a rule for the pattern. Then draw the next two figures.

- The pattern uses large and small circles. The large and small circles alternate: 1 large, 1 small, 1 large, 1 small, etc.
- The large dot on each pair of circles alternates between being on the top of each set of circles and being on the right side of each set of circles.
- So, the next two figures would be:

Write a rule for the pattern. Then draw the next two figures.

- The pattern uses large triangles and small triangles.
- Each large triangle is followed by 3 small triangles. Then the pattern starts over again: one large triangle, three small triangles, etc.
- So, the next two figures would be:

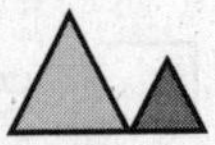

Write a rule for the pattern. Then draw the next two figures.

1. 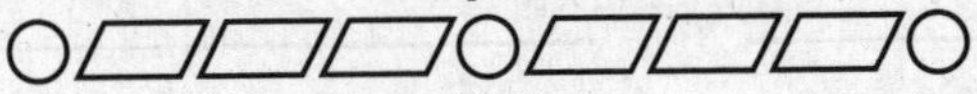

2. 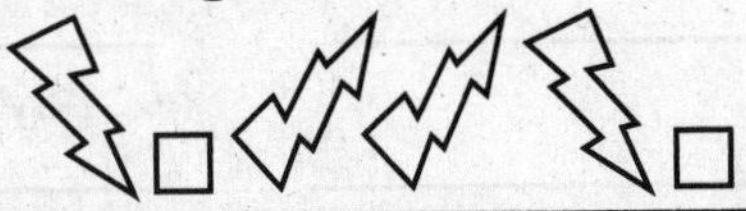

3. 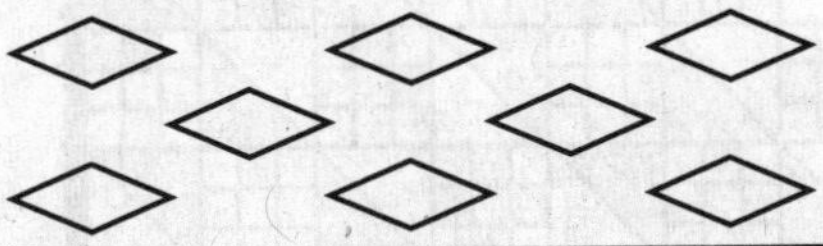

4. 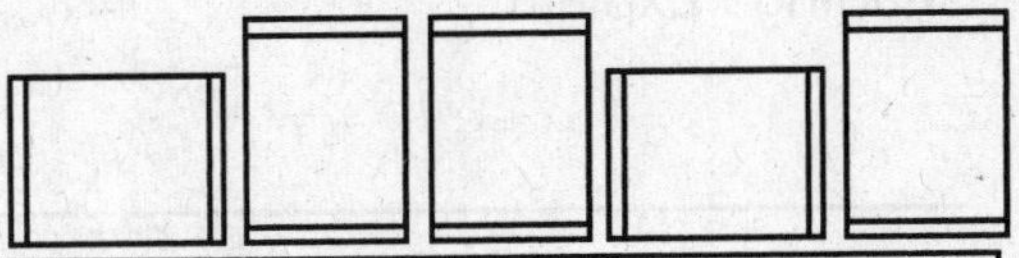

5. 

6. 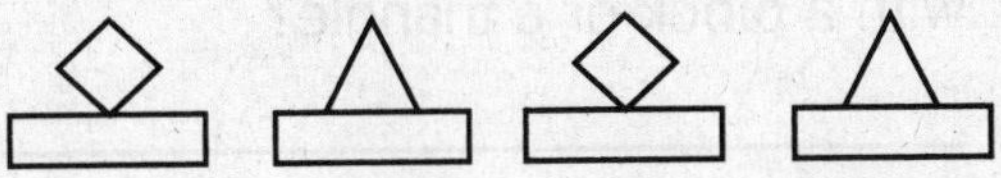

Name____________________

Geometric Patterns

Write a rule for the pattern. Then draw the next two figures in your pattern.

1. 

____

____

2. 

____

____

3. 

____

____

4. 

____

____

Write a rule for the pattern. Then draw the missing figure in your pattern.

5. 

6. 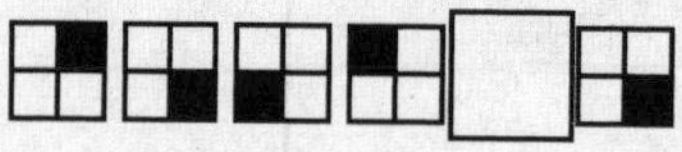

7. 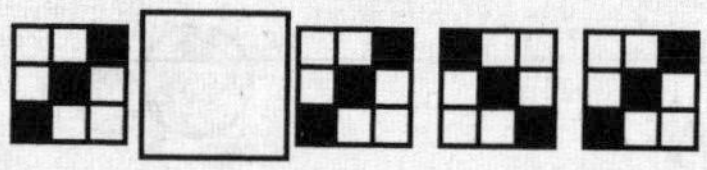

____

____

Problem Solving and Test Prep

USE DATA For 8–9, use the quilt.

8. Does the rule for the pattern include shading? Explain.

____

____

9. If you remove the border and add a row at the bottom, will that row start with a block or a triangle?

____

10. In Exercise 6, what will be the tenth figure in the pattern?

A 

B 

C 

D 

11. In Exercise 2, if the white arrow continues to rotate, what will be the fifteenth figure in the pattern?

A 

B 

C 

D 

Name____

Faces, Edges, and Vertices

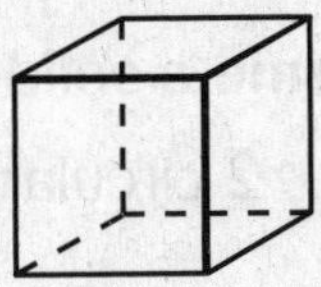

- The **face** is a polygon that is a flat surface.
- The **edge** is where two faces meet.
- A **vertex** is where 3 or more edges meet.
- The **base** is the flat surface on which a solid figure rests.

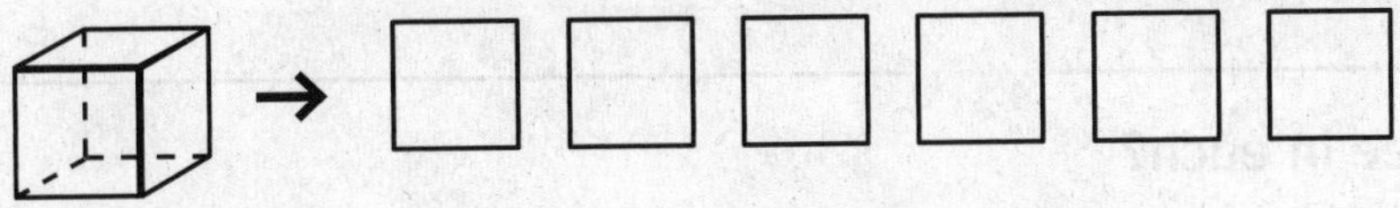

A **cube** has 6 faces.

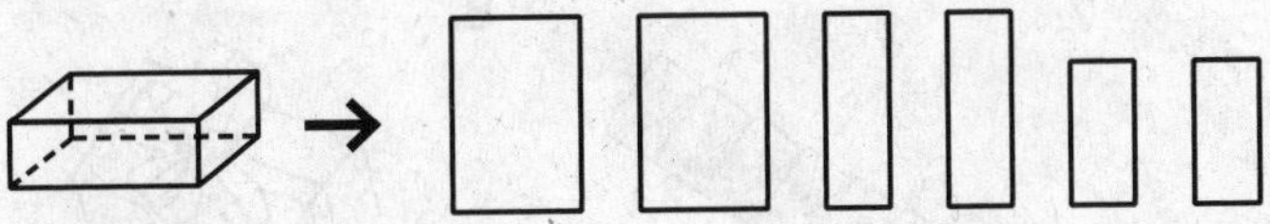

A **rectangular prism** has 6 faces.

A **triangular pyramid** has 4 faces.

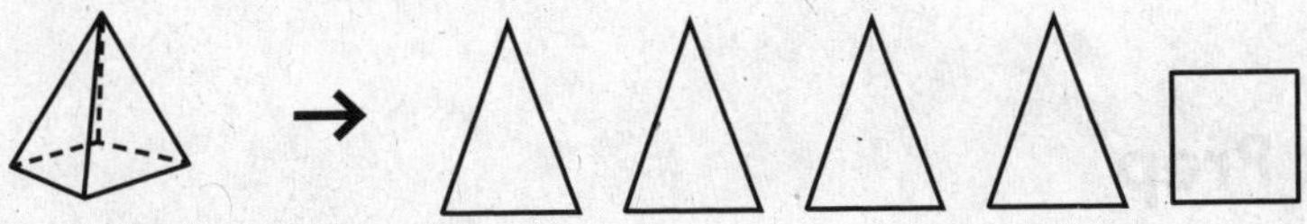

A **square pyramid** has 5 faces.

A **rectangular pyramid** has 5 faces.

Name a solid figure that has fewer than 6 faces.

A triangular prism, square pyramid, and rectangular pyramid all have fewer than 6 faces.

Name a solid figure that is described.

1. 5 faces ____________

2. 4 faces ____________

3. all square faces ____________

4. 8 edges ____________

5. 4 vertices ____________

6. all rectangular faces ____________

MG 3.6 Visualize, describe, and make models of geometric solids (e.g., prisms, pyramids) in terms of the number of shapes of faces, edges, and vertices; interpret two-dimensional representations of three-dimensional objects; and draw patterns (of faces) for a solid that, when cut and folded, will make a model of the solid.

Name ____________________

Faces, Edges, and Vertices

Name a solid figure that is described.

1. 2 circular bases

2. 6 square faces

3. 1 rectangular and 4 triangular faces

4. 1 circular base

Which solid figure do you see in each?

5.

6.

7.

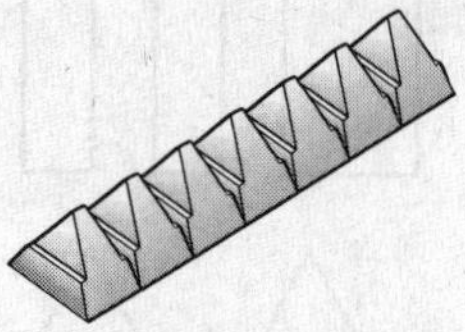

8.

Problem Solving and Test Prep

For 9–10, look at the edges of the rectangular prism.

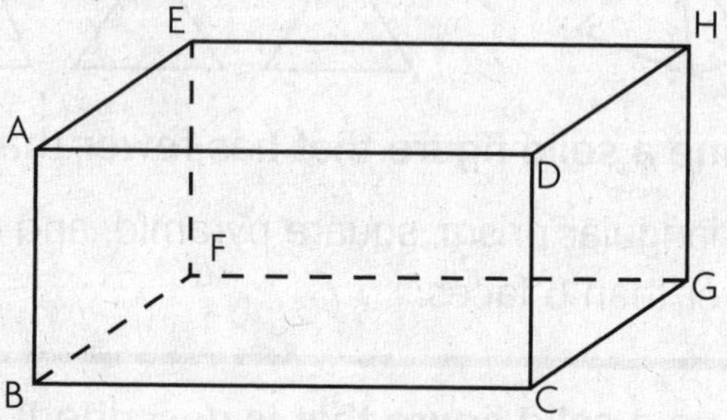

9. Name a pair of parallel line segments.

10. Name a pair of perpendicular line segments.

11. Which solid figure has more edges, a rectangular prism or a triangular prism? How many more?

12. What is the relationship between the number of faces and the number of edges of a triangular pyramid?

13. Which of the figures below has only one circular base?

A cone
B sphere
C cylinder
D square prism

Name____

Lesson 22.2

Patterns for Solid Figures

A net is a pattern that can be folded to make a three-dimensional figure.

Draw a net that can be cut to make a model of each solid figure.

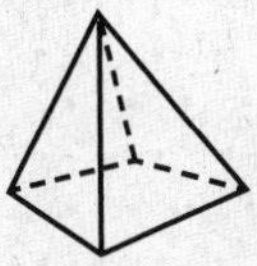

* A square pyramid has 5 faces.

The base of a square pyramid is a square.
Draw the base.

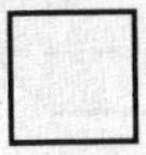

The other faces of a square pyramid are triangles.
Draw the other 4 faces of the solid figure.

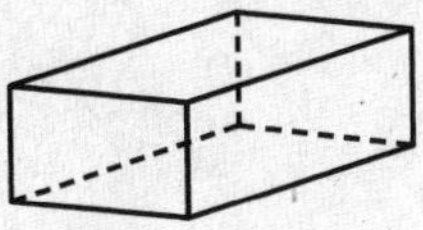

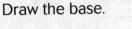

* A rectangular prism has 6 faces.

The base of a rectangular prism is a rectangle.
Draw the base.

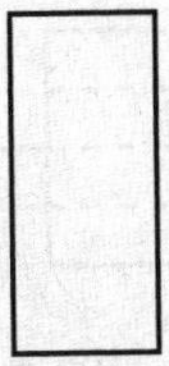

Draw the other 5 faces.

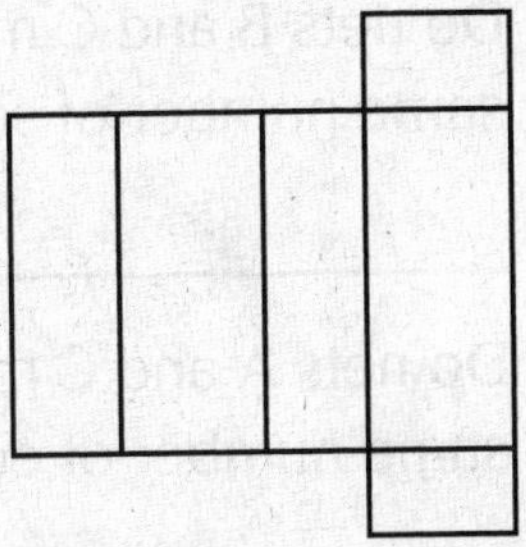

Draw a net that can be cut to make a model of each solid figure.

1.

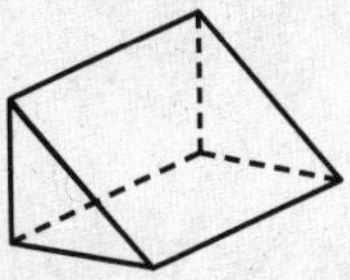

2.

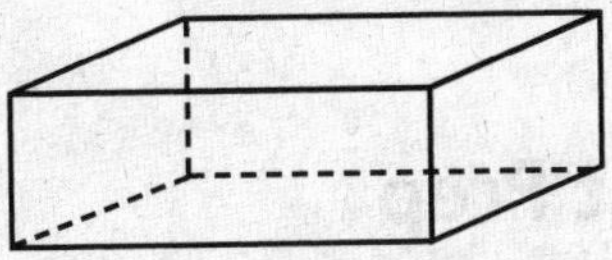

3.

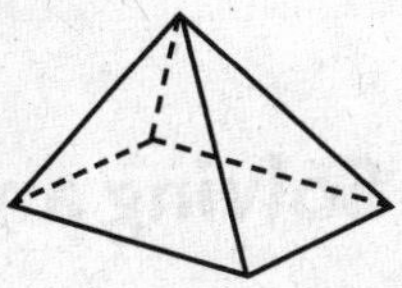

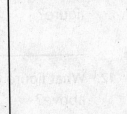

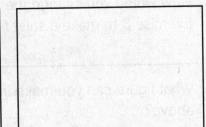

MG 3.6 Visualize, describe, and make models of geometric solids (e.g., prisms, pyramids) in terms of the number and shape of faces, edges, and vertices; interpret two-dimensional representations of three-dimensional objects; and draw patterns (of faces) for a solid that, when cut and folded, will make a model of that solid.

Name_______________________________________

Lesson 22.2

Patterns for Solid Figures

Draw a net that can be cut to make a model of the solid figure shown.

1.

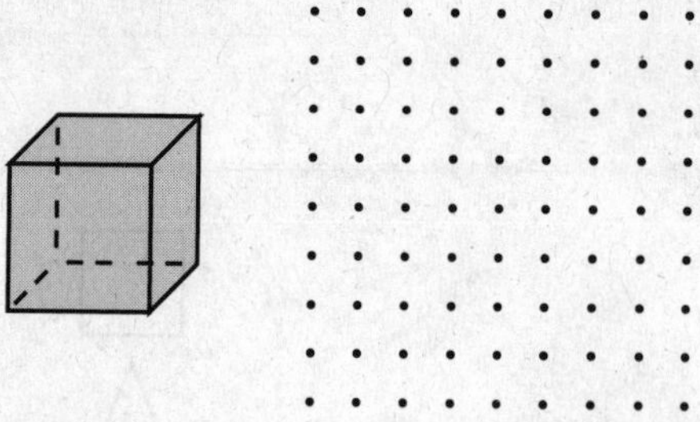

2.

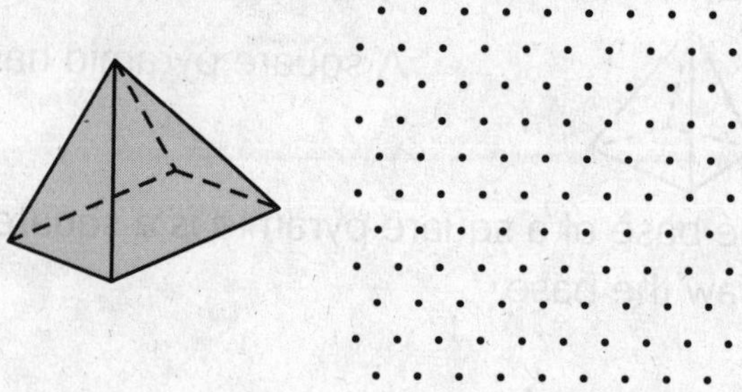

Would the net make a rectangular prism? Write yes or no.

3.

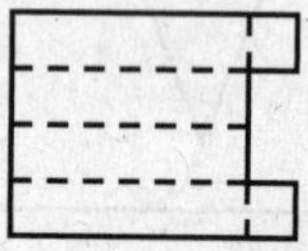

____

4.

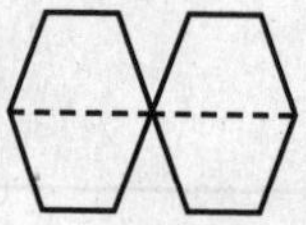

____

5.

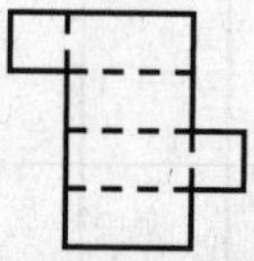

____

6.

____

For 7–8, use the nets.

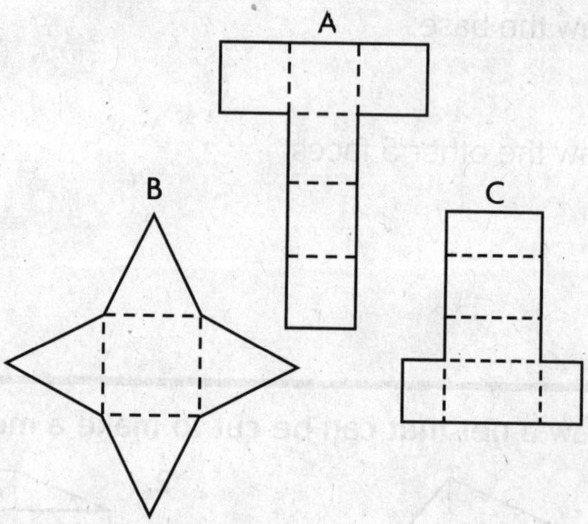

7. Do nets B and C make figures with the same number of sides?

____

8. Do nets A and C make figures with the same number of edges? Explain.

____

Problem Solving and Test Prep

9. How would you change the figure in Exercise 3 to make a solid figure?

____

10. Can the net in Exercise 6 make a solid figure?

____

11. What figure can you make from net A above?

____

12. What figure can you make from net B above?

____

Name ________________________________

Different Views of Solid Figures

You can identify solid figures by the way they look from different views. For example, you can look at the object from the top, from the front, and from the sides.

Name the solid figure that has the following views.

top view front view side view

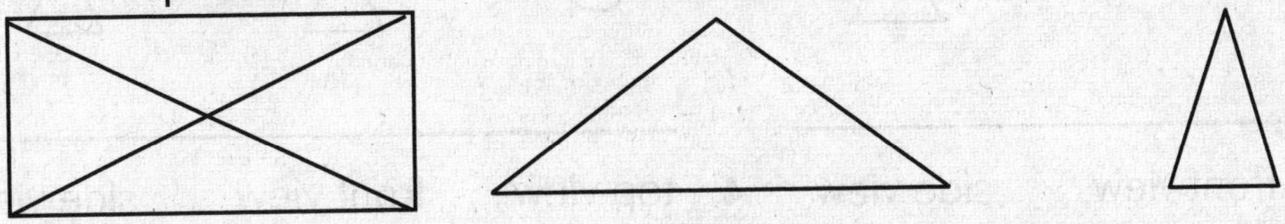

- The top view shows that the figure has a rectangular base. The crossed lines show that the sides come together to form a point.
- The front and side views show that the faces look like a triangles.
- A solid figure with a rectangular base and triangular faces is a rectangular pyramid.

So, the figure is a rectangular pyramid.

Name the solid figure that has the following views.

1. top view front view side view

____

2. top view front view side view

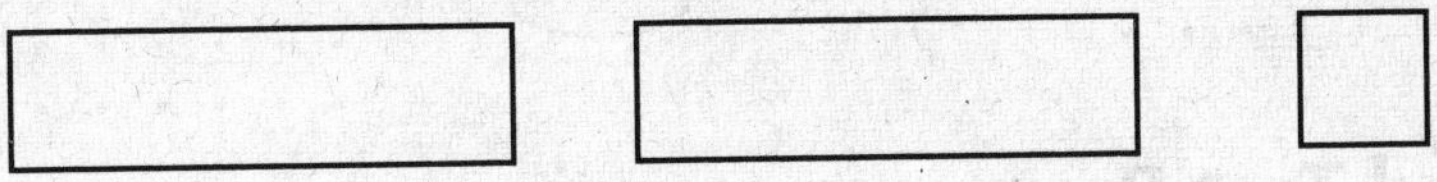

____

3. top view front view side view

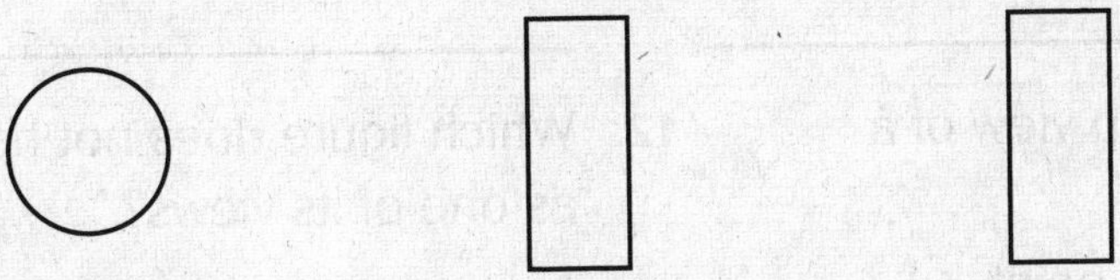

____

Name ______________________________

Different Views of Solid Figures

Name the figure that has the following views.

1. top view front view side view

____

2. top view front view side view

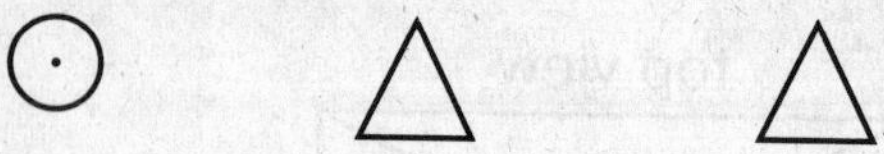

____

3. top view front view side view

____

4. top view front view side view

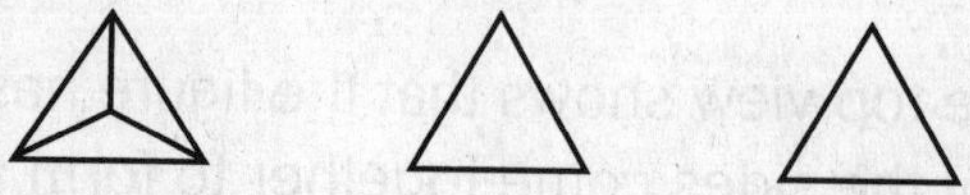

____

Draw the top, front, and side views of each solid figure.

5.

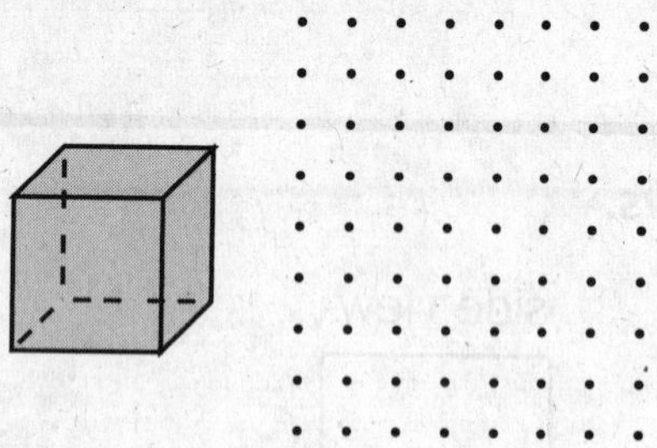

6.

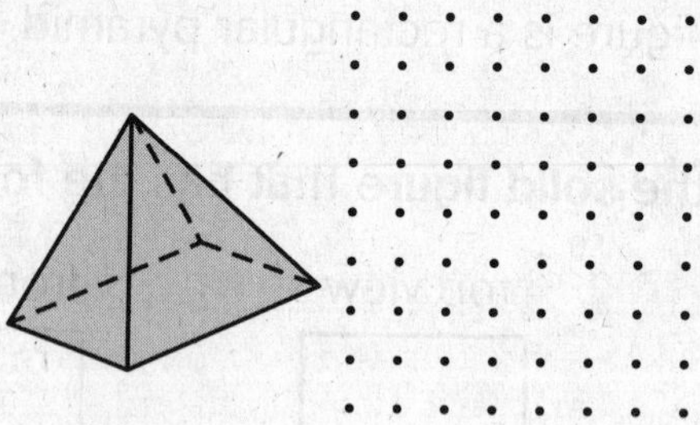

7.

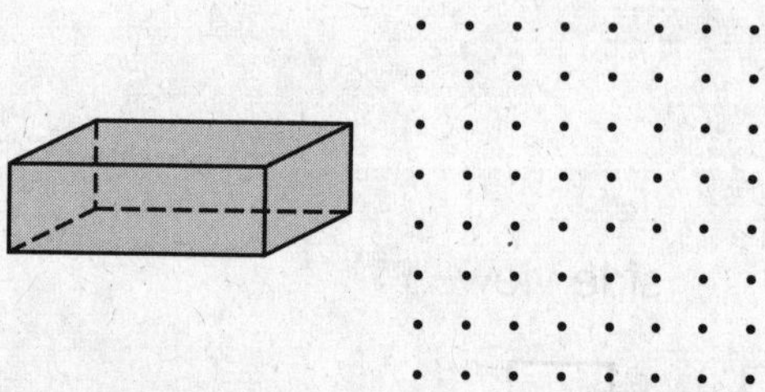

8.

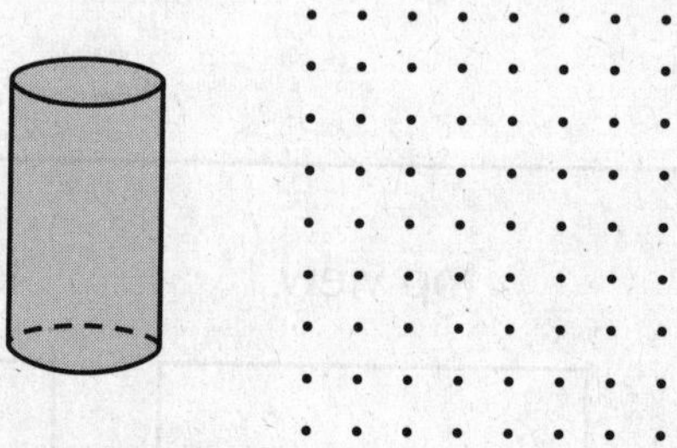

Problem Solving and Test Prep

9. What solid figures have a circle on two views?

____

____

10. What solid figures have a triangle on at least one of its views?

____

____

11. Which figure is the top view of a cube?

A square **C** point
B cylinder **D** triangle

12. Which figure does not have a triangle as one of its views?

A cone **C** cylinder
B triangular pyramid **D** triangular prism

Name ____________________

Problem Solving Workshop Strategy: Making a Model

Alicia used the fewest possible cubes to make a building whose views are shown below. How many cubes did Alicia use?

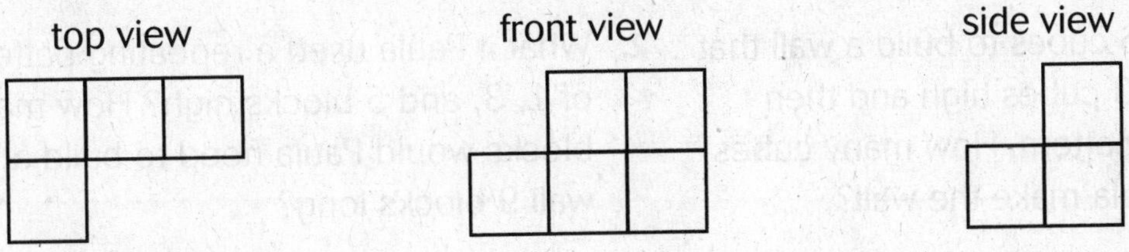

Read to Understand

1. What are you asked to find?

____

Plan

2. How can you use the make-a-model strategy to help solve the problem?

____

Solve

3. Use centimeter cubes to build the top view. Then add more cubes to create the front view and the side view. How many cubes did Alicia use?

____

Check

4. How can you check your answer to the problem?

____

____

Make a model to solve.

5. James uses centimeter cubes to make a rectangular prism. The prism is 7 cubes wide, 2 cubes high and 2 cubes long. How many cubes does James use?

____

6. Pam uses blocks to make a square. Each side of the square is 4 blocks long and one block high. Then Pam pulls out 4 blocks in the center of the square. How many blocks are left?

____

Name ______________________________

Problem Solving Workshop Strategy: Make a Model

Problem Solving Strategy Practice

Make a model to solve.

1. Paula has 36 cubes to build a wall that is 1, 2, and 3 cubes high and then repeats the pattern. How many cubes long can Paula make the wall?

____

2. What if Paula used a repeating pattern of 1, 3, and 5 blocks high? How many blocks would Paula need to build a wall 9 blocks long?

____

3. John has 66 cubes. He gives 21 to Mark and then builds a staircase beginning with 1 cube, then 2, and so on. How tall will John's staircase be?

____

4. How many cubes would John need to build the next step of his staircase?

____

Mixed Strategy Practice

5. Sandra and Jan have a total of 88 cubes, half of which are blue. Jan uses 34 to make a wall and Sandra uses 25 to make a building. What is the least number of blue cubes they could use?

____

6. Mrs. Lutie left home and went to the bank. Then she drove 18 miles to the dentist, 9 miles for groceries, 8 miles to pick up her kids, and 3 miles back home. If Mrs. Lutie drove a total of 45 miles, how far was it from home to the bank?

____

7. Pose a Problem Change the numbers in Exercise 6. Make a new problem about Mrs. Lutie's errands.

____

____

____

8. How many ways can you arrange 12 cubes in more than one row? Name the ways.

____

____

____

Name ____________________

Spiral Review

For 1–5, estimate.
Then find the product.

1. $\begin{array}{r} 95 \\ \times\ 11 \\ \hline \end{array}$

2. $\begin{array}{r} 618 \\ \times\ 49 \\ \hline \end{array}$

3. $\begin{array}{r} 904 \\ \times\ 89 \\ \hline \end{array}$

4. $18 \times \$5.66$

5. 37×292

For 8–10, choose the best type of graph or plot for the data. Explain your choice.

8. How long it took Marie to fill up her pool

____

____

9. How Mikela spent $20 at the mall

____

____

10. The average inches of snow in Ben's backyard each month

____

____

For 6–7, find the area and perimeter of each figure. Then draw another figure that has the same perimeter but a different area.

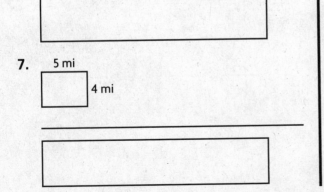

6.

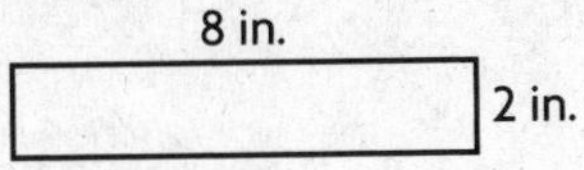

____

7.

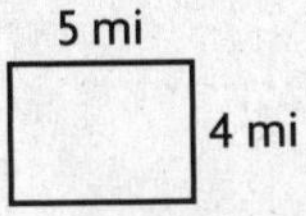

____

For 11–12, use $A = lw$ to find the area.

11.

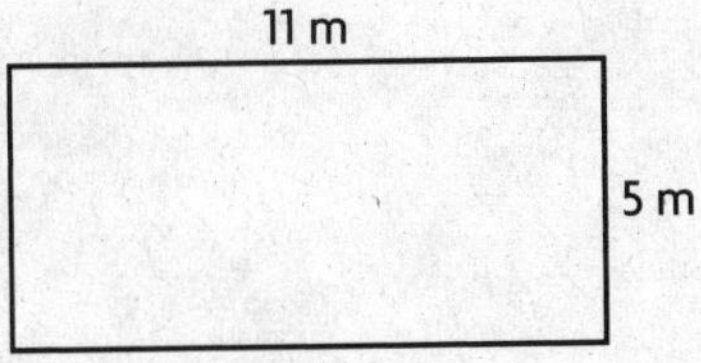

____

12.

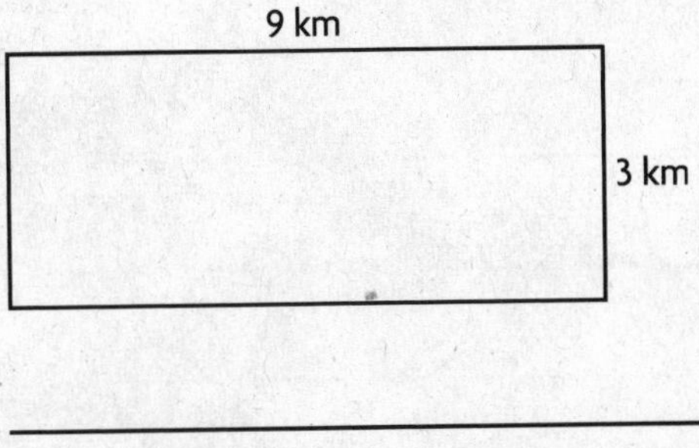

____

Name____ **Lesson 23.1**

Customary Measurements

You can use customary units to measure length, weight, and capacity.

Length: Customary units of length include inch (in.), foot (ft), yard (yd), and mile (mi). The length of $\frac{1}{2}$ of your thumb is about 1 inch long. A license plate is about 1 foot long. A baseball bat is about 1 yard long. The distance you can walk in 20 minutes is about a mile.

Weight: Customary units of weight include ounce (oz), pound (lb), and ton (T). Five new pencils weigh about 1 oz. Four sticks of butter weigh about 1 lb. A small car weights about 1 T.

Capacity: Customary units of capacity include cup (c), pint (pt), quart (qt), and gallon (gal). 1 pint is 2 cups. 4 quarts make 1 gallon.

Choose the more reasonable measurement.

2 c or 2 qt

6 oz or 6 lb

- Study the coffee mug. 2 qt is half a gallon. A coffee mug cannot hold that much, so 2 qt is too much. 2 c is a more reasonable estimate.
- Study the grapes. 5 lb is about as much as 20 sticks of butter. 5 grapes could not weigh as much as 20 sticks of butter, so 6 lb is too much. 6 oz is a more reasonable estimate.

Choose the more reasonable measurement.

1.  1 oz or 1 lb ____________

2. 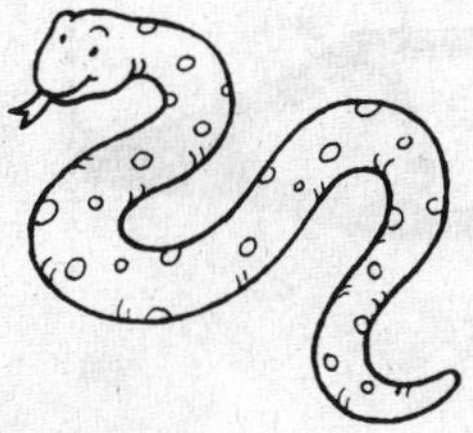 3 ft or 3 in. ____________

3.  3 qt or 3 gal ____________

4.  5 oz or 5 lb ____________

5. 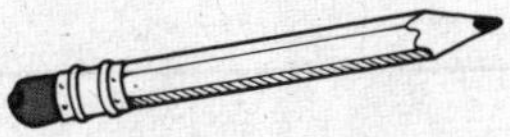 7 ft or 7 in. ____________

6.  1 gal or 1 pt ____________

Name ______________________________ **Lesson 23.1**

Customary Measurements

Circle the most reasonable measurement.

1.

70 ft or 70 mi

2.

2 c or 2 gal

3.

180 lb or 180 T

Write an equation you can use to complete each table. Then complete each table.

4. ______________________

Inches, *n*	48	60	72	84	96
Feet, *f*	4	5			

5. ______________________

Gallons, *g*	2	4	6	8	10
Pints, *p*	16				

Estimate to the nearest inch. Then measure to the nearest $\frac{1}{2}$ and $\frac{1}{4}$ inch.

6.

____

7.

____

Problem Solving and Test Prep

8. How many more cups are in a gallon than cups in a quart?

____

9. How many more feet are in 1 mile than feet in 1,000 yards?

____

10. How many inches are in 6 ft?

A 12 in. **C** 108 in.

B 72 in. **D** 144 in.

11. Walt's stew pot holds 2 gallons. How large is Walt's stew pot in cups?

____

Name________________________________ **Lesson 23.2**

Metric Measurements

- Metric units can be used to measure length, weight, and capacity.
- Metric units of **length** include millimeter (mm), centimeter (cm), decimeter (dm), meter (m), and kilometer (km). A door is about 1 m wide. A dime is about 1 mm thick.
- Metric units of mass include the gram (g) and kilogram (kg). A dollar's mass is about 1 g. A small textbook's mass is about 1 kg.
- Metric units of capacity include milliliter (mL) and liter (L). A regular-sized water bottle holds about 1 L.

Choose the more reasonable measurement.

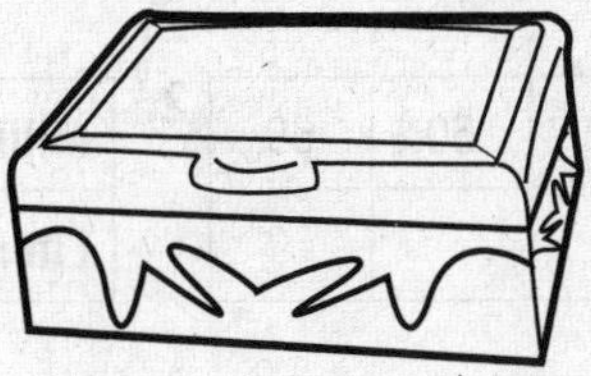

4 mm or 4 dm

- Millimeters are only the width of a dime. A jewelry box is much longer than 4 mm. So, 4 dm is the more reasonable estimate.

Write an equation you can use to complete the table. Then complete the table.

Centimeters, c	1,100	1,000	900	800	700
Meters, m	■	10	■	■	■

- The chart shows that 1,000 cm = 10 m. You know that $1{,}000 \div 100 = 10$. So the formula is $m = c \div 100$, and the complete table would be:

Centimeters, c	1,100	1,000	900	800	700
Meters, m	11	10	9	8	7

Choose the more reasonable measurement.

1. 

50 g or 50 kg ____________

2. 

2 cm or 2 m ____________

3. 

8 mL or 8 L ____________

4. 

1 mm or 1 cm ____________

Write an equation you can use to complete the table. Then complete the table.

5. ____________________

Decimeters, dm	60	50	40	30	20
Meters, m	■	■	4	■	■

6. ____________________

Kilograms, kg	4,000	5,000	6,000	7,000	8,000
Grams, g	4	■	6	■	■

AF 1.5 Understand that an equation such as $y = 3x + 5$ is a prescription for determining a second number when a first number is given.

Name____

Metric Measurements

Write an equation you can use to complete each table. Then complete each table.

1.

Meters, *m*	20	30	40	50	60
Decimeters, *d*		300			

____

2.

Milliliters, *mL*	4,000	6,000	8,000	10,000	12,000
Liters, *L*		6			

____

Estimate to the nearest centimeter. Then measure to the nearest half centimeter and millimeter.

3.

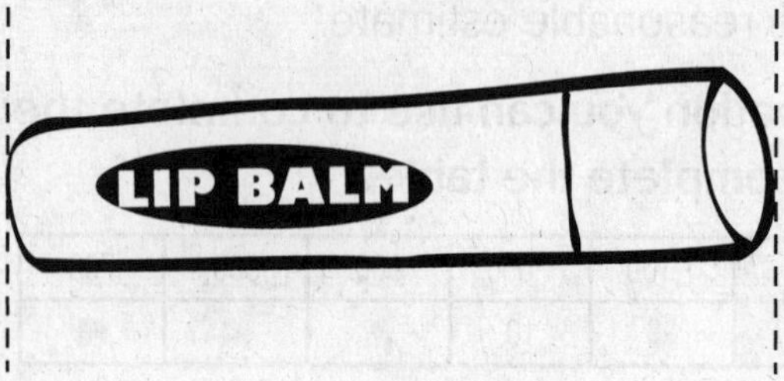

____

4.

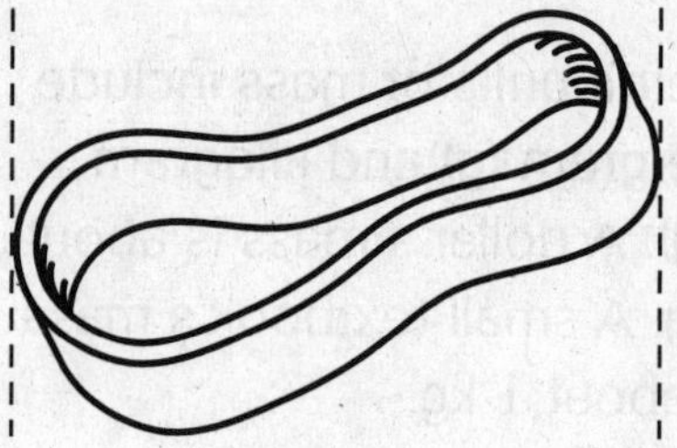

____

Order the measurements from greatest to least.

5. $\frac{1}{2}$ km; 700 m; 80,000 mm; 1 km

____

6. 3 kg; 3,100 g; 2 kg; 5,000 g

____

7. 3,000 mL; $2\frac{1}{2}$ L; 2 L; 1,600 mL

____

Problem Solving and Test Prep

USE DATA For 8–9, use the table.

8. How many Cockatoos like Max would it take to have a combined mass of 4.4 kilograms?

____

9. How many millimeters long is one of the porcupine's quills?

____

Animals at the San Diego Zoo

Animal	Fact
Pocahontas the Porcupine	Quills are about 30 centimeters long.
Dotti and Tevi the Clouded Leopard Sisters	Each is about 1.5 meters long.
Max the Salmon Crested Cockatoo	Has a mass of about 550 grams.
Tembo the African Elephant	Can hold about 14 liters of water in her trunk.

10. Orville's model airplane is $4\frac{1}{2}$ decimeters long. How many millimeters long is it?

A 4,500 **C** 45

B 450 **D** $4\frac{1}{2}$

11. Trina is going to school. Her book bag weighs 7 kilograms. How many grams does it weigh?

____

Name_______________________________________ **Lesson 23.3**

Estimate and Measure Perimeter

Perimeter is the distance around a figure. You can use string and a ruler to find the perimeter of an object.

Use string to estimate and measure the perimeter of your desk.

- Take a long piece of string and wrap it all the way around your desk. Make sure that the string is tight against the sides of your desk. If the string is too loose, your estimate and measurement will be less accurate.
- Mark or cut the string to show the perimeter of the desk.
- Lay the string on the ground so that it is straight. Now estimate: about how many inches long is the string?
- Measure the string with a ruler. Record the perimeter in inches.
- A typical school desk has a perimeter of about 72 inches.

Find the perimeter of the figure at the right.

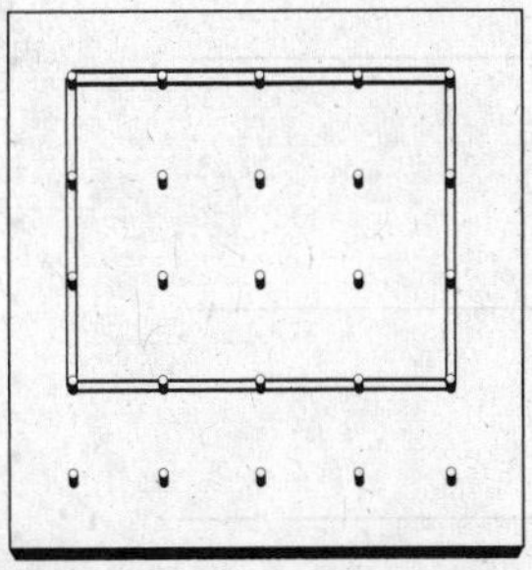

- The shape is a rectangle. Count the number of units on each side of the rectangle.
- 4 units + 4 units + 3 units + 3 units = 14 units.
- So, the figure has a perimeter of 14 units.

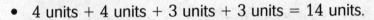

Use string to estimate and measure the perimeter of each object.

1. a textbook

2. a folder

3. a computer screen

______________ ______________ ______________

Find the perimeter of each figure.

4.

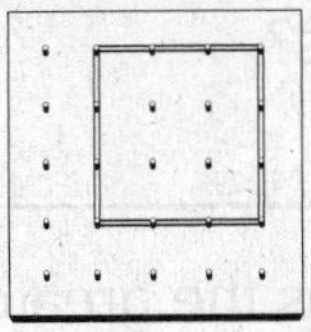

5.

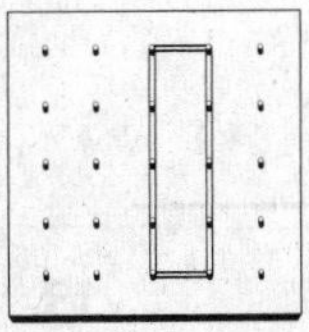

6.

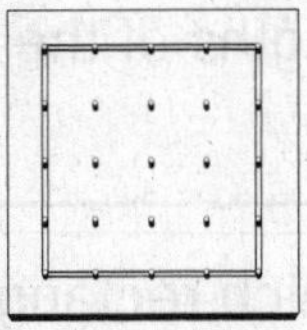

______________ ______________ ______________

7.

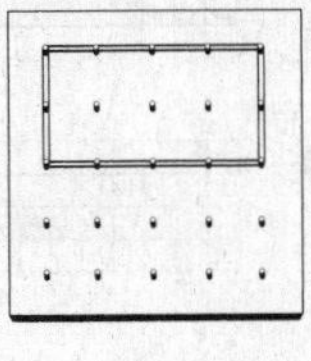

8.

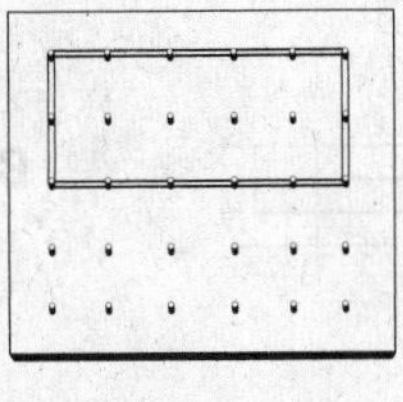

9.

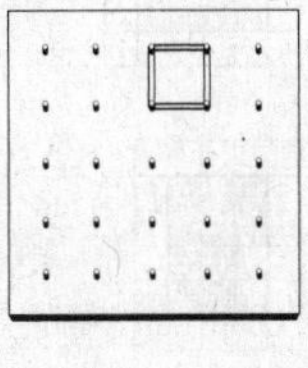

______________ ______________ ______________

Name________________________________ **Lesson 23.3**

Estimate and Measure Perimeter

Use string to estimate and measure the perimeter of each object.

1. this workbook ____________ ____________
2. the doorway to your bedroom ____________ ____________
3. the face of a TV ____________ ____________
4. the door of your refrigerator ____________ ____________

Find the perimeter of each figure.

5. ____________
6. ____________
7. ____________
8. ____________
9. ____________

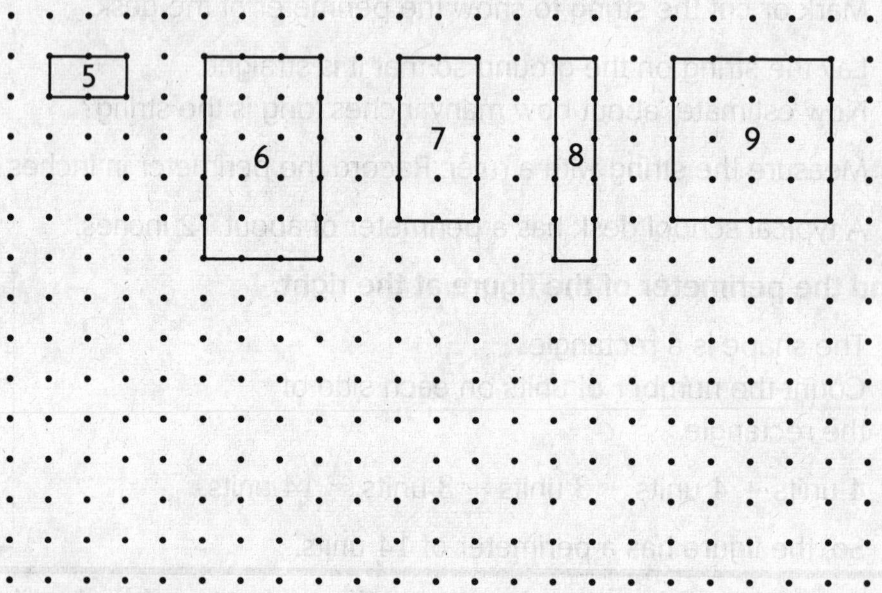

Problem Solving and Test Prep

For 10–11, use the dot paper above.

10. Draw and label a square with a perimeter of 8 units. What are the lengths of the sides?

____

11. Draw and label a square with a perimeter of 16 units. What are the lengths of the sides?

____

12. Which rectangle has the greatest perimeter?

A 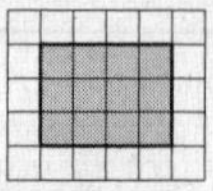

B 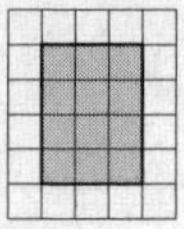

C 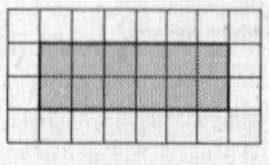

D 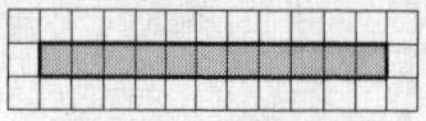

13. Which rectangle has the greatest perimeter?

A 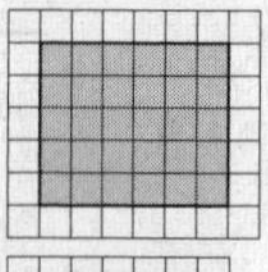

B 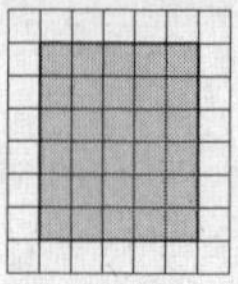

C 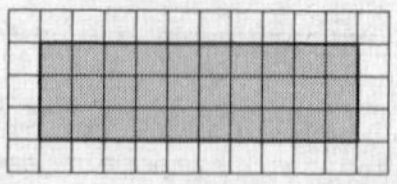

D 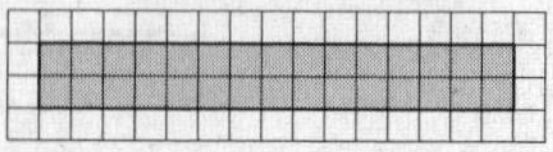

Name____

Algebra: Find Perimeter

You can find the perimeter of a figure by adding up the lengths of the sides, or by using a formula.

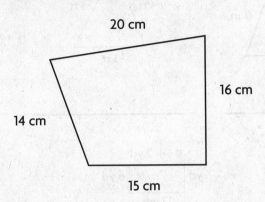

Find the perimeter.

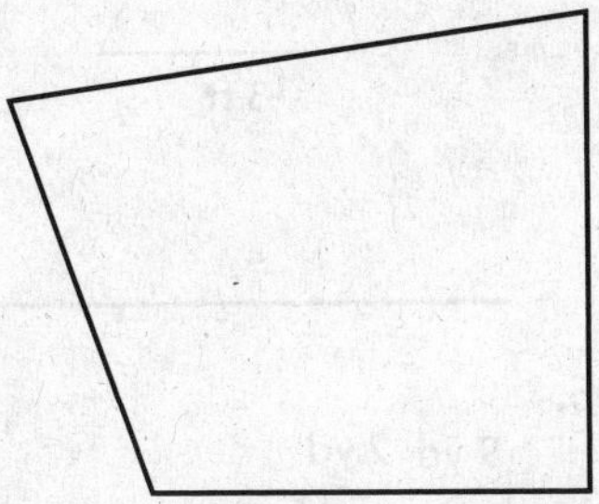

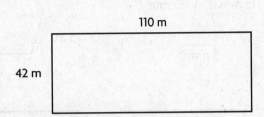

- Add up the lengths of all of the sides.
- The figure has sides that are 20 cm, 16 cm, 15 cm, and 14 cm long.
- $20 + 16 + 15 + 14 = 65$.
- So, the perimeter of the figure is 65 cm.

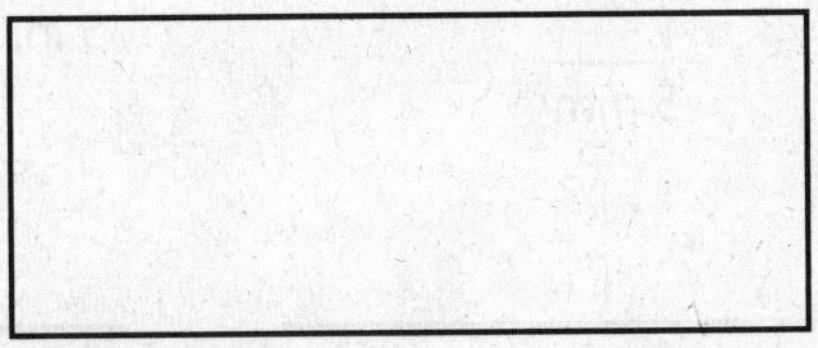

- Use the formula $P = (2 \times l) + (2 \times w)$. This means $Perimeter = (2 \times length + 2 \times width)$.
- Substitute the measurements for the length and width into the formula.

$$P = (2 \times l) + (2 \times w)$$
$$P = (2 \times 42) + (2 \times 110)$$
$$P = (84) + (220)$$
$$P = 304$$

- So, the perimeter of the figure is 304 m.

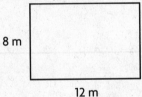

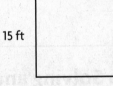

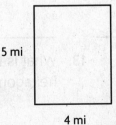

Find the perimeter.

1.

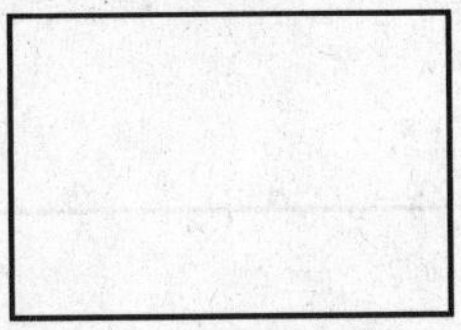

____

2.

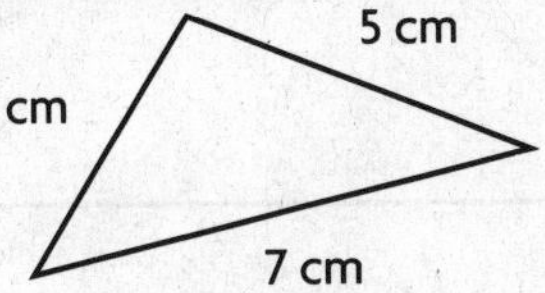

____

3.

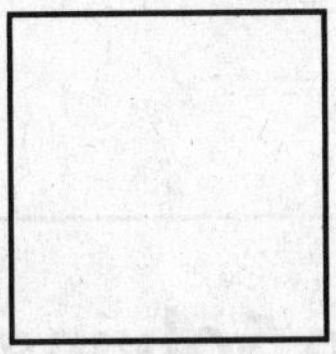

____

4.

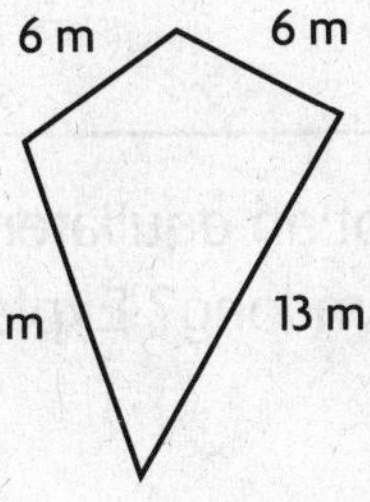

____

5.

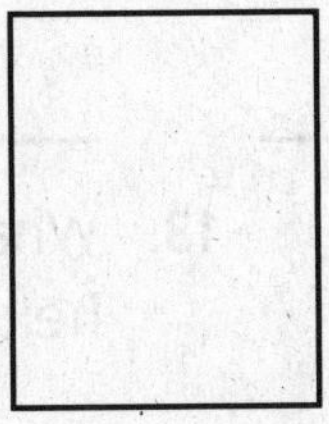

____

6.

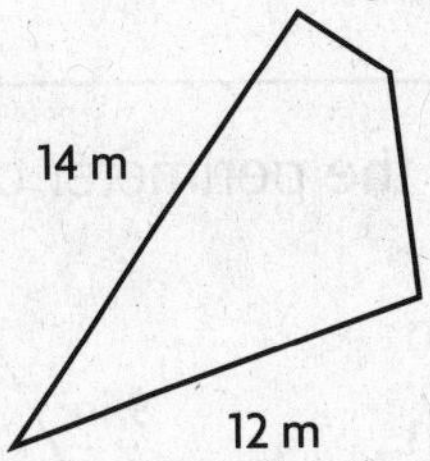

____

Name______________________________

Algebra: Find Perimeter

Find the perimeter.

1.

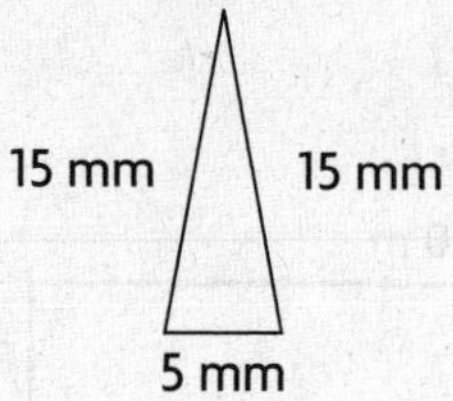

____

2.

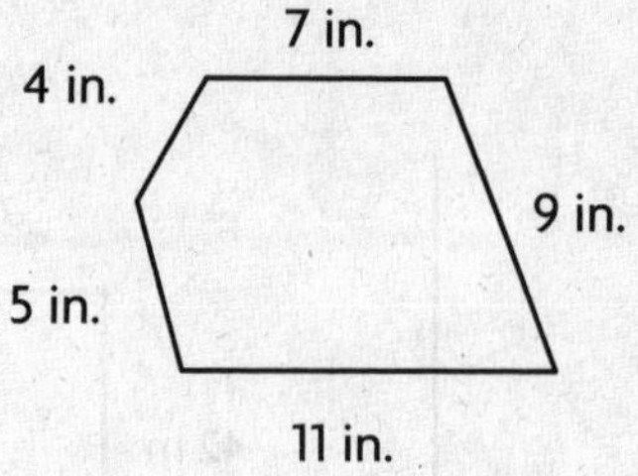

____

3.

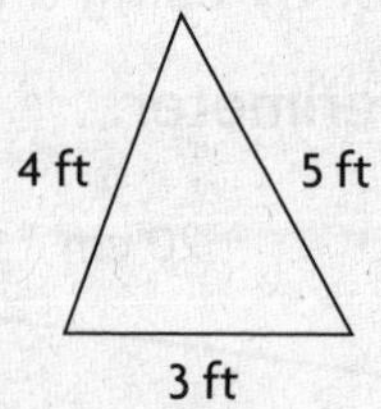

____

4.

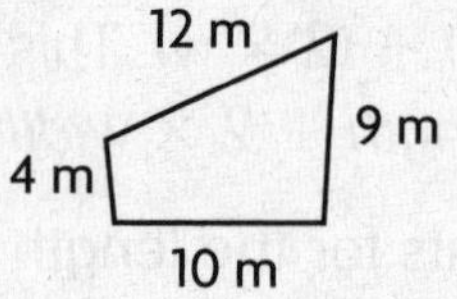

____

5.

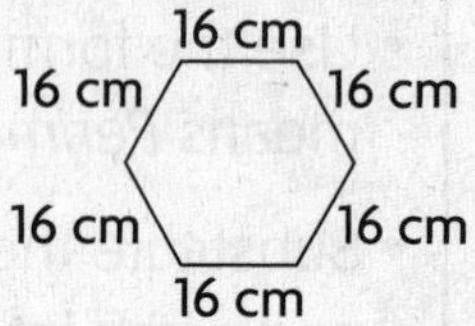

____

6.

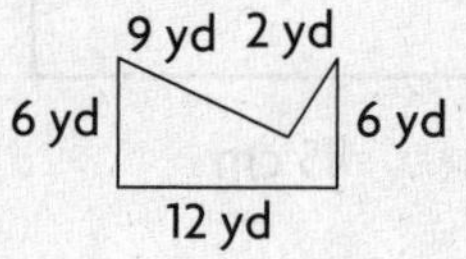

____

Use a formula to find the perimeter.

7.

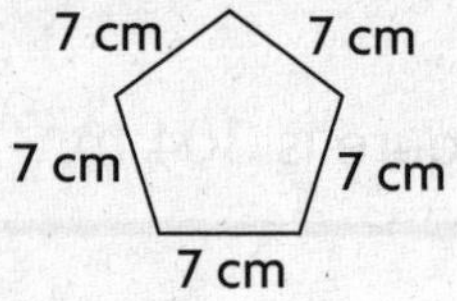

____

8.

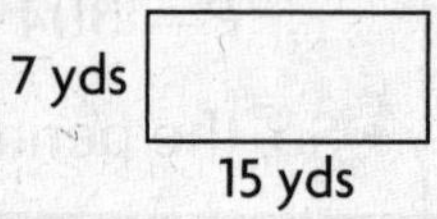

____

9.

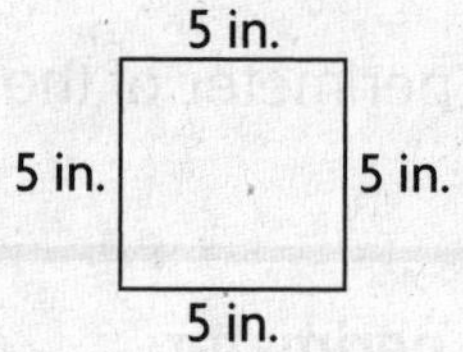

____

Problem Solving and Test Prep

10. Reasoning The perimeter of an isosceles triangle is 30 in. Its base is 8 in. How long are each of the other two sides?

____

11. Reasoning The perimeter of a rectangle is 46 ft. The width is 10 ft. What is the length?

____

12. What is the perimeter of this figure?

A 18 in.

B 27 in.

C 36 in.

D 45 in.

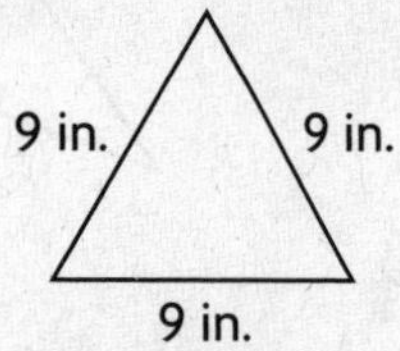

13. What is the perimeter of an equilateral hexagon with sides 6 cm long? Explain.

____

____

Name________________________________

Problem Solving Workshop Skill: Use a Formula

Stacy's bike path forms a pentagon. One side is 700 meters, one side is 650 meters, one side is 730 meters, and one side is 670 meters. The perimeter of the path is 3 km. What is the length of the fifth side of the bike path?

1. What are you asked to find?

____

2. What do you know about the bike path?

____

____

____

3. What measurement do you need to convert before you can solve the problem? Use a formula to make this conversion.

____

4. Think about the information you now have about the bike path. Place this information into the formula below to solve the problem.

$P = a + b + c + d + e$

■ = 700 + 650 + ___ + ___ + ___

5. What is the length of the fifth side of the bike path?

____

6. How can you check that your answer is correct?

____

____

Use a formula to solve.

7. The school playground is shaped like a trapezoid. The north side is 70 feet, the south side is 85 feet, and the west side is 60 feet. The perimeter of the playground is 295 feet. What is the length of the playground's east side?

____

8. Mary's quadrilateral shape yard is 26 meters on one side, 25 meters on another side, and 31 meters on a third side. The perimeter of the yard is 109 meters. How many meters long is the fourth side of the yard?

____

Name____

Problem Solving Workshop Skill: Use a Formula

Problem Solving Skill Practice

Use a formula to solve.

1. Stacy's backyard is 50 ft by 95 ft. She wants to put in a privacy fence like she saw at the Palms Marketplace. How many feet of fencing will Stacy need for the perimeter of her backyard?

dog run: 10 ft by 80 ft

sandbox: 13 ft by 50 ft

Zen garden 15 ft by 25 ft

____

For 2–3, use the diagram.

2. Stacy's dog runs one time around the perimeter of the dog run. How many feet does Stacy's dog run?

____

3. Stacy wants to place edging around the sandbox and the Zen garden. How many feet of edging will Stacy need in all?

____

Mixed Applications

4. Use the diagram above. How much more fencing does Stacy need to fence the sandbox than the Zen garden?

____

Bart's Building Supplies

Supply	Cost
How-to Book	$15
Outdoor Siding	$8/foot
Fencing	$15/yard

USE DATA For 5–6, use the table.

5. Bethany bought a how-to book and 80 feet of siding. How much did she spend?

____

6. Mr. Daley spent $195 for 3 How-to Books and fencing. How many yards of fencing did he buy?

____

Name ____________________ **Lesson 23.6**

Estimate Area

Area refers to how many **square units** (sq un) are needed to cover a surface. A square unit is 1 unit long and 1 unit wide.

Estimate the area of the figure. Each unit stands for 1 sq m.

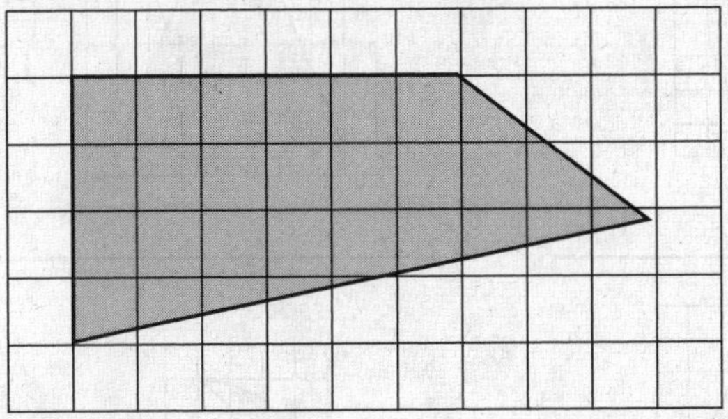

- Count the squares.
- There are 17 full squares.
- There are 3 almost-full squares.
- There are 6 half-full or almost half-full squares. $6 \div 2 =$ about 3 full squares
- Add up all of the squares you counted.

  $17 + 3 + 3 = 23.$

- So, the area of the figure is about 23 square meters.

Draw an oval on grid paper. Then estimate the area.

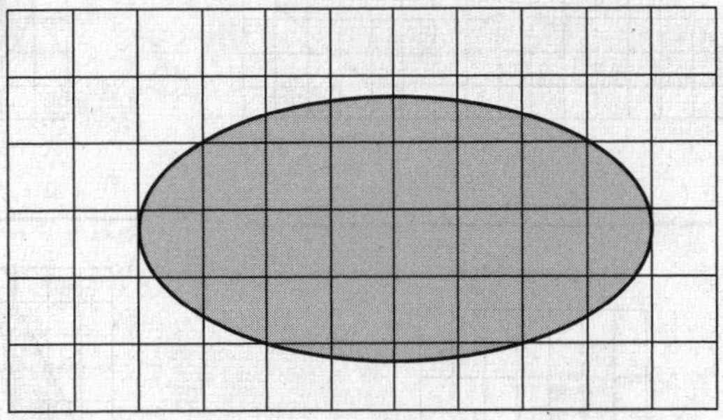

- There are 16 full squares.
- There are 7 almost-full squares.
- There are 6 half-full or almost half-full squares. $6 \div 2 =$ about 3 full squares

  $16 + 7 + 3 = 26.$

- So, the area of the oval is about 26 square units.

Estimate the area of each figure. Each unit stands for 1 sq m.

1.

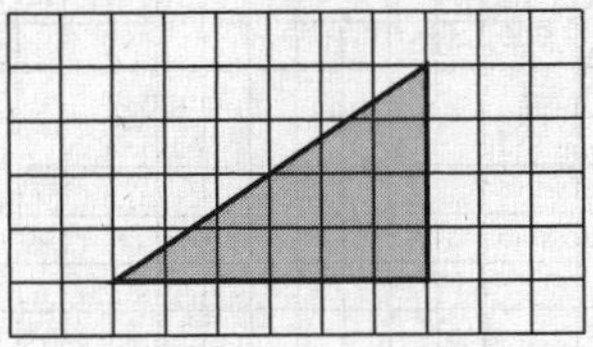

____

2.

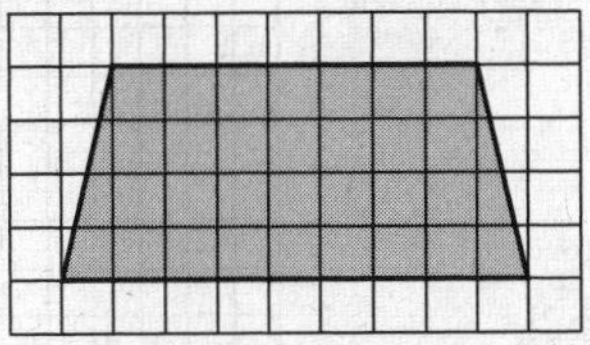

____

3.

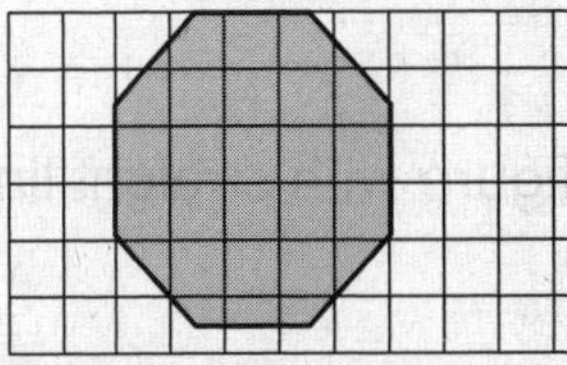

____

Draw each figure on grid paper. Then estimate the areas.

4. trapezoid

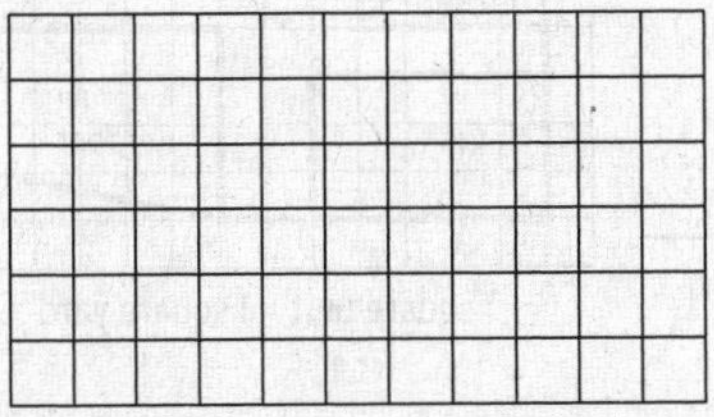

5. a figure with two straight lines and two curved lines

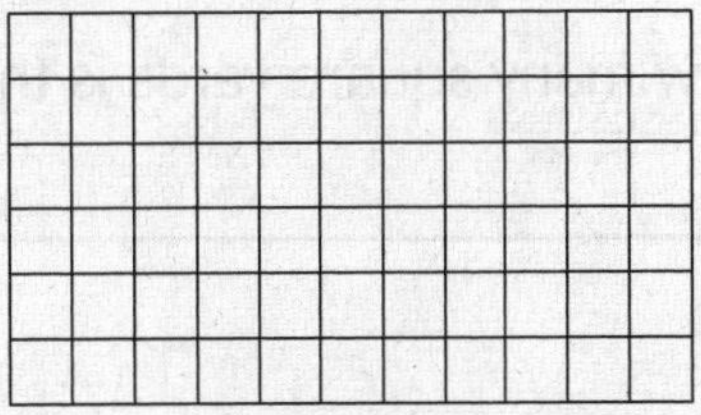

Name____

Estimate Area

Estimate the area of each figure. Each unit stands for 1 sq m.

1.

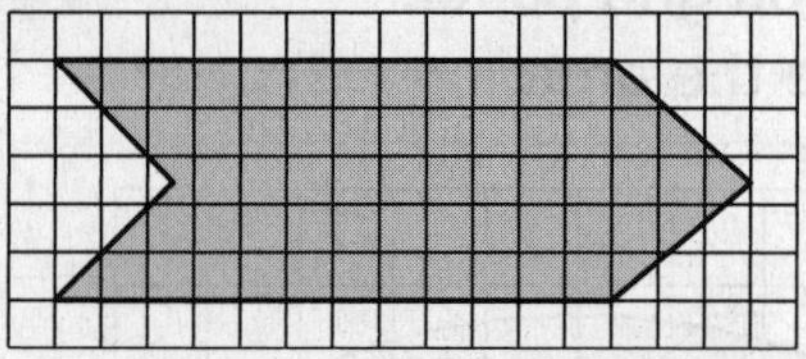

____

2.

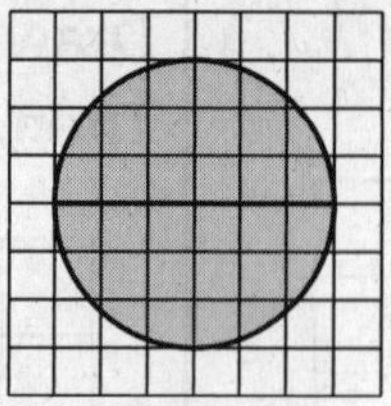

____

3.

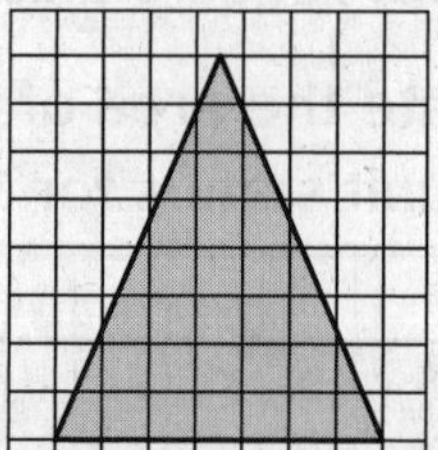

____

4.

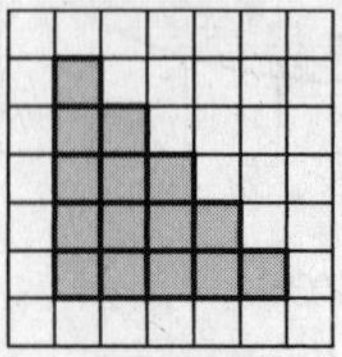

____

5.

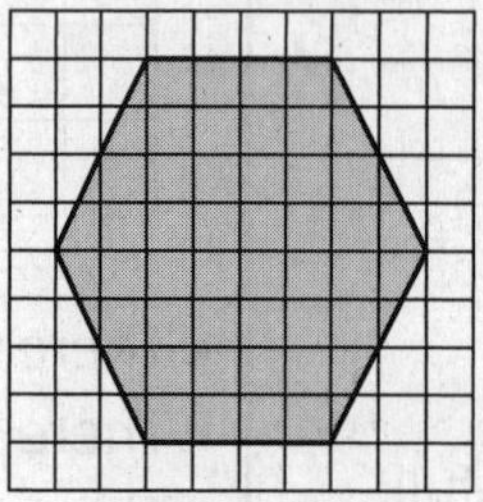

____

6.

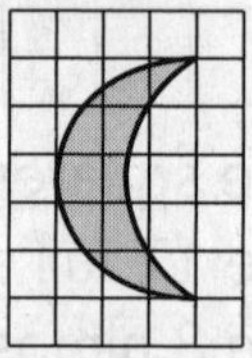

____

Draw each figure on the grid paper at the right. Then estimate the areas.

7. hexagon

8. right triangle

9. figure with straight lines

10. figure with curved and straight lines

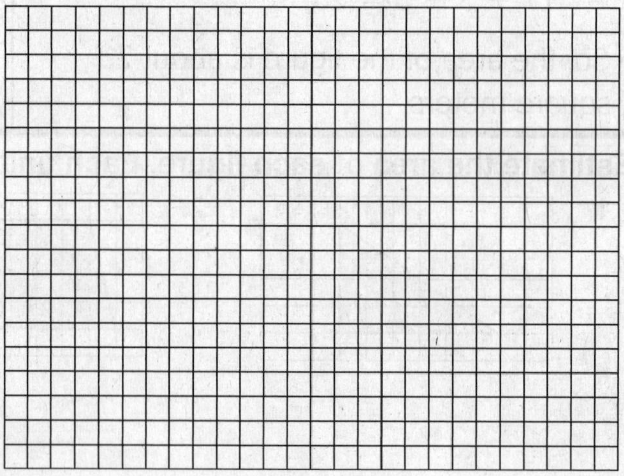

USE DATA For 11–12, use the diagram.

11. About how many square yards is the hallway?

____

12. About how many square yards is the closet?

____

Floor Plan

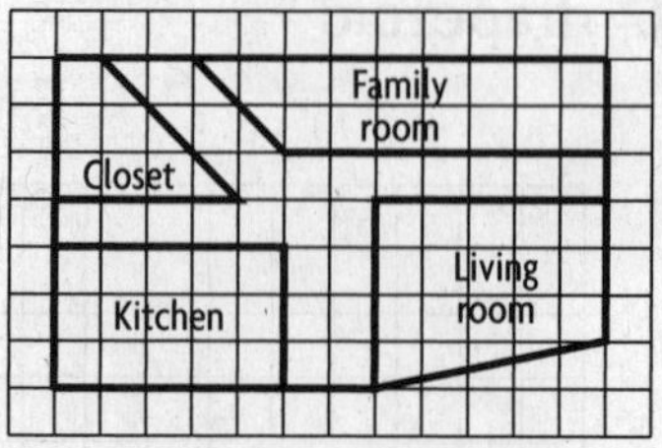

1 square unit = 1 square yard

Spiral Review

For 1–5, divide. You may wish to use counters or draw a picture to help.

1. $16 \div 3$

____

2. $95 \div 2$

____

3. $6\overline{)38}$

____

4. $7\overline{)52}$

____

5. $9\overline{)84}$

____

For 8–10, list all the possible outcomes of each experiment.

8. tossing a quarter

____

9. spinning the pointer of a spinner with a red, yellow, green, and blue section

____

10. tossing a quarter and spinning the same pointer

____

____

____

For 6–7, label each of the following on circle *J*.

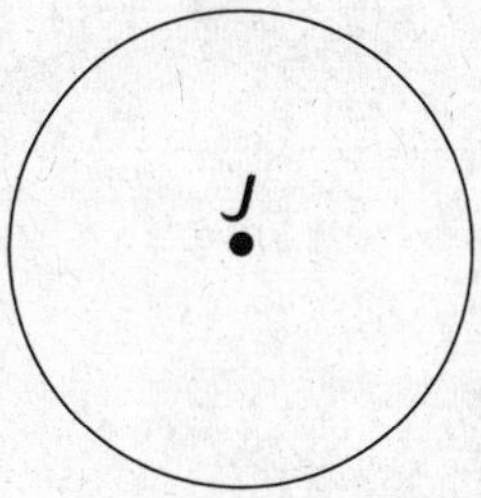

6. radius *JP*
7. diameter *LM*

For 11–16, find the value of each expression if $x = 6$ and $y = 3$.

11. $y + 9$

____

12. $11 + (x - 5)$

____

13. $(y + 8) - 2$

____

14. $(x - y) + 10$

____

15. $(x + 15) - y$

____

16. $36 - (x + y)$

____

Name____ **Lesson 23.7**

Algebra: Find Area

You can use multiplication and formulas in order to find the area of a figure. You must know the figure's length and width.

Find the area. Use multiplication.

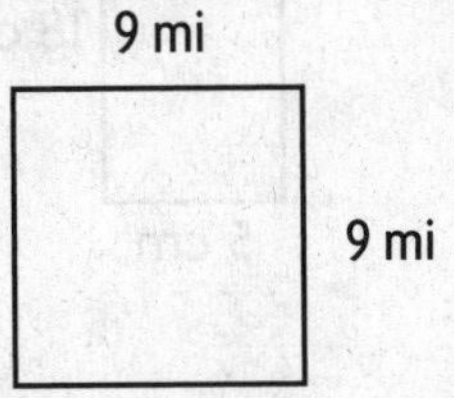

- Find the figure's length. The length is the distance across the figure. The length of the figure is 9 miles.
- Find the figure's width. The width is the distance from the top of the figure to the bottom of the figure. The width of the figure is 9 miles.
- Multiply the length by the width.

  $9 \times 9 = 81.$

- So, the area of the figure is 81 square miles.

Find the area. Use a formula.

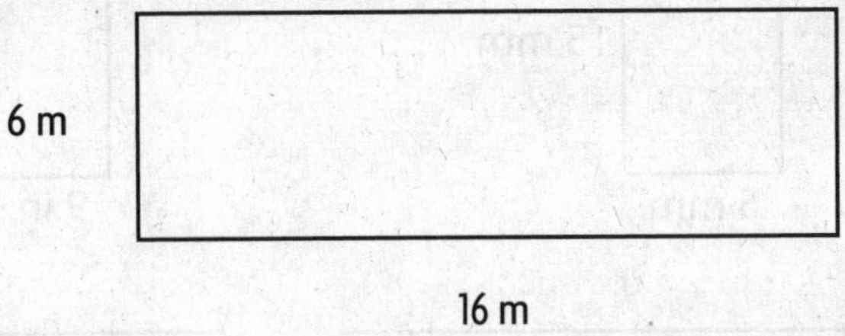

- Use the formula $A = l \times w$. This means to multiply the figure's length by its width.
- The figure is 16 m long and 6 m wide.

  $A = l \times w$

  $A = 6 \times 16$

  $A = 96$

- So, the figure has an area of 96 square m.

Find the area.

1. 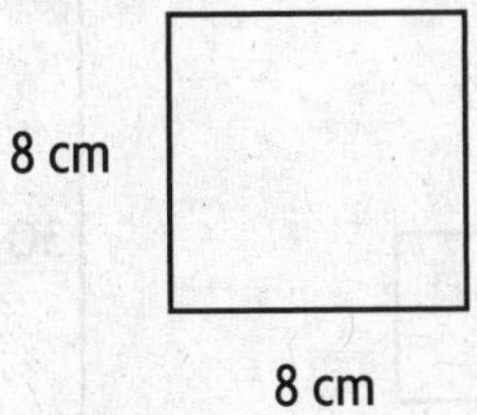

____

2. 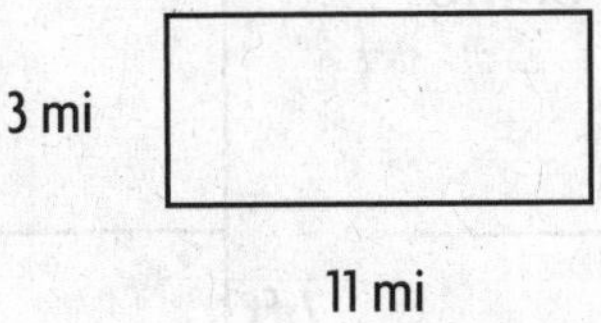

____

3. 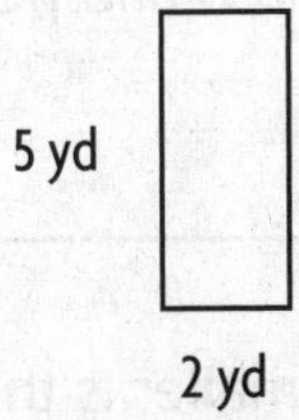

____

4. 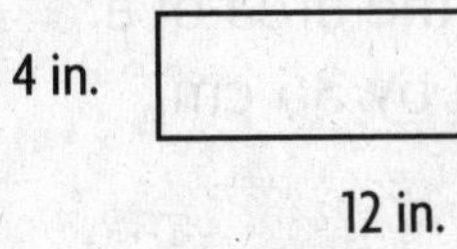

____

5. 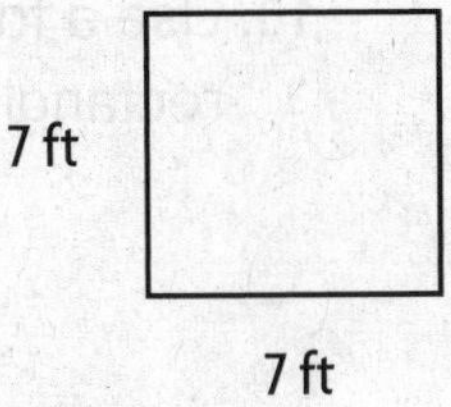

____

6. 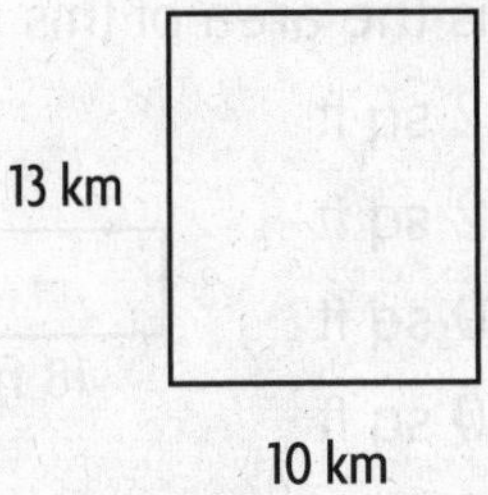

____

MG 1.4 Understand and use formulas to solve problems involving perimeters and areas of rectangles and squares. Use those formulas to find the areas of more complex figures by dividing the figures into basic shapes.

Algebra: Find Area

Find the area.

1.
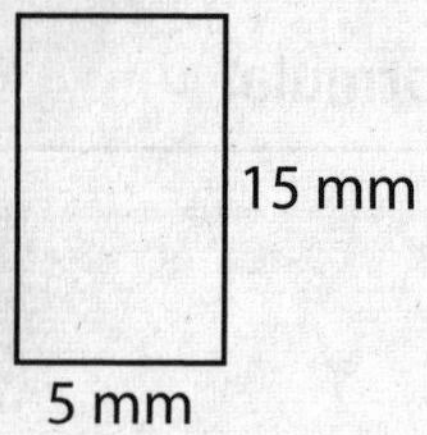

____

2.
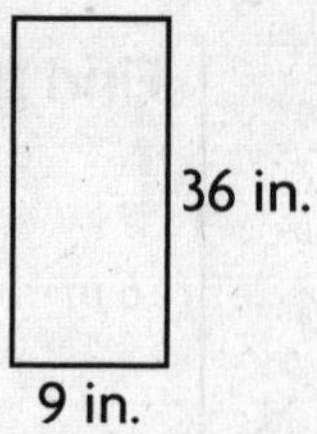

____

3.
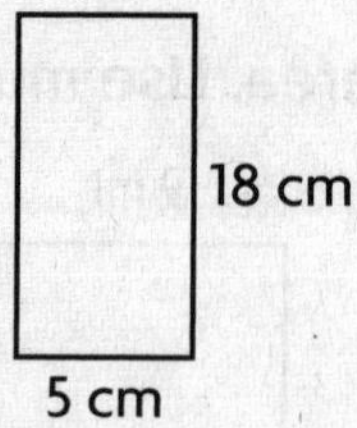

____

Find the area and perimeter.

4.
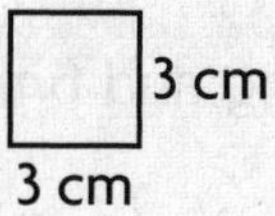

____

____

5.
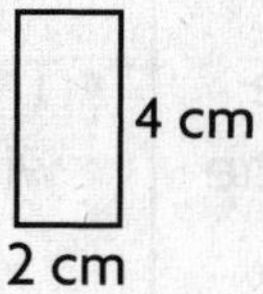

____

____

6.
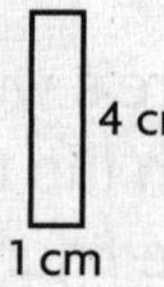

____

____

Problem Solving and Test Prep

For 7–8, use the diagram.

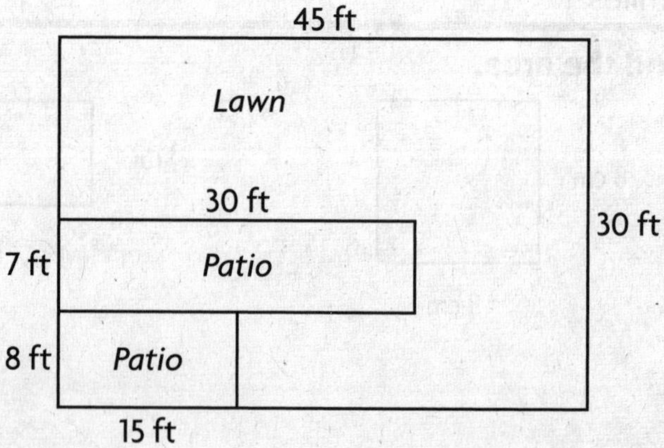

7. What is the area and perimeter of the entire patio?

____

8. How much smaller is the area of the patio than the area of the lawn?

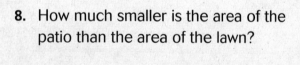

____

9. What is the area of this figure?

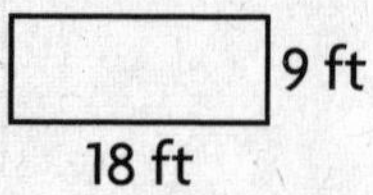

A 152 sq ft

B 162 sq ft

C 180 sq ft

D 200 sq ft

10. Use a formula to find the area of a rectangle that is 7 cm by 35 cm.

____

Name ____________________

Problem Solving Workshop Strategy: Solve a Simpler Problem

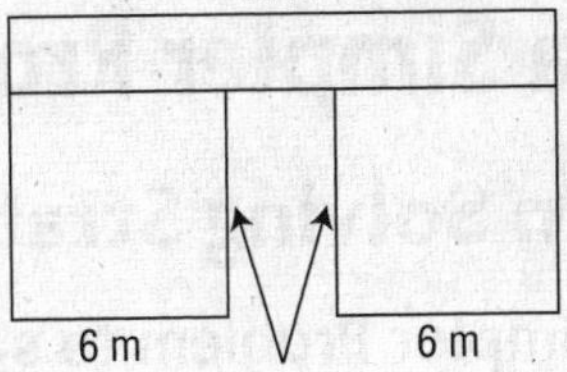

George is competing in a sandcastle competition. The diagram shows part of the competition arena. The competition staff decides to put a fence around the outside part of the arena shown. How much fencing is needed?

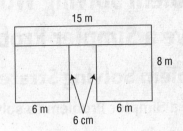

Read to Understand

1. What are you asked to find?

____

Plan

2. Why do you need to use the Solve a Simpler Problem strategy to solve this problem?

____

____

Solve

3. Solve the simpler problem. Then solve the main problem. Show your work below. How much fencing is needed?

$15 = 6 + ■ + a$ $a = ■$. So, ______________ is ■ m.

Perimeter of arena = ■ + ■ + ■ + ■ + ■ + ■ + ■ + ■

____

Check

4. How did this strategy help you solve the problem?

____

Solve a simpler problem to solve.

5. What is the perimeter of the figure?

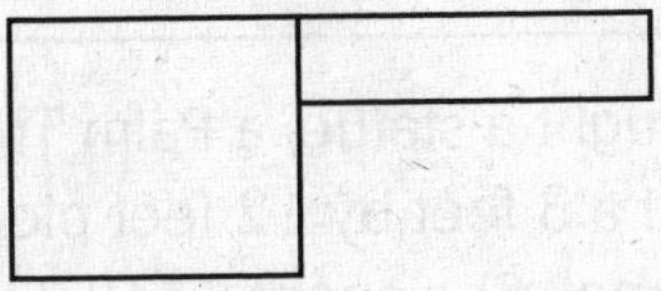

____

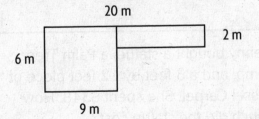

6. What is the total area of the figure?

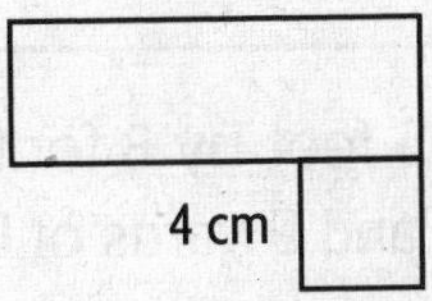

____

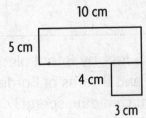

MG 1.4 Understand and use formulas to solve problems involving perimeters and areas of rectangles and squares. Use those formulas to find the areas of more complex figures by diving the figures into basic shapes.

Name______________________________

Problem Solving Workshop Strategy: Solve a Simpler Problem

Problem Solving Strategy Practice

Solve a Simpler Problem to solve.

For 1–3, use the diagram.

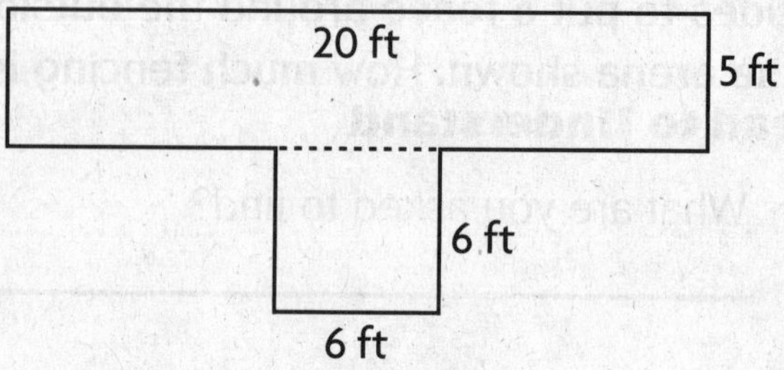

1. Workers will put sod on the meadow shown at the right. How many square feet of sod do they need?

____

2. How many feet of fencing will be needed to enclose the meadow?

____

3. What if the square were 12 ft by 12 ft? How much greater would the area of the entire meadow be?

____

Mixed Strategy Practice

4. Look at the diagram above. What if the meadow had 2 more 6 ft by 6 ft squares, one on each side of the existing square. How much fencing would be needed to enclose the entire meadow?

____

Vic's Souvenir Shoppe	
Supply	**Cost**
Palm Tree Lamp	\$25
Sissel Carpet	\$8/square foot
Border Fencing	\$15/yard
5-Foot Mango Light String	\$9

USE DATA For 5–8, use the table.

5. Justine wants to string Mango lights from Vic's Souvenir Shop around a 7-foot square patio. How many strings of lights will she need?

____

6. Wally bought 3 strings of Mango lights, a Palm Tree Lamp, and 3 yards of Border Fencing. How much change would he get from a \$100 bill?

____

7. Grant bought a 5 feet by 8 feet piece of Sissel Carpet and 9 yards of Border Fencing. How much did he spend?

____

8. Jenny bought a statue, a Palm Tree lamp, and a 3 feet by 12 feet piece of Sissel Carpet. She spent \$348. How much did the statue cost?

____

Name____

Relate Perimeter and Area

The perimeter and area of a figure are related. Sometimes when you change the perimeter, the area stays the same. Similarly, the perimeter of a figure may stay the same when you change the area.

Find the area and perimeter of the square below. Then draw another figure that has the same perimeter but a different area.

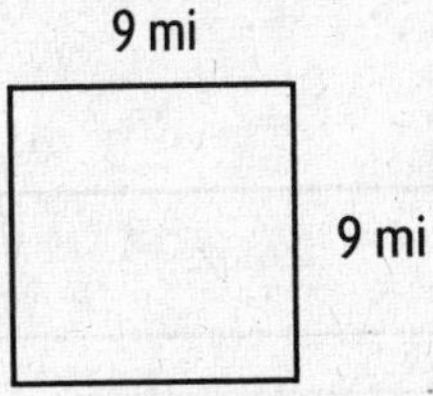

- $9 + 9 + 9 + 9 = 36$
  The figure's perimeter is 36 mi.
- $9 \times 9 = 81$
  The figure's area is 81 sq mi.
- The figure you draw must have a perimeter of 36, and an area greater than or less than 81.

Try this figure:

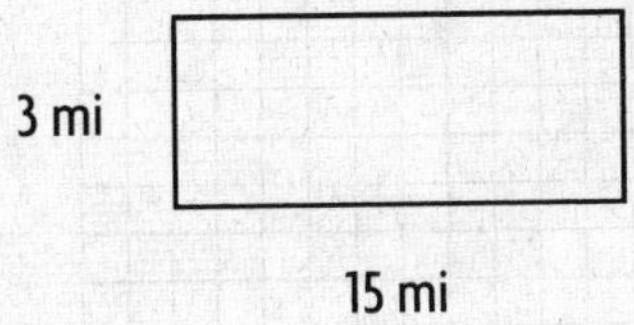

- $15 + 3 + 15 + 3 = 36$
  The figures perimeter is also 36 mi.
- $15 \times 3 = 45$.
  The figure's area is different. It is 45 sq mi.
- So, a rectangle that has sides of 15, 15, 3, and 3 has a perimeter of 36 miles but an area of 45 sq mi., and is a correct solution.

Find the area and perimeter of each figure. Then draw another figure that has the same perimeter but a different area.

1. 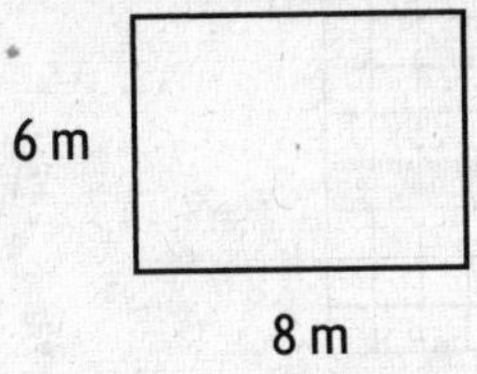

2. 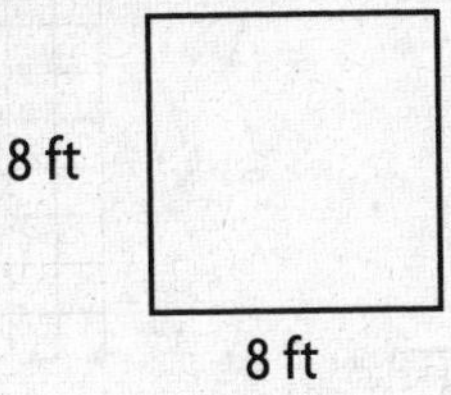

3. 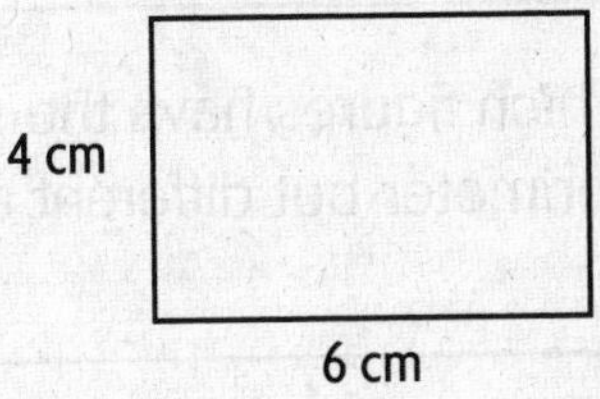

______________ ______________ ______________

Name____

Relate Perimeter and Area

Find the area and perimeter of each figure. Then draw another figure that has the same perimeter but a different area.

1.

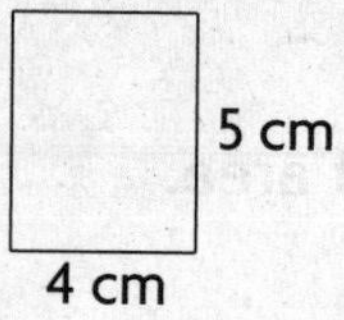

2.

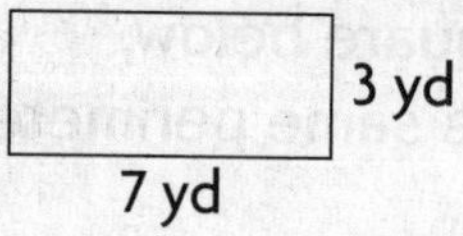

3.

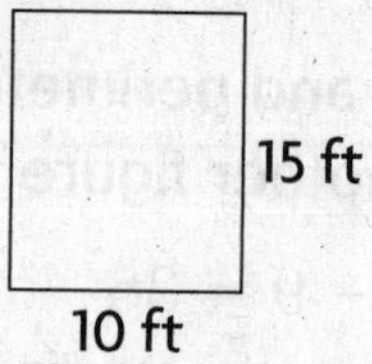

______________________ ______________________ ______________________

______________________ ______________________ ______________________

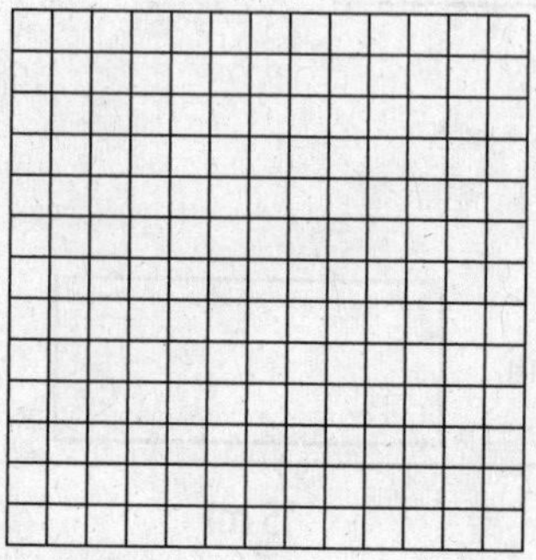

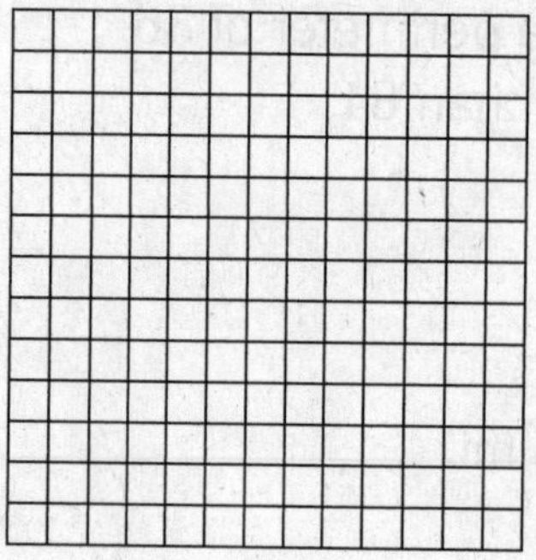

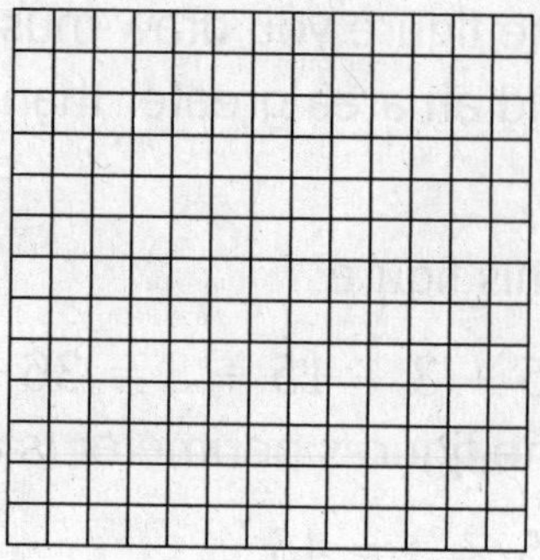

Problem Solving and Test Prep

For 4–5, use figures a–c.

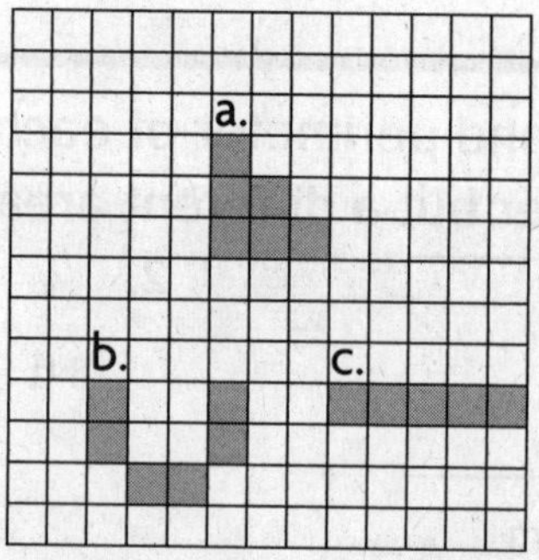

4. Which figures have the same area but different perimeters?

____

5. Which figures have the same perimeter but different areas?

____

6. The rectangles below have the same area. Which has the greatest perimeter?

A 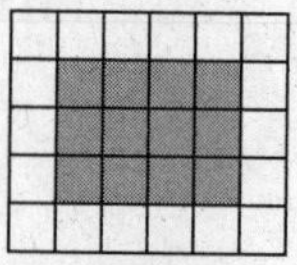

C 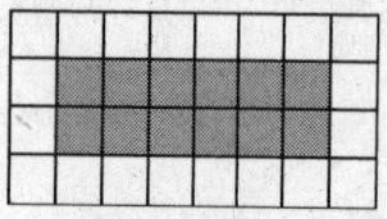

B 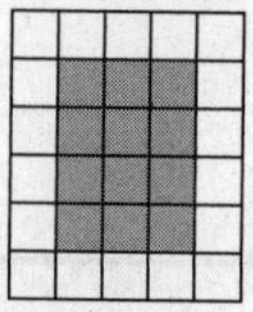

D 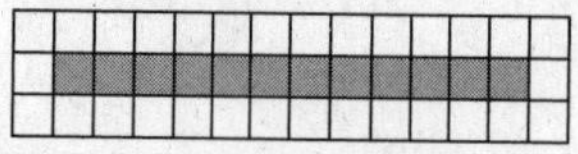

7. The rectangles below have the same perimeter. Which has the greatest area?

A 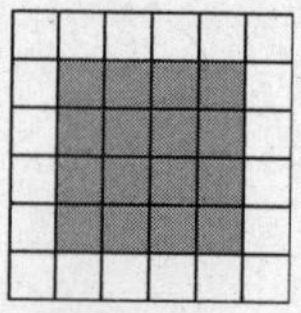

C 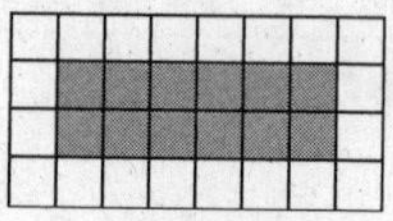

B 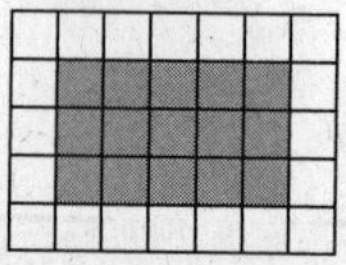

D 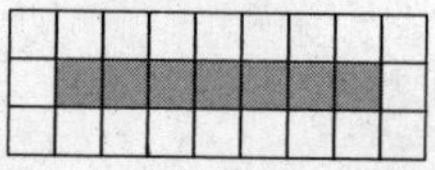

Name____________________

List All Possible Outcomes

Use the table.
List all the possible outcomes for the experiment.

Sally's Experiment
Roll a Number Cube and Toss a Coin

	1	2	3	4	5	6
Heads	//		///		//	//
Tails	/	//	/	//		////

- Make an organized list.
- Write the possible outcomes for the coin: heads, tails.
- Write the possible outcomes for the number cube: 1, 2, 3, 4, 5, 6.
- Pair the possible outcomes of the coin with the possible outcomes of the number cube. *Heads, 1; Heads, 2; Heads, 3; Heads, 4; Heads, 5; Heads, 6; Tails, 1, Tails, 2; Tails, 3; Tails, 4; Tails, 5; Tails, 6.*
- So, the possible outcomes for tossing the coin and rolling the number cube are *Heads, 1; Heads, 2; Heads, 3; Heads, 4; Heads, 5; Heads, 6; Tails, 1, Tails, 2; Tails, 3; Tails, 4; Tails, 5; Tails, 6.*

Use the picture below. List all the possible outcomes of the experiment.

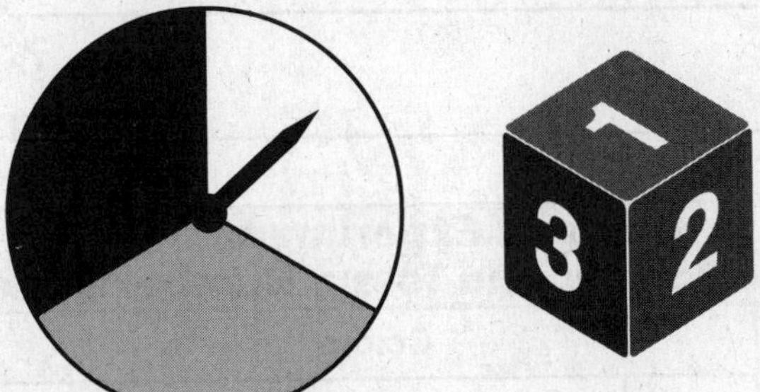

1. How many possible outcomes are there?

____

____

____

____

For 2 – 3, use the table below.

Todd's Experiment
Spin the Pointer and Toss a Coin

	1	2	3	4	5
Heads	///	/		//	////
Tails	//	//	/		/

2. How many possible outcomes are there?

____

3. How many times did the outcome *Tails, 5* occur?

____

Name ______________________________

List All Possible Outcomes

USE DATA For 1–4, use the pictures. List all possible outcomes for each experiment.

1. spinning the pointer

____

____

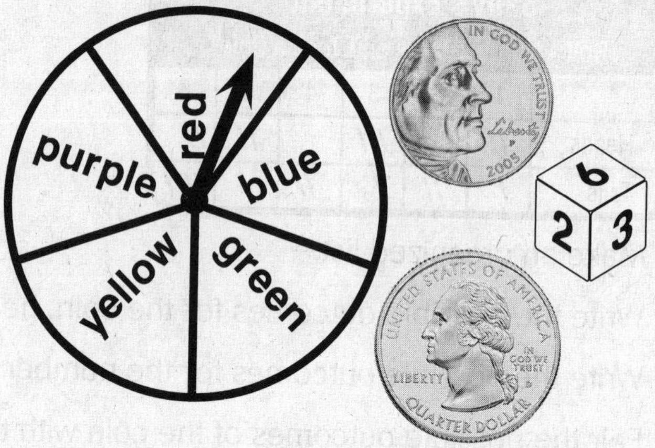

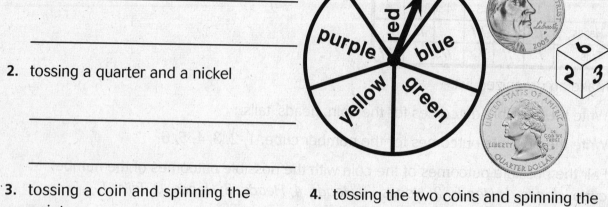

2. tossing a quarter and a nickel

____

____

3. tossing a coin and spinning the pointer

____

____

____

____

4. tossing the two coins and spinning the pointer

____

____

____

____

USE DATA For 5–8, use the table.

5. How many times did the outcome *Green, 5* occur? __________

6. How many times did the outcome *Yellow, 4* occur? __________

7. List all the possible outcomes of the experiment.

____

____

____

____

Ahmed's Experiment
Spin the Pointer and Toss a Number Cube

Number Cube	Colors				
	Red	Blue	Green	Yellow	Purple
1	III		I	IIII	II
2		IIII	III	I	I
3	II	III			I
4	IIII	I	III		III
5	I		IIII	III	I
6	II	IIII		II	I

8. How many possible outcomes did Ahmed have by spinning the pointer and tossing the cube? __________

Name____

Problem Solving Workshop Strategy: Make an Organized List

Yuri is playing a game using a coin and Spinner 1. He tosses the coin and spins the pointer. List all the possible outcomes.

Read to Understand

1. What are you asked to find?

____

Plan

2. How can making an organized list help you solve this problem?

____

Solve

3. Make an organized list to solve the problem. Show your work below.

heads, 1	heads, 2	____ , ___	____ , ___
____ , 1	____ , ___	____ , 3	tails, 4

4. How many possible outcomes were there for this problem?

____

Check

5. Does the answer make sense for the problem? Explain.

____

Make an organized list to solve.

6. Tom spins the pointer and tosses a number cube with sides labeled 1–6. List all the possible outcomes.

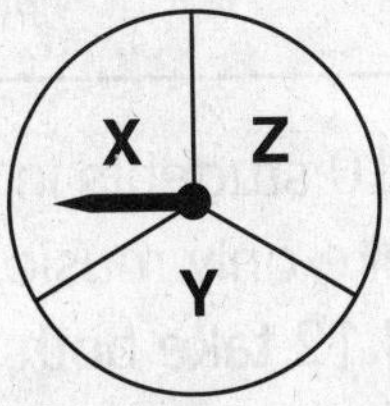

____

____

7. Mary spins the pointer below and also tosses a coin. How many possible outcomes are there?

____

Name______________________________

Problem Solving Workshop Strategy: Make an Organized List

Problem Solving Strategy Practice

USE DATA For 1–3, use the spinners. Make an organized list to solve.

1. Franco made these spinners for a school carnival game. What are the possible outcomes?

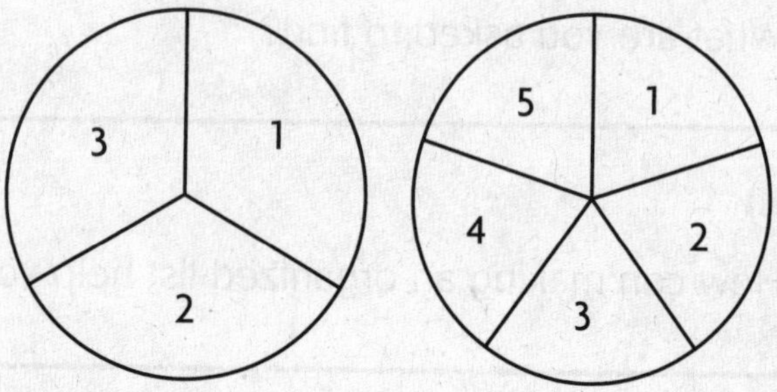

2. To win, Gloria must spin both pointers for a total more than 6. Name the ways Gloria can win.

3. Patty can win if she spins both pointers for a total of more than 5. Name the ways Patty can win.

Mixed Strategy Practice

4. Pedro is making cards for a game. Each type of card will be a different color. The suits will be hearts and flags. In each suit, there will be 3 sets: numbers, letters, and symbols. How many colors will there be?

5. **Open Ended** You probably made an organized list to solve Exercise 4. What is another strategy you could use to solve it? Explain.

6. Jorge's father has driven his car 103,240 miles. His mother has driven hers 69,879. How much further has his father driven?

7. There are 110 students in fourth grade. Thirty-two take only music, 25 take only art, and 12 take both. How many students do not take art or music?

Spiral Review

For 1–2, complete the factor tree to find the prime factors.

1. 16

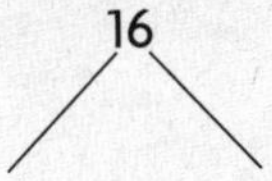

2. 81

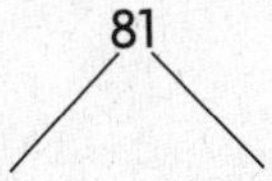

For 3–5, tell whether the rays on the circle show a $\frac{1}{4}$, $\frac{1}{2}$, $\frac{3}{4}$, or full turn. Then identify the number of degrees the rays have been turned.

3.

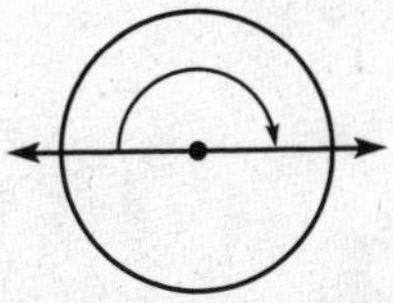

____

4.

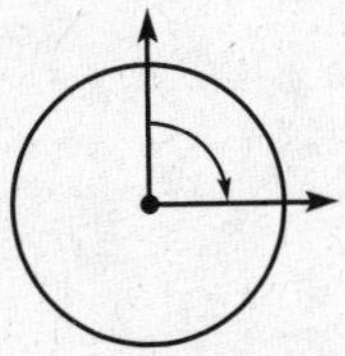

____

5.

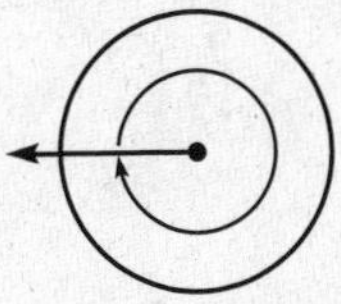

____

For 6–8, use the spinner to find the probability of each event. Write the answers as fractions.

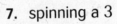

6. spinning a 2

____

7. spinning a 3

____

8. spinning a 4

____

For 9–10, write an equation. Then work backward to solve.

9. The Lion Preserve has many lions. This month, 3 were transferred to a local zoo. Later in the month, 4 were added, making 10 lions in all. How many lions were there at the beginning of the month?

____

____

10. Tina received 4 toy cars for her birthday. She gave away 5. She now has 9 toy cars. How many cars did Tina have before her birthday?

____

____

Make Predictions

You can predict the likelihood of some events. Some events are **certain**, which means they will always happen. Some events are **impossible**, which means they will never happen. An event is **likely** if it has a greater than even chance of happening. An **unlikely** event has a less than even chance of happening. Two events can be **equally likely**, which means that they have the same chance of happening.

Tell whether the event is *likely, unlikely, certain,* or *impossible.*

Tossing a number greater than 1 on a cube labeled 1 to 6.

- The number cube has the numbers 1, 2, 3, 4, 5, and 6. Five of these numbers are greater than 1, so there is a better than even chance of tossing a 2, 3, 4, 5, or 6.
- So, the event is *likely*.

Spinning a multiple of 4 on a spinner with 4 equal parts labeled 4, 8, 12, and 16.

- Each of the numbers on the spinner is a multiple of 4. Therefore, the pointer will always land on a multiple of 4.
- Events that will always happen are certain.
- So, the event is *certain*.

Tell whether the event is *likely, unlikely, certain,* or *impossible.*

1. Pulling a red marble from a bag that contains 6 green marbles, 4 white marbles, and 2 yellow marbles.

____

2. Tossing a number greater than 2 on a number cube labeled 1 to 6.

____

3. Spinning an odd number on a spinner with four equal parts labeled 3, 5, 7, and 9.

____

4. Pulling a white marble from a bag that contains 12 green marbles, 2 white marbles, and 14 yellow marbles.

____

5. Tossing a number less than 6 on a number cube labeled 1 to 6.

____

6. Spinning a multiple of 6 on a spinner with five equal parts labeled 21, 24, 31, 35, and 47.

____

Name_______________________________________

Make Predictions

Tell whether the event is *likely, unlikely, certain,* or *impossible.*

1. Having the pointer land on blue on a spinner with equal sections of red, yellow, and green

____

2. Tossing the number 2 on a cube numbered from 1 to 6

____

3. Pulling a red tile from a bag that contains 6 red, 2 white, and 1 blue tile

____

4. Having a pointer land on red on a spinner that is all red

____

For each experiment, tell whether Events A and B are *equally likely* or *not equally likely*. If they are not equally likely, name the event that is more likely.

5. Experiment:
Toss a cube numbered 1–6.
Event A: tossing an odd number
Event B: tossing an even number

____

6. Experiment:
Spin the pointer
Event A: blue
Event B: yellow

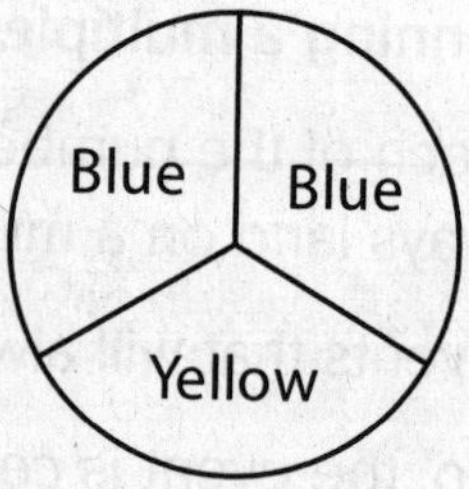

____

Problem Solving and Test Prep

USE DATA For 7–10, use the spinner.

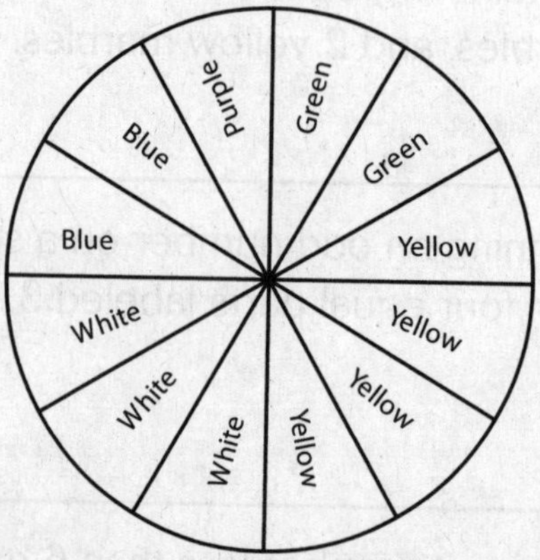

7. Which two events are equally likely?

____

8. Which event is most likely?

____

9. Which event is impossible?

A brown
B blue
C purple
D green

10. Which event is least likely?

A yellow
B blue
C purple
D green

Name________________________________

Lesson 24.4

Probability as a Fraction

You can use numbers to express the **mathematical probability** of an event happening. If an event will never happen, its mathematical probability is 0. If the event is certain to happen, its mathematical probability is 1. All other mathematical probabilities are written as fractions. The numerator tells the number of favorable outcomes, and the denominator tells the number of possible outcomes.

Use the spinner. Write the probability as a fraction. Find the probability of not spinning gray.

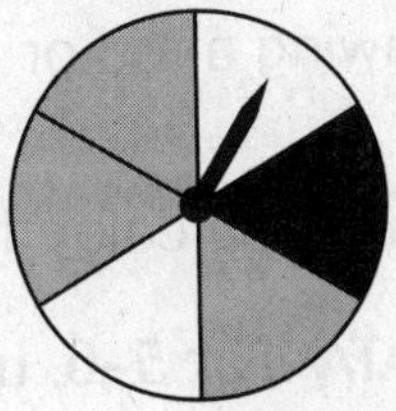

- There are 2 white sections, 3 gray sections, and 1 black section. So there are 6 possible outcomes.
- A favorable outcome would be to land on a section that is not gray. So there are 3 favorable outcomes: landing on 2 white sections and on 1 black section.
- Plug this information into the fraction.

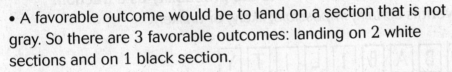

$$\text{probability of not gray} = \frac{\text{favorable outcomes (2 whites, 1 black)}}{\text{possible outcomes (3 gray, 2 white, 1 black)}} = \frac{3}{6}$$

- So, the probability of not gray is $\frac{3}{6}$, or $\frac{1}{2}$.

For 1–3, use the spinner. Write the probability as a fraction.

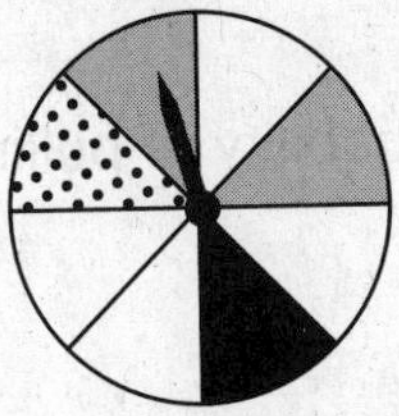

1. spinning white

____

2. not spinning gray

____

3. spinning gray or spotted

____

For 4–6, use the spinner. Write the probability as a fraction.

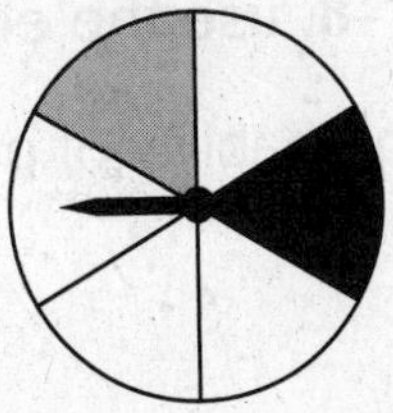

4. not spinning white

____

5. spinning gray or white

____

6. not spinning white or black

____

Name ____________________

Lesson 24.4

Probability as a Fraction

USE DATA For 1–4, use the equal-sized marbles. Write the probability as a fraction.

1. drawing a white marble ____________

2. drawing an orange marble ____________

3. drawing a red or a yellow marble ____________

4. drawing a color that is not green ____________

USE DATA For 5–6, use the equal-sized cards. Write the probability as a fraction. Then, tell whether each event is *certain*, *impossible*, *likely*, or *unlikely*.

| P | R | O | B | A | B | I | L | I | T | Y |

5. pulling an L

____

6. pulling a B or an I

____

Problem Solving and Test Prep

USE DATA For 7–8, use the equal-sized cards above.

7. What is the probability of pulling a C, F, or E?

____

8. What is the probability of pulling an A, P, R, O, L, or T?

____

9. All the marbles are the same size. What is the probability of pulling a green marble?

A $\frac{1}{12}$

B $\frac{1}{4}$

C $\frac{1}{2}$

D $\frac{3}{4}$

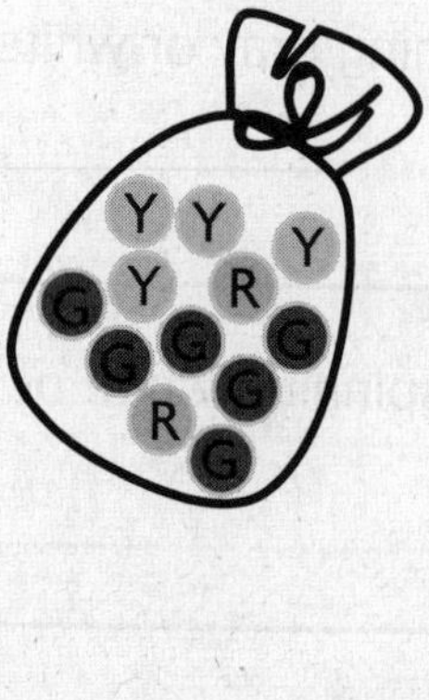

10. What is the likelihood of pulling a pink tile from a bag of pink tiles? Explain.

____

____

____

Name____ **Lesson 24.5**

Experimental Probability

Toss a cube labeled 1 to 6 thirty times. Record the outcomes in a tally table. Write the experimental probability of rolling 1 as a fraction.

Results						
Outcome	1	2	3	4	5	6
Tally	////	𝍸 //	///	////	//	𝍸 𝍸

- The number 1 was tossed 4 times. There were a total of 30 tosses.
- $\dfrac{\text{number of times tossed}}{\text{total number of tosses}} = \dfrac{4}{30}$
- So, the experimental probability of tossing 1 is $\frac{4}{30}$, or $\frac{2}{15}$.

What is the experimental probability of not spinning black? What is the mathematical probability?

Maryellen's Results				
Outcomes	White	Gray	Black	Spotted
Tally	𝍸 𝍸 /	𝍸 ///	𝍸 ///	𝍸 //// 𝍸

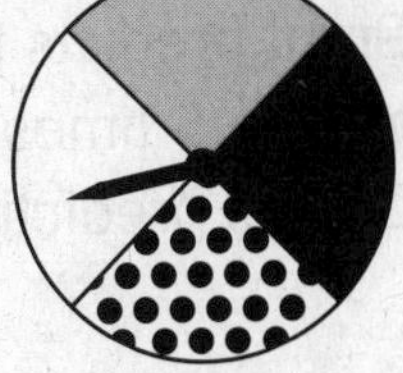

- experimental probability:

$\dfrac{\text{number of times not spinning black}}{\text{total number of spins}}$ $\dfrac{33}{41}$

- mathematical probability:

$\dfrac{\text{favorable outcomes}}{\text{total possible outcomes}} = \dfrac{3}{4}$

- So, the experimental probability of not spinning shaded is $\frac{33}{41}$, and the mathematical probability is $\frac{3}{4}$.

Complete the exercise below.

1. Flip a coin 20 times. Record the outcomes in a tally table. Write the experimental probability of flipping heads as a fraction.

Outcome	Heads	Tails
Tally		

____

Use the number cube and the table.

2. What is the experimental probability of tossing 4? What is the mathematical probability?

Results						
Outcome	1	2	3	4	5	6
Tally	///	///	/	////	//	///

____

Name ____

Experimental Probability

1. Toss a coin 20 times. Record outcomes in the tally table. Write as a fraction the experimental probability of heads.

Tally Table		
Outcome	Heads	Tails
Tally		

____

2. **Reasoning** Grant plans to pull a marble from the bag, return it, and then choose another one 30 times. Grant predicts that he will pull a yellow marble 5 times. Do you agree with Grant's prediction? Why or why not?

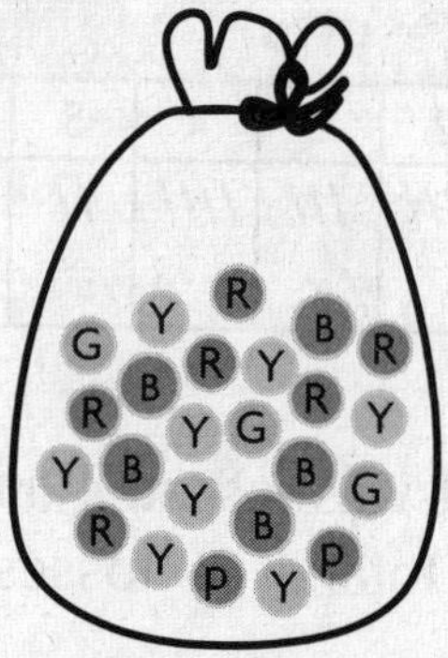

____

3. Toss two coins thirty times. Make a tally table to record the outcomes. How close do you think your experimental probability is to the mathematical probability?

Tally Table				
Outcomes	Coin 1 Heads	Coin 1 Tails	Coin 2 Heads	Coin 2 Tails
Tally				

____

USE DATA For 4–6, use the spinner and the table.

4. What is the experimental probability of spinning blue? What is the mathematical probability?

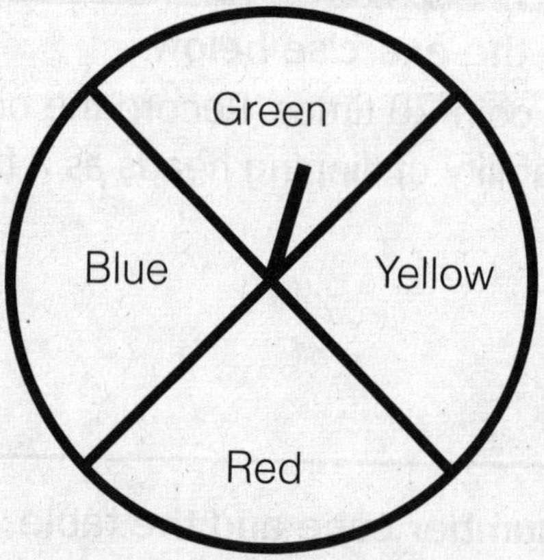

____

5. What is the experimental probability of not spinning blue? What is the mathematical probability?

____

6. How does the experimental probability of spinning green or yellow compare to the mathematical probability of spinning them?

Maryellen's Results				
Outcomes	Blue	Red	Green	Yellow
Tally	卌 卌	卌 |||	卌 |||	卌卌||||

____

Name ____________________

Tree Diagrams

A **tree diagram** is an organized list you can use to find all the possible outcomes of an event.

Make a tree diagram to solve.

Tara tosses a coin and tosses a cube labeled 1 to 6. How many outcomes show tossing heads and an even number?

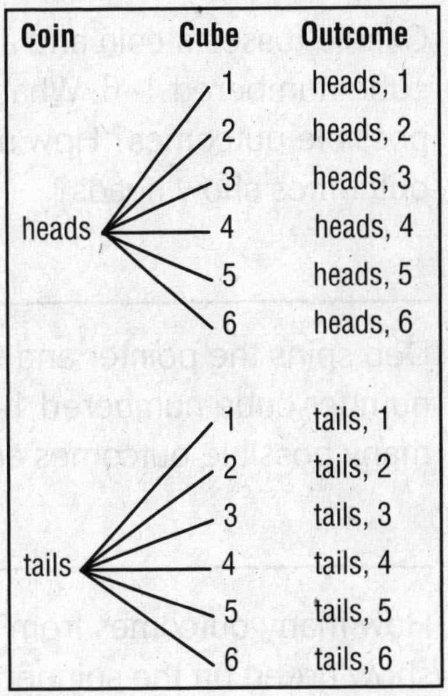

- List all the possible coin flip outcomes in the left column.
- List all the possible number cube outcomes in the center column.
- Draw lines to connect the coin flip outcomes to the number cube outcomes. List each outcome in the third column.
- Find all the outcomes that show heads with an even number: heads–2, heads–4, and heads–6.
- So, there are a total of 3 outcomes for tossing heads and an even number.

Make a tree diagram to solve.

1. Tina spins a spinner with 3 equal–sized sections: 1 red, 1 white, and 1 blue. She also tosses a coin. How many outcomes show spinning red and tossing tails?

____

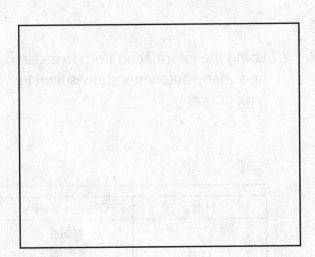

2. Jack tosses a coin and a number cube labeled 7–12. How many outcomes show tossing tails and a number less than 9?

____

SDAP 2.1 Represent all possible outcomes for a simple probability situation in an organized way (e.g., tables, grids, tree diagrams).

Name____

Tree Diagrams

USE DATA Make a tree diagram to solve. For 1–3, use the pictures.

1. Charlie tosses a coin and a number cube numbered 1–6. What are the possible outcomes? How many outcomes show heads?

____

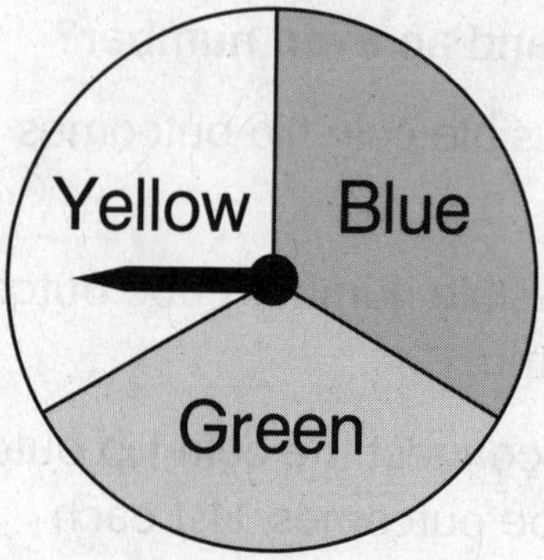

2. Deb spins the pointer and tosses a number cube numbered 1–6. How many possible outcomes are there?

____

3. How many outcomes from Exercise 2 show green on the spinner?

____

For 4–5, choose one of each. Find the number of possible outcomes by making a tree diagram.

4. Event choices
Events: sports, play, movie
Day: Saturday, Sunday

____

5. Footwear choices
Shoes: navy, black, brown
Socks: white, stripes, tan

____

Problem Solving and Test Prep

6. Nora tosses a coin and spins a pointer with pink, yellow, brown, and orange sections. List all possible outcomes.

____

____

7. Using the information from Exercise 6, how many outcomes show spinning pink or orange?

____

8. Higgins the Clown has to choose one hat and one clown suit. How many different outcomes are possible? Use the table at the right.

____

Hats	Clown Suits
red yellow blue	gold orange purple white green silver